# Wine Economics

# Wine Economics

Stefano Castriota
Foreword by Orley Ashenfelter
Translated by Judith Turnbull

The MIT Press
Cambridge, Massachusetts
London, England

The open access edition of this book was made possible by generous funding from Arcadia—a charitable fund of Lisbet Rausing and Peter Baldwin.

The translation of this work has been funded in part by these organizations:

SEPS—Segretariato Europeo per le Pubblicazioni Scientifiche

Via Val d'Aposa 7—40123 Bologna—Italy, seps@seps.it—www.seps.it

Consorzio di Tutela Barolo Barbaresco Alba Langhe e Dogliani

www.langhevini.it

This book was originally published as *Economia del Vino* in 2015 by Egea S.P.A.

This book was set in Sabon LT by Westchester Publishing Services. Printed and bound in the United States of America.

Library of Congress Cataloging-in-Publication Data

Title: Wine economics / Stefano Castriota ; foreword by Orley Ashenfelter; translated by Judith Turnbull.
Other titles: Economia del vino. English
Description: Cambridge, Massachusetts : MIT Press, [2020] | Includes bibliographical references and index.
Identifiers: LCCN 2020003044 | ISBN 9780262044677 (hardcover)
Subjects: LCSH: Wine industry. | Wine and wine-making—Economic aspects.
Classification: LCC HD9370.5 .C37513 2020 | DDC 338.4/76632—dc23
LC record available at https://lccn.loc.gov/2020003044

10  9  8  7  6  5  4  3  2  1

# Contents

# Foreword

It is my distinct pleasure to introduce you to the English-language version of Stefano Castriota's extraordinary book on the economics of wine. Combining insights from economics with carefully selected and relevant sets of data, this volume is essential reading for anyone interested in the economic aspects of the consumption, production, or distribution of wine.

I had just one question when I first previewed the original book in Italian: when would we have an English version? We finally have the version I wanted, but we have more too. By adding extensive new material, the book has become even more useful. Although the basics of the economic aspects of wine may be similar across space and even over time, regulatory regimes often differ extensively across (and even within) countries as well as over time. By incorporating a detailed discussion of regulatory regimes, Castriota has enhanced the value of this book for many readers.

To some people it may come as a surprise to learn that in the last decades, a group of economists, statisticians, psychologists, and agronomists has created an academic subject that encompasses "wine economics." Judged by the usual standards of the academy, wine economics has become a respectable field of economics. There are several academic associations devoted to research about the economics of wine, including the American Association of Wine Economists (AAWE), of which I am currently president. These organizations have lively annual meetings where often over one hundred papers are presented. And there are several journals in the field, including the *Journal of Wine Economics*, which is published for the AAWE in partnership with the Cambridge University Press. The *JWE* is now in its fourteenth volume, and it has reached an audience far beyond the academy. Published articles now receive wide coverage in newspapers and magazines.

Castriota has provided the reader with a look at the general nature of the subject of wine economics but with an eye to its applicability in matters of public policy. He sets the stage for this discussion with a broad description of the evolution of wine consumption and production worldwide. Many readers will be surprised to learn that the consumption of wine, beer, and spirits has tended to converge across countries over time so that total per capita alcohol consumption has also tended to converge.

Much of this book is devoted to the role that government plays, whether for better or for worse, in the market for wine. There are several different aspects of wine that lead to a natural interest in government regulation. One important issue is the abuse of alcohol and what role the state should play to ameliorate such problems. Castriota provides a careful review of the health economics literature so as to inform a sensible public policy approach. Another important research area that Castriota has himself made important contributions is the role of wine region appellations, typically designated by government regulation, in creating value for both consumers and producers. Finally, Castriota does not shy away from the discussion of the role that governments have played in impeding the smooth adjustment of wine prices in response to changes in demand and supply.

One extremely useful distinction that Castriota treats with care is the difference between homogeneous commodity wines and wines that aspire to have distinguishable characteristics. In the usual jargon of economics, homogeneous wines are sold in a perfectly competitive market where producers cannot affect their prices while wines with distinguishable characteristics operate in imperfectly competitive markets where suppliers have pricing power. The characteristics that make wines imperfect substitutes are not entirely known, and to some extent, as with any branded product, subjective matters may play an important role in determining the consumer's perception of the quality of a wine. This area of research is one that makes "wine economics" an especially interesting subject. Although differences in wine quality no doubt exist, it can be extremely difficult to assess these differences in practice. The result is that wine prices can diverge dramatically from what could be justified by the costs of production. With such dramatic divergences between price and cost the role of expert opinion becomes of special interest. The result is that a great deal of research in wine economics is about the assessment of the role of experts in price determination.

This fine book deserves a wide audience. It will be of interest to a broad cross-section of producers, investors, and consumers in the world of wine.

**Orley Ashenfelter**
Joseph Douglas Green 1895 Professor of Economics
Princeton University

# Acknowledgments

I would like to thank Julian Alston, Kym Anderson, Orley Ashenfelter, Leonardo Becchetti, Tito Boeri, Federico Boffa, Paolo Buonanno, Giacomo Calzolari, Jean-Marie Cardebat, Davide Castellani, Maurizio Ciaschini, Alessandro Corsi, Marco Costanigro, Daniele Curzi, Marco Delmastro, Arthur Fishman, Pierre Fleckinger, Olivier Gergaud, Elisa Giuliani, Philipp Kircher, Mikel Larreina, Alessandro Olper, Marika Santoro, Giancarlo Spagnolo, Steven Stillman, Karl Storchmann, Mirco Tonin, David Yanagizawa-Drott, and Angelo Zago for the exchange of opinions and continuous enrichment of ideas; the Italian Sommelier Association (AIS) for interesting lessons and excellent textbooks; and members of all the sommelier associations—AIS, the Italian Federation of Sommeliers, Hotels, Restaurants (FISAR), and the Italian Sommelier Foundation (FIS)—for continuous comparisons and tastings at Vinitaly and around Italy.

# Introduction

Several reasons, other than my personal interests, have led me to write a handbook of wine economics. First and foremost, there is the fact that no well-structured monograph with a comprehensive and interdisciplinary approach ranging from industrial to welfare economics, from economic policy to political economy, from management to finance, and from medicine to law and crime exists. Second is the growing interest in the subject—and not only in wine producing countries—as can be seen by the foundation of the American Association of Wine Economists (AAWE) (set up by Orley Ashenfelter of Princeton University) and the European Association of Wine Economists (EuAWE) as well as the launching of many academic journals on wine economics (e.g., *Journal of Wine Economics*, *Journal of Wine and Business Research*). Third is the publication of articles on the topic in the most prestigious cross-cutting economics journals (for example, *American Economic Review* and the *Economic Journal*) when previously they had appeared only in agronomic journals (for example, *American Journal of Agricultural Economics*, *Journal of Agricultural Economics*, *Agricultural Economics*, etc.). Further, many universities have activated modules or even courses on wine economics, and over the last few years we have been witnessing a growing interest in everything to do with nutrition.

This is an academic book, but it hopes to appeal to a wider public. It intends, therefore, to serve as a useful tool of study for students and consultation for researchers and professionals by limiting technical terminology while at the same time ensuring rigor in the literature review and in the use of data. The book is divided into two parts: the first (chapters 1–5) presents the mechanisms involved in the functioning of the wine market while the second (chapters 6–8) focuses on the measures taken by public authorities to regulate the market and correct failures. The hope is to contribute to the spread and study of the economics of wine as well as stimulating governments, trade associations, and businesses to take concrete and effective action to encourage the growth of the sector and of the wine-making culture in their countries.

The first chapter provides an overview of the world market in a historical perspective. It illustrates the trends and the determinants of both the demand (consumption) and supply (production and export) of wine in the world and highlights the dynamics, developments, and happenings that have taken place in the market in the course of time. Particular attention is given to the so-called "wine war"—that is, the clash between the traditional wine-producing countries (Mediterranean Europe) and the countries of the New World (especially the United States, Australia, New Zealand, Chile, Argentina, and South Africa)—with an analysis of the strengths and weaknesses of the two groups.

In the second chapter we discuss issues concerning consumption and the quality and price of wine. Drawing heavily on health economics literature, the first section looks at the variables that affect the consumption of alcoholic beverages, from genetic predisposition to individual characteristics and from social and economic to environmental factors. The next two sections focus on the determinants of the quality and price of wine and in particular the variables that interest an economist rather than a wine-making expert.

The third chapter deals with competition and the profitability of wineries. It begins by applying the concepts to the three forms of market which are relevant to the wine sector (perfect competition, monopoly, and monopolistic competition). It then goes on to apply Michael Porter's five forces model to identify the elements that feed competition in the wine market. The last section analyzes the profitability of wineries.

The fourth chapter highlights the differences between the various types of companies, classified according to the objectives pursued and their legal status. This topic is particularly relevant as big multinationals and conglomerates dominate in the New World countries while there is a prevalence of small family businesses and cooperatives in the Old World countries.

The fifth and last chapter of the first part deals with the theme of finance in the world of wine. It first analyzes the risks and returns on investment in wine in comparison with other assets and considers the purchase of prestigious bottles and shares in wine-making companies separately. This is followed by an evaluation of the advantage of including wine in an investment portfolio in terms of diversification, risk, and return. The last section is a discussion of the use of derivatives in the wine sector.

The second part of the book addresses the question of market regulation by public authorities, be they national (like the US Congress or single states) or supranational (such as the European Union). The sixth chapter illustrates market failures caused by information asymmetries that have prompted national legislators (first in France in Bordeaux in 1855 and later at a national level in 1935) and supranational legislators (the European Union) to create a pyramidal product classification system based on four levels and that have encouraged companies to group together in consortia to create collective brands (appellations). The chapter then goes on to review the main scientific studies on the corporate reputation of wineries (individual reputation), appellations of origin (collective reputation), and hierarchy within the classification system (institutional reputation).

The seventh chapter analyzes both the positive and negative externalities of the production and consumption of wine (and alcohol in general). Externalities are either advantages or disadvantages to third parties created by the activity of an agent and for which the latter does not receive or pay a price. Unlike the case of information asymmetries, here it is not the imperfection of the market but rather its total absence that justifies public intervention in externalities. Among the positive externalities included is the effect of moderate alcohol consumption on health and social well-being. As for the negatives, detailed analyses have been made of the consequences of alcohol abuse on health and risky behavior, such as unprotected sex, the likelihood of having a fatal accident, and the incidence of criminal actions. The chapter ends with a discussion of the main tools used by public authorities to counter harmful or illegal conduct caused by alcohol abuse.

The eighth chapter describes the economic policies adopted in different parts of the world to influence the supply of wine. The objectives of regulation vary dramatically across countries and over time. In Europe, where the consumption of wine is deeply rooted in the culture and the wine business gives employment to a significant number of workers, the common agricultural policy (CAP) has been heavily influenced by France and its priority is to ensure producers' rather than consumers' welfare. Over the last decades the European Union has tried to adapt the supply to declining internal demand, thus ex officio reestablishing an equilibrium. The main focus is on planting rights, incentives for grubbing up vineyards, subsidies for "crisis distillation," and so on. Unlike the reasons given in the two previous chapters, the regulations imposed here by authorities to influence production do not arise from the imperfection or absence of the market—since the latter in the long run is able to automatically rebalance demand and supply—but rather from a number of other needs that will be explained in due course.

The United States, instead, has been shaped by Anglo-Saxons and Germans who have imposed their beer and spirits culture. Further, in many areas strong religious—especially Protestant—temperance movements fought the consumption of alcohol, leading to Prohibition from 1920 to 1933. When the Cullen-Harrison Act finally amended the Volstead Act, thereby repealing Prohibition, the Twenty-First Amendment allowed bans to be maintained at the state and county levels. Further, the new law established a very complex three-tier system whereby producers had to sell to distributors who had to sell to retailers who could finally sell to consumers. In most states, the direct-to-consumers shipment of alcoholic beverages is not officially forbidden. However, state or even county laws impose a number of constraints and administrative tasks that make it easier for a typical consumer to arrange the interstate shipment of a gun than a Californian bottle of wine (Colman, 2008, p. 2).

The conclusions identify the six key variables necessary to survive the so-called "wine war" and to ensure open and efficient markets, along with the need to protect public health: (1) product quality, (2) changes to the tax system, (3) marketing and a clear classification of wines, (4) competition and support to small wineries, (5) economies of scale and competitive prices, and (6) promotion of a wine culture among consumers.

# 1

## The World Wine Market

Wine is one of the most civilized things in the world
—A phrase attributed to Ernest Hemingway (1899–1961)

The first chapter is divided into three sections. The first section describes the evolution of wine production, consumption, and export in time and across countries. The second describes more in detail the "wine war" between New World and Old World countries, from the "Paris judgment" in 1976 to the present day. This section will discuss the heavy investments in terms of both quantity and quality made by New World countries and the strategy adopted by Old World countries to reclaim the historical dominant position they used to hold. The last section highlights the strengths and weaknesses of the two groups; these concepts underlie the policy conclusions at the end of the book.

### 1.1 Production, Consumption, and Export of Wine

### 1.1.1 Production of Wine

Wine is the beverage resulting exclusively from the partial or complete alcoholic fermentation of fresh grapes, whether crushed or not, or of grape must. Its actual alcohol content shall not be less than 8.5% vol. Nevertheless, taking into account climate, soil, vine variety, special qualitative factors or traditions specific to certain vineyards, the minimum total alcohol content may be able to be reduced to 7% vol. by legislation particular to the region considered (definition 18/73; International Organisation of the Vineyard and the Wine [OIV], 2017).

Only grapes belonging to the *Vitis vinifera* species or coming from a cross between this species and others of the *Vitis genus* (as for example, *Vitis labrusca* and *Vitis rupestris*) may be used to produce wine. However, the European Union insists that wine can be produced only with *Vitis vinifera* since this provides a higher quality product.

The vine is a very resistant plant that can be cultivated between the thirtieth and the fiftieth parallel in the Northern and Southern Hemispheres and at an altitude between sea level and one thousand meters, though global warming is slowly extending these boundaries (e.g., to southern England and Belgium). Red grapes prefer warm climates while white grapes can withstand colder temperatures.

It is impossible to say with certainty who first produced wine and where. It is quite plausible that the discovery of wine making happened by chance with the spontaneous fermentation of grapes left in a bowl. As for the place, the first traces of grape and wine production dated back to periods between 10,000 and 5,000 BCE and have been found in the Middle East in the area amid the Caucasus, eastern Turkey, and Iran. Many scholars believe that vine cultivation for wine making goes back to 4,000 BCE, and they mark the place on the slopes of Mount Ararat where, according to the Bible, Noah's ark ran aground. Vine cultivation spread from Mesopotamia to the rest of the world in various stages, first reaching Egypt and Greece and then later taken to the southern coasts of Italy (*Magna Graecia*, "Big Greece"), France, and Spain by the Greeks and Phoenicians. In Italy the *vitis vinifera sylvestris*, commonly used throughout Europe, was cultivated by the Etruscans before the tenth century BCE, well before Greek domination introduced the *vitis vinifera sativa*, which was found throughout the Middle East (Buono and Vallariello, 2002).

With the birth of the Roman Empire, viticulture spread to the provinces of northern Europe, replacing beer as the favorite drink—or at least among the higher social classes. In Germany, however, the preference for beer remained deeply rooted in popular culture, probably due to the Germanic influence on the Celts (Colen and Swinnen, 2010). In *De Bello Gallico* Julius Caesar reports, with reference to the Nervii and the Germanic people,

> that there was no access for merchants to them; that they suffered no wine and other things tending to luxury to be imported; because, they thought that by their use the mind is enervated and the courage impaired: that they were a savage people and of great bravery: that they upbraided and condemned the rest of the Belgae who had surrendered themselves to the Roman people and thrown aside their national courage: that they openly declared they would neither send ambassadors, nor accept any condition of peace. (Book 2, Chapter 15)
>
> Merchants have access to them rather that they may have persons to whom they may sell those things which they have taken in war, than because they need any commodity to be imported to them. Moreover, even as to labouring cattle, in which the Gauls take the greatest pleasure, and which they procure at a great price, the Germans do not employ such as are imported, but those poor and ill-shaped animals, which belong to their country; these, however, they render capable of the greatest labor by daily exercise. ... They on no account permit wine to be imported to them, because they consider that men degenerate in their powers of enduring fatigue, and are rendered effeminate by that commodity. (Book 4, Chapter 2) (Caesar, 1869)

Although Germany today is an important producer and consumer of wine, beer still remains the most popular drink for the reason mentioned above. However, the fall of the Roman Empire led to such a rapid decline in wine production throughout Europe that it nearly came to an end. The knowledge and practice of viticulture was largely kept alive by monasteries because they used wine for the Eucharist. At this time, therefore, wine was produced and consumed primarily in the Mediterranean.

With the discovery of the Americas, viticulture was exported to the New World countries that had the right climatic and territorial conditions while wine production in North Africa resumed with the beginning of French colonialism. Although Europeans had introduced viticulture to other continents, wine production was almost entirely concentrated in the western Mediterranean; wine production in other countries was negligible until the mid-twentieth century. In the second half of the century, however, the geography of wine underwent unprecedented changes. New World producers (especially Argentina, Australia, Chile, New Zealand, the United States, and South Africa) broke into the market while the well-established North African producers, such as Algeria and Tunisia, suffered a drastic downsizing after the raising of customs barriers and the end of French colonialism.

Figure 1.1a shows the trend of world wine production from 1961 to 2014. It was relatively stable over the period, ranging between twenty-one and thirty-seven million tonnes, even though the world population rose from 3.0 to 7.2 billion individuals.[1] Production peaked in the late 1970s and early 1980s before falling in the following decades and then leveling off. As a result, the marked imbalance that had emerged between demand and supply was reduced. In Italy (see figure 1.1b), the economic and demographic growth of the postwar period led to an enormous expansion followed by a rapid reduction in the 1980s. In relative terms the share of table grapes in the total amount of grapes produced has increased and now stands at around 20 percent (figure 1.1c).

If we compare figures 1.2a and 1.2b, which show the main countries' shares of world wine production in hectoliters in 1961 and 2014 respectively, we can see that the

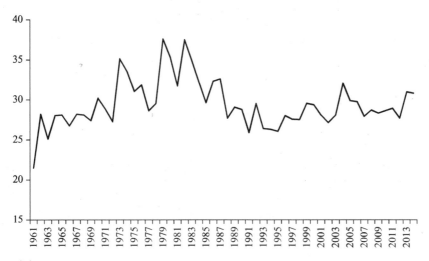

**Figure 1.1a**
World wine production (million tonnes).
*Source*: Author's calculations using data from the US Food and Agriculture Organization (FAO).

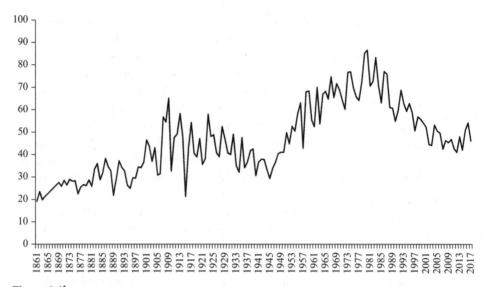

**Figure 1.1b**
Italian wine production (million hectoliters).
*Source*: Author's calculations using data from the Italian National Institute of Statistics (ISTAT).

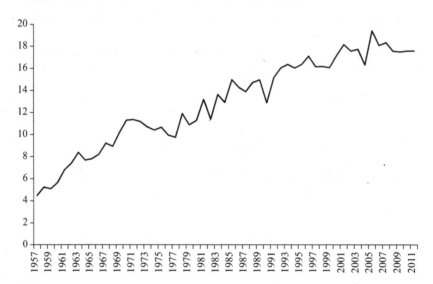

**Figure 1.1c**
Table grapes as share of total grapes produced in Italy (%).
*Source*: Author's calculations using data from ISTAT.

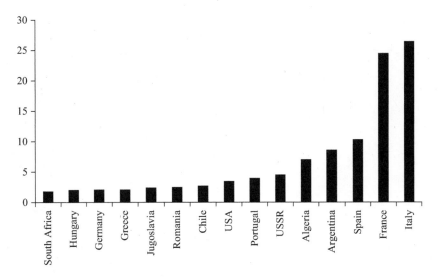

**Figure 1.2a**
Share of world production (% of total quantities), 1961.
*Source*: Author's calculations using data from FAO.

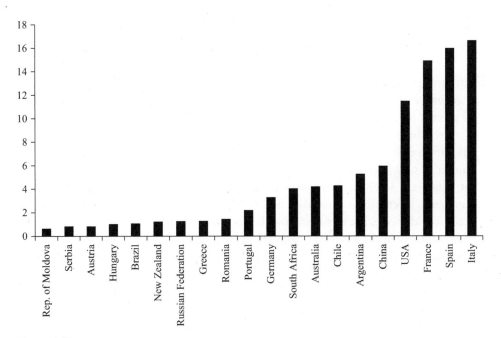

**Figure 1.2b**
Share of world production (% of total quantities), 2014.
*Source*: Author's calculations using data from FAO.

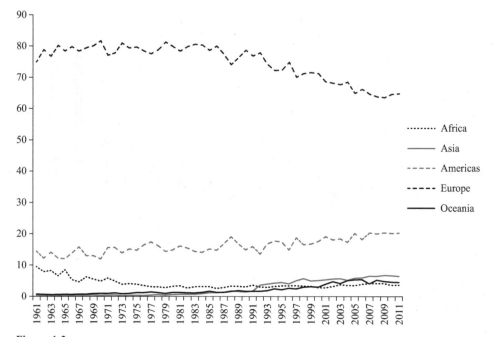

**Figure 1.3**
Wine production as % share of total hectoliters, by continent.
*Source*: Author's calculations using data from FAO.

first three positions are held by Italy, France, and Spain (with Spain overtaking France in recent years), even though the total volume of the three producers has fallen from 60 percent to less than 50 percent. There has been a progressive decline in production in all southern Mediterranean countries: over the last fifty years Algeria's share has fallen from 6.3 to 0.1 percent of world production, Morocco from 1 to 0.1 percent, and Tunisia from 0.7 to 0.07 percent. Europe's loss of volume becomes even more apparent in figure 1.3, which shows wine production in hectoliters as a percentage of total production by continent. Until the mid 1980s Europe consistently produced 80 percent of the world's wine, but this share had dropped to 65 percent by 2011. Production has grown notably in the Americas, from about 12 percent to 20 percent, and in Asia and Oceania, reaching 6.5 percent and 4.6 percent respectively. Africa has seen its share fall from 10 to 4 percent. This has happened for the reason mentioned above, despite strong growth in production in South Africa that has tripled in the last half century.

### 1.1.2    Consumption of Wine
On the demand side, the world per capita consumption of alcoholic drinks among persons of age fifteen and over converted into terms of pure alcohol stood at 6.2 liters in 2010. More than a quarter of alcoholic beverages are produced illegally

at home or without being registered (WHO, 2014). Consumption of this kind of alcohol is particularly dangerous because it can contain impurities or other substances that are toxic for the body (e.g., methanol).

As will be explained in detail in chapter 2, there are marked differences in the consumption levels of the various continents, with the highest levels being recorded not only in the Northern Hemisphere but also in Argentina and Oceania. Intermediate levels of consumption can be found in South Africa and the Americas while the lowest levels are observed in northern and sub-Saharan Africa, in the eastern Mediterranean, and southern Asia. Alcohol consumption is strongly influenced by per capita income and religion, explaining the low consumption levels of most of the latter areas since the Islamic religion explicitly forbids its consumption. As to income, it is correlated positively with total consumption and negatively with home-produced or illegal alcoholic beverages (see WHO, 2011, figure 1).

In dynamic terms (figure 1.4), wine consumption grew to over 280 million hectoliters after World War II, but at the end of the 1980s it suffered a sharp contraction due to the decline recorded in European Mediterranean countries. After falling to just over 220 million hectoliters in the early 1990s, consumption started to grow again in northern European countries and the rest of the world and exceeded 240 million in the five-year period 2006–2010, even though it declined slightly afterward.

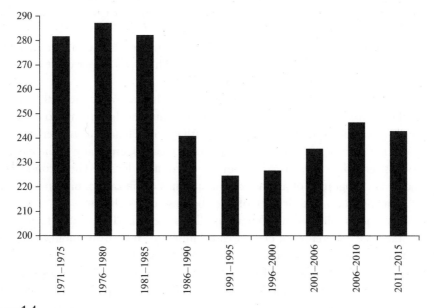

**Figure 1.4**
World wine consumption (million hectoliters).
*Source*: Author's calculations using data from the International Organisation of Vine and Wine (OIV).

A process of convergence in consumption has gradually taken place, involving both the number of liters of pure alcohol consumed (and hence total consumption regardless of the product) and the preferences for the different alcoholic beverages available on the market. Figures 1.5a and 1.5b show the evolution of consumption in liters of alcohol per capita per year in West European and New World countries. The Old World here is understood as France, Germany, Italy, Portugal, and Spain, and the New World as Argentina, Australia, Chile, China, New Zealand, the Russian Federation, South Africa, and the United States (this classification will also be maintained in the following tables and figures).[2]

In the last fifty years, per capita consumption has been progressively moving toward similar values both within European and among New World countries. The same is true when all the countries are considered together.[3] What is most striking is that, contrary to what is often believed, the countries with the highest consumption per capita are not in the north of Europe but in the Mediterranean area with France in first place, although it is fast converging toward the values of its neighbors. This process is even more evident if we look at figure 1.6 that shows average consumption for groups of countries.[4]

The most important point for the present discussion is the breakdown of the annual per capita pure alcohol consumption for wine, beer, and spirits to identify preferences and classify each country as "wine, beer, or spirits drinking." Wine accounts for the largest part of alcohol consumed in Argentina, Chile, and in some West European countries. Spirits are the favorite drink in eastern Europe and in a large part of Asia while beer is ranked first in most of northern Europe, the rest of the Americas, Oceania, and much of Africa. The category "other alcoholic drinks" comes top in sub-Saharan Africa where, however, the levels of per capita consumption are very low (see WHO, 2011, figure 2).

What factors influence the consumption of alcoholic beverages and make countries a wine-, beer-, or spirits-drinking country? The first answer that springs to mind is the climate. Wine can be produced only in areas with specific climatic characteristics. Since transportation costs significantly affect the price of goods, especially in the past, each country has tended to consume what could be produced locally. The second element influencing the geography of alcohol consumption is colonization. Britain has always been a great producer and drinker of beer because of its climate, so it was natural that its ex-colonies followed suit. Migratory flows are a third important factor. In the United States the preference for beer, already "imposed" by British colonists, was reinforced by migratory flows from Germany, Ireland, and the Netherlands. Beer is also the most consumed beverage in former Spanish and Portuguese colonies in Central and South America, even though the two colonial powers were historically wine drinkers. The only exceptions are Argentina, Chile, and Uruguay, and the difference in consumption between these three countries and other South American ex-colonies can be explained, first by the unfavorable climatic conditions for viticulture in the equatorial

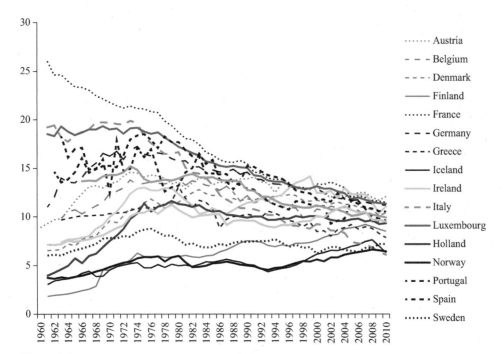

**Figure 1.5a**
Annual total per capita alcohol consumption, Western Europe (liters).
*Source*: Author's calculations using data from the World Health Organization (WHO).

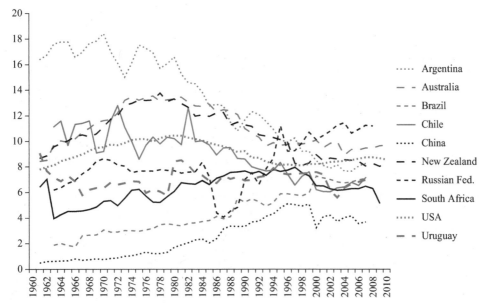

**Figure 1.5b**
Annual total per capita alcohol consumption, New World (liters).
*Source*: Author's calculations using data from WHO.

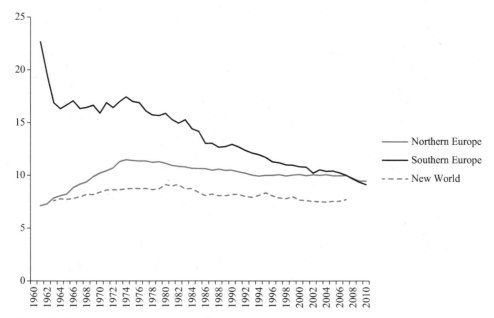

**Figure 1.6**
Annual total per capita alcohol consumption, Europe and New World (liters).
*Source*: Author's calculations using data from WHO.

area and second by their larger communities of European migrants from Mediterranean countries.[5] Lastly, religion has conditioned not so much the preference than as the absolute levels of alcohol consumed; indeed, in some areas of Muslim influence they have reached such a low as to render the classification of a country by this criterion meaningless. The same holds true in the United States where state or county laws can forbid the sale of alcohol (in areas known as "dry" states or counties), and this type of regulation is primarily based on moral and religious objections (Marks, 2015, p. 129).

Figure 1.7 shows the 2014 ranking of countries with the highest per capita consumption of wine expressed in liters of pure alcohol. France leads with 6.4 liters, followed by Croatia and Portugal with almost six, and then by Slovenia, Italy, and Moldova. Luxembourg and Switzerland come next, and although they are climatically unfit for wine production, they have very high per capita income and are surrounded by countries with a great wine-making tradition; this has led to a move away from beer. Denmark and Belgium, which also have high income levels as well as a cold climate, are ranked ninth and eleventh. Spain, with its 2.1 liter per capita of alcohol, is ranked thirty-third, and countries that would be expected to be beer consumers, like Belgium, the United Kingdom, the Netherlands, Germany, and Sweden, follow below. The so-called BRICs (Brazil, Russia, India and China) still have

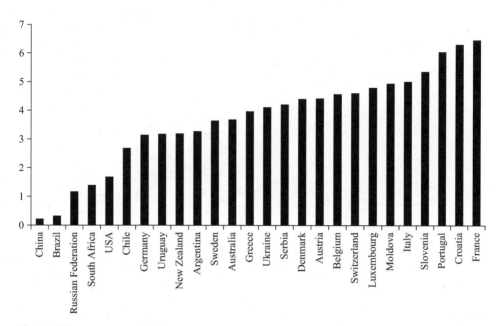

**Figure 1.7**
Annual total per capita wine consumption in liters of pure alcohol, 2014.
*Source*: Author's calculations using data from WHO.

very low per capita consumption levels and are therefore not among the top fifteen countries, but given their strong growth rates of recent years, they are expected to become increasingly important markets for companies in the Old and New World.

If we look at how the share of wine in the total consumption of alcoholic beverages is evolving, we can see that the percentages are converging to similar values. Figure 1.8 shows the share of pure alcohol per capita per year in terms of wine in West European countries. In northern Europe, the alcohol content attributable to wine is increasing, whereas the opposite is happening in Mediterranean Europe.

The same phenomenon can be observed in countries in the New World (figure 1.9). This trend is even more evident in figures 1.10 and 1.11, which show the average shares of wine in the total in northern and southern Europe and the standard deviation of the shares in Western European countries and major wine-producing countries (New and Old World), respectively. The wine share is converging toward 40–45 percent in northern and southern Europe. Given the constant growth in one area and the decline in the other, it will be interesting to see in the coming decades whether the roles will be reversed, with traditionally wine-drinking countries preferring beer and vice versa.

The tastes and habits of consumers can, in fact, change in time. Some important examples are Spain, which up to a few decades ago preferred wine and has now

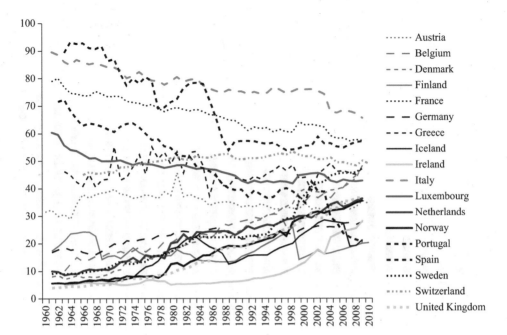

**Figure 1.8**
Share of annual per capita pure alcohol attributable to wine (%), Western Europe.
*Source*: Author's calculations using data from WHO.

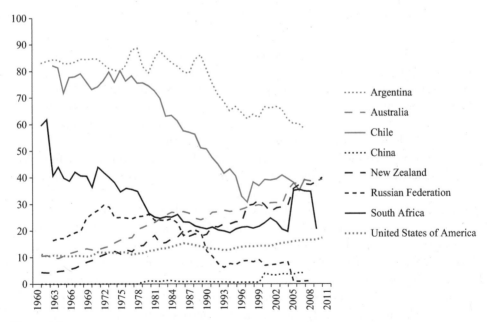

**Figure 1.9**
Share of annual per capita pure alcohol attributable to wine (%), New World.
*Source*: Author's calculations using data from WHO.

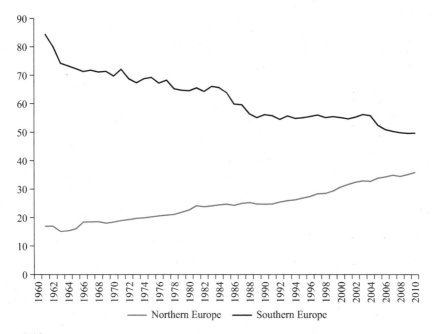

**Figure 1.10**
Share of annual per capita pure alcohol attributable to wine (%), northern and southern Europe.
*Source*: Author's calculations using data from WHO.

become beer drinking. Similarly, over a fifteen-year period Russia and Poland have replaced vodka with beer and can be classified as "drinkers of beer" while Denmark and Sweden today belong to the wine-drinking countries (see table 1.1a).[6] Replicating the same charts for the share of beer and spirits on the total produces quite similar results. The United States, traditionally a beer-drinking country, has been gradually replacing this beverage with wine and spirits (see figures 1.12a and 1.12b) over the last several years. Table 1.1b shows the number of countries which prefers each alcoholic beverage.

Colen and Swinnen (2010) and Aizenman and Brooks (2008) confirmed the converging consumption patterns in an econometric analysis conducted on a large sample of countries. Colen and Swinnen used data on 104 countries over a period of thirty-five years (from 1970 to 2005) and showed how the share for beer drinking increased in both traditional wine-producing and emerging countries and dropped in those countries that traditionally drink beer. There is, therefore, a clear sign of convergence in consumption patterns captured by specific variables and econometric methodologies net of other determinants and disturbance factors. Aizenman and Brooks came to the same conclusion by analyzing a sample of thirty-eight countries over the period 1963–2000.

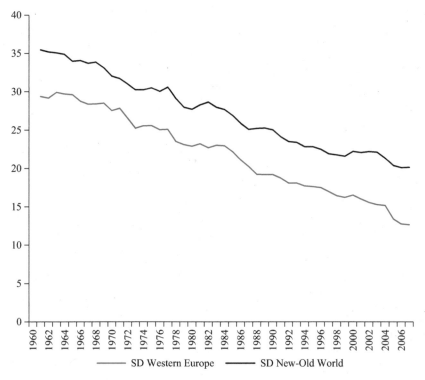

**Figure 1.11**
Standard deviation of the annual per capita share of pure alcohol attributable to wine, Western European and New/Old World countries.
*Source*: Author's calculations using data from WHO.

Alcohol consumption is influenced by addiction (Becker and Murphy, 1988; Fehr and Zych, 1998; Grossman, Chaloupka, and Sirtalan, 1998), by models of consumption inherited from previous generations (internal habits; see Sundaresan, 1989; Detemple and Zapatero, 1991), and by imitating peer behavior (external habits; see Abel, 1990; Campbell and Cochrane, 1999). While the first two factors tend to slow down the process of convergence, the third can encourage changes in preferences, as has happened in Russia over the last few years with the replacement of vodka with beer (Dekonink and Swinnen, 2012). The convergence process also appears to be faster in groups of countries with a higher level of integration (Aizenman and Brooks, 2008).

### 1.1.3    Export of Wine
The abolition of barriers and customs duties and a fall in transportation costs have encouraged trade integration between countries—even those that are quite a distance from each other. At the beginning of the nineteenth century, for example, when a batch of wine was sent from Strasbourg to the Dutch border, it had to go through

**Table 1.1a**

Consumption of pure alcohol by type of beverage (%), year 2010.

| Country | Beer | Wine | Spirits | Other | Main |
|---|---|---|---|---|---|
| Australia | 44 | 36.7 | 12.5 | 6.8 | Beer |
| Austria | 50.4 | 35.5 | 14 | 0 | Beer |
| Belgium | 49.2 | 36.3 | 14.4 | 0.1 | Beer |
| Brazil | 59.6 | 4 | 36.3 | 0.1 | Beer |
| Bulgaria | 39.3 | 16.5 | 44.1 | 0.1 | Spirits |
| Canada | 51.2 | 22 | 26.8 | 0 | Beer |
| Chile | 29.9 | 40.7 | 29.4 | 0 | Wine |
| China | 27.8 | 3 | 69.2 | 0 | Spirits |
| Croatia | 39.5 | 44.8 | 15.4 | 0.2 | Wine |
| Czechia | 53.5 | 20.5 | 26 | 0 | Beer |
| Denmark | 37.7 | 48.2 | 14.1 | 0 | Wine |
| Estonia | 41.2 | 11.1 | 36.8 | 10.9 | Beer |
| Finland | 46 | 17.5 | 24 | 12.6 | Beer |
| France | 18.8 | 56.4 | 23.1 | 1.7 | Wine |
| Germany | 53.6 | 27.8 | 18.6 | 0 | Beer |
| Greece | 28.1 | 47.3 | 24.2 | 0.4 | Wine |
| Hungary | 36.3 | 29.4 | 34.3 | 0 | Beer |
| Iceland | 61.8 | 21.2 | 16.5 | 0.5 | Beer |
| India | 6.8 | 0.1 | 93.1 | 0 | Spirits |
| Indonesia | 84.5 | 0.1 | 15.3 | 0 | Beer |
| Ireland | 48.1 | 26.1 | 18.7 | 7.1 | Beer |
| Israel | 44 | 6.2 | 49.5 | 0.3 | Spirits |
| Italy | 23 | 65.6 | 11.5 | 0 | Wine |
| Japan | 19.2 | 4.1 | 52 | 24.7 | Spirits |
| Latvia | 46.9 | 10.7 | 37 | 5.4 | Beer |
| Lithuania | 46.5 | 7.8 | 34.1 | 11.6 | Beer |
| Luxembourg | 36.2 | 42.8 | 21 | 0 | Wine |
| Mexico | 75.7 | 1.5 | 22.2 | 0.5 | Beer |
| The Netherlands | 46.8 | 36.4 | 16.9 | 0 | Beer |
| New Zealand | 38.2 | 33.9 | 15.2 | 12.5 | Beer |
| Norway | 44.2 | 34.7 | 19 | 2.1 | Beer |
| Poland | 55.1 | 9.3 | 35.5 | 0 | Beer |
| Portugal | 30.8 | 55.5 | 10.9 | 2.8 | Wine |
| Romania | 50 | 28.9 | 21.1 | 0 | Beer |
| Russian Federation | 37.6 | 11.4 | 51 | 0 | Spirits |

(*continued*)

**Table 1.1a** (continued)

| Country | Beer | Wine | Spirits | Other | Main |
|---|---|---|---|---|---|
| Slovakia | 30.1 | 18.3 | 46.2 | 5.5 | Spirits |
| Slovenia | 44.5 | 46.9 | 8.6 | 0 | Wine |
| South Africa | 48.1 | 17.8 | 16.7 | 17.4 | Beer |
| Spain | 49.7 | 20.1 | 28.2 | 1.8 | Beer |
| Sweden | 37 | 46.6 | 15.1 | 1.4 | Wine |
| Switzerland | 31.8 | 49.4 | 17.6 | 1.2 | Wine |
| Thailand | 27 | 0.4 | 72.6 | 0 | Spirits |
| Turkey | 63.6 | 8.6 | 27.9 | 0 | Beer |
| Ukraine | 40.5 | 9 | 48 | 2.6 | Spirits |
| United Arab Emirates | 10.3 | 2.9 | 86.7 | 0 | Spirits |
| United Kingdom | 36.9 | 33.8 | 21.8 | 7.5 | Beer |
| United States | 50 | 17.3 | 32.7 | 0 | Beer |
| Uruguay | 30.6 | 59.9 | 9.5 | 0 | Wine |
| Venezuela | 75.6 | 0.8 | 23.4 | 0.2 | Beer |
| Vietnam | 97.3 | 0.6 | 2.1 | 0 | Beer |

*Source*: Author's calculations using data from WHO.

thirty-one checkpoints and tolls (Robinson, 1998, p. 308). The creation of free trade agreements—for example, the European Common Market in 1957 and the North American Free Trade Agreement in 1992—has seen a progressive elimination of barriers and tariffs between neighboring and non-neighboring countries.[7]

As for transportation costs, the spread of steamships and the development of the railways in the nineteenth century played a crucial role in the early years of globalization (North, 1958) while the invention of containers and innovations in the aeronautics industry revolutionized transportation modes in the second half of the twentieth century (Levinsohn, 2006). This substantiates a key forecast of international trade theory that trade integration reduces the correlation between production sites and consumption because goods from faraway countries are easily available at moderate prices. While Aizenman and Brooks (2008) found that the per capita consumption of wine in 1963 was largely attributable to latitude and grape production, this correlation was much weaker in 2000.

The decrease in transportation costs and the increase in the volume of exports play a key role in this convergence process. Unlike production, the volume of the world's wine exports is constantly growing as can be seen in figure 1.13; they have, in fact, increased almost four times over, from 2.7 million tonnes in 1961 to 10.9 in 2016. The combination of stagnant production and growing exports has led to a

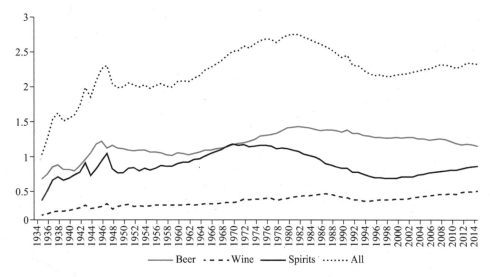

**Figure 1.12a**
Per capita consumption of alcohol by type of beverage in the United States (gallons).
*Source*: Author's calculations using data from the US National Institute on Alcohol Abuse and Alcoholism (NIAA).

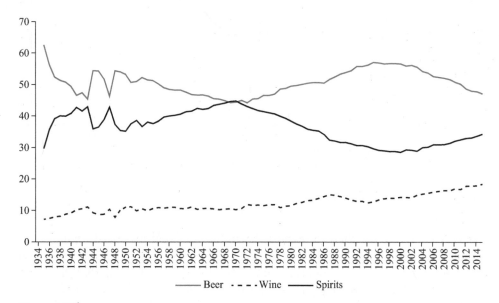

**Figure 1.12b**
Share of per capita consumption of alcohol by type of beverage in the United States (%).
*Source*: Author's calculations using data from NIAA.

**Table 1.1b**
Favorite beverage, 2010.

| Beverage | No. of countries that prefer this beverage |
| --- | --- |
| Beer | 83 |
| Wine | 17 |
| Spirits | 60 |
| Other | 21 |
| Total | 181 |

*Source*: Author's calculations using data from WHO.

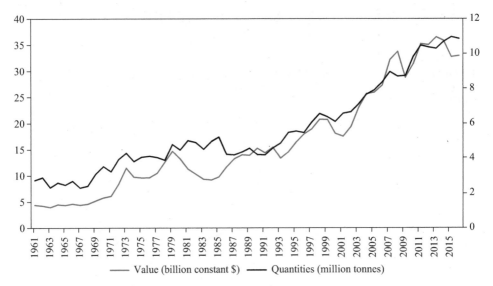

**Figure 1.13**
World wine exports, billion $ constant 2017 terms (left axis) and million tonnes (right axis).
*Source*: Author's calculations using data from FAO.

sharp increase in the share of exported wine in the total produced (figure 1.14). While less than 10 percent of the total wine production was destined for export in the 1960s, this figure now stands at more than 30 percent, showing that the export sector is becoming increasingly important for wineries all over the world.

There are essentially three reasons for such a sudden increase in wine exports. First, a reduction in consumption in the main wine-producing countries has driven domestic companies in Mediterranean Europe to seek out markets for their surplus product overseas. The second is linked to an increased demand for wine by countries

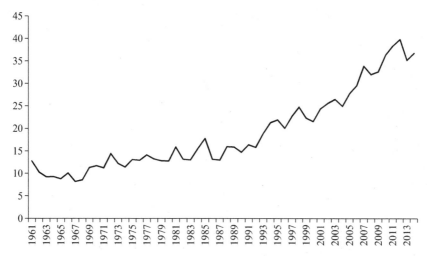

**Figure 1.14**
Share of world wine exported (as % of total quantities).
*Source*: Author's calculations using data from FAO.

that traditionally drink beer or other alcoholic beverages after the globalization of consumption patterns, and not just in the agricultural sector. This is an old process that had already been observed by Gide (1907) in his analysis of the French wine market crisis from the end of the ninetieth century to the beginning of the twentieth century; he believed it was caused more by a lack of demand as a consequence of substituting beer for wine rather than by excess supply. The third can be attributed to the reduction or elimination of customs duties and barriers as a result of international agreements and to reduced transportation costs.

Transporting beverages is expensive because the goods are mainly made up of water and are therefore bulky and heavy (Colen and Swinnen, 2010). The lower the value of the drink, the less cost-effective transportation costs will be since the burden of these costs against the drink's price can become unsustainable. For this reason, breweries have expanded their operations abroad mainly through mergers and acquisitions or through on-site production by licensing rather than through exports from the country of origin. In the case of wine, however, exporting is indispensable; climate conditions make it impossible to produce in some areas of the world, and there may be a total or partial ban on producing wine using grapes coming from areas planted outside the boundaries of a protected appellation area.

The exchange of goods between neighboring countries takes place mainly via road, rail, and pipeline (e.g., oil) and almost exclusively by sea and air between distant countries (Hummels, 2007). The choice of transportation is influenced by a number of factors, including the value and weight of the goods. Almost all raw materials, heavy

Table 1.2
Export of Italian wine according to mode of transportation (%).

| Share of wine | | | | | | |
| --- | --- | --- | --- | --- | --- | --- |
| Quantity | Sea | Air | Rail | Road | Other | Total |
| Germany | | | | | | |
| 1999 | 1.0 | 0.0 | 6.3 | 92.7 | 0.0 | 100 |
| 2012 | 0.0 | 0.0 | 2.4 | 37.6 | 60.0 | 100 |
| France | | | | | | |
| 1999 | 60.1 | 0.0 | 3.1 | 36.8 | 0.0 | 100 |
| 2012 | 2.6 | 0.0 | 0.5 | 44.1 | 52.8 | 100 |
| Great Britain | | | | | | |
| 1999 | 9.9 | 0.0 | 9.9 | 80.2 | 0.0 | 100 |
| 2012 | 0.2 | 0.0 | 0.1 | 56.7 | 43.0 | 100 |
| **Value** | | | | | | |
| Germany | | | | | | |
| 1999 | 0.9 | 0.0 | 1.7 | 97.4 | 0.0 | 100 |
| 2012 | 0.0 | 0.0 | 0.8 | 36.6 | 62.6 | 100 |
| France | | | | | | |
| 1999 | 45.1 | 0.0 | 2.2 | 52.7 | 0.0 | 100 |
| 2012 | 1.5 | 0.0 | 0.5 | 40.5 | 57.4 | 100 |
| Great Britain | | | | | | |
| 1999 | 9.0 | 0.0 | 8.3 | 82.7 | 0.0 | 100 |
| 2012 | 0.1 | 0.0 | 0.1 | 53.7 | 46.0 | 100 |

*Source*: Author's calculations using data from Eurostat.

goods, and low value goods are transported by ship and fuels via pipeline while air transportation represents a frequent mode for light and high value-added products (e.g., electronics). The willingness of consumers to pay for fast air transportation depends on the incidence of this cost on the final price of the good and the value assigned to the speed and punctuality of delivery. These considerations also apply to wine, a heavy commodity. The cost of air transport is justified in economic terms only for a niche of very valuable products and for a small circle of wealthy buyers.

Table 1.2 shows Italian wine exports to three main European trading partners—Germany, France, and Great Britain—broken down according to transport modes. Air transportation does not account for even 0.1 percent of the total either in quantity or in value for any of the three countries considered. The 1999 data shows that

**Table 1.3**
Share (%) of EU-27 wine exported by air.

| Year | Quantity | | Value | |
|------|------|-------|------|-------|
| | USA | China | USA | China |
| 1999 | 0.9 | 0.2 | 3.1 | 1.0 |
| 2000 | 0.9 | 0.1 | 3.1 | 1.5 |
| 2001 | 0.8 | 0.5 | 2.9 | 4.7 |
| 2002 | 0.8 | 0.9 | 2.7 | 4.4 |
| 2003 | 0.7 | 0.8 | 2.2 | 4.1 |
| 2004 | 0.6 | 0.7 | 1.8 | 4.0 |
| 2005 | 0.5 | 0.2 | 1.9 | 5.5 |
| 2006 | 0.5 | 0.6 | 2.0 | 8.1 |
| 2007 | 0.5 | 0.6 | 2.0 | 7.2 |
| 2008 | 0.4 | 0.6 | 1.4 | 7.6 |
| 2009 | 0.2 | 0.6 | 1.3 | 6.7 |
| 2010 | 0.2 | 0.5 | 1.1 | 9.1 |
| 2011 | 0.1 | 0.6 | 0.8 | 8.9 |
| 2012 | 0.1 | 0.3 | 1.1 | 5.8 |

*Source*: Author's calculations using data from Eurostat.

they were mostly carried by road (to Germany and Great Britain), followed by sea (especially to France). It is not possible to analyze the evolution of the operators' choices on the basis of the 2012 data because the mode of transportation is no longer recorded for much of the wine (and is therefore classified as "Other"). There is, however, no particular reason to expect changes from the 1999 shares.

Table 1.3 shows the share of EU-27 wine exports to China and the United States transported by air. Expressed as a percentage of the tonnes, this ratio never reaches even 1 percent, and for the United States it appears to be in sharp decline. When expressed as quantity, however, it can be seen that wine is transported almost entirely by sea. If expressed in value, instead, the situation changes. First of all, the percentage of wine transported by air is from eleven (United States) to nineteen (China) times higher when it is expressed in value compared to quantity. Second, while the share expressed in value and quantity is falling for the United States, there is a fluctuating trend for China with peaks in 2006 and 2010, though it is still growing. This is probably due to an increase in the value of imported wines, reflecting the exponential growth of the Chinese economy and the evolution of consumer tastes. What emerges clearly, however, is that the transportation of wine over long distances takes place almost exclusively by ship because of the high costs while air transportation is reserved for a niche of valuable products.

Costs have fallen primarily for air transportation as a result of technological innovations, especially in the 1950s and 1960s (Hummels, 2007), though they continued to fall in subsequent decades, albeit at a lower rate. Other modes of transportation were also affected by innovations and improvements that cut costs, though periodically external shocks (e.g., oil) reduced the savings that had been achieved. Golub and Tomasik (2008) studied transport costs in twenty-one Organisation for Economic Cooperation and Development (OECD) countries over the 1973–2005 period and, contrary to Hummels' results (2007), did not find a downward trend in shipping costs. As far as wine is concerned, it is not possible to study the evolution of costs for all these modes because the only information available is about maritime transportation.

Table 1.4 shows the costs expressed as a percentage of the value (ad valorem) for some combinations of exporting and importing countries. Costs depend on the distance between the two countries, the value of the commodity exchanged, and a set of other factors. Shipping costs, as a percentage of the value of the wine, noticeably diminished over the fifteen-year period, helping to explain the constant growth of

**Table 1.4**
Cost of sea transport as % of value of wine (ad valorem).

| Importer<br>Exporter | USA<br>EU | USA<br>Australia | USA<br>Chile | USA<br>South Africa | EU<br>USA | China<br>EU | China<br>USA |
|---|---|---|---|---|---|---|---|
| 1991 | 7.14 | 15.3 | 13.18 | 11.64 | – | – | – |
| 1992 | 6.57 | 10.26 | 11.52 | 6.63 | – | – | – |
| 1993 | 6.85 | 8.68 | 10.04 | 7.36 | 8.69 | 3.5 | 15.21 |
| 1994 | 7.13 | 7.88 | 10.27 | 6.13 | 8.28 | 6.34 | 14.1 |
| 1995 | 7.01 | 7.17 | 10.38 | 7.28 | 7.74 | 3.15 | 17.62 |
| 1996 | 6.6 | 5.53 | 9.66 | 6.99 | 8.88 | 8.66 | 12.62 |
| 1997 | 6.53 | 4.88 | 8.62 | 7.82 | 9.09 | 10.7 | 12.71 |
| 1998 | 5.81 | 4.35 | 7.5 | 7.29 | 9.25 | 10.85 | 11.72 |
| 1999 | 5.09 | 3.97 | 6.49 | 5.29 | 6.58 | 5.71 | 6.28 |
| 2000 | 5.54 | 4.24 | 7.16 | 7.56 | 2.75 | 5.53 | 6.54 |
| 2001 | 5.64 | 4.69 | 7.29 | 10.28 | 2.42 | 4.6 | 7.29 |
| 2002 | 5.32 | 5.95 | 7.23 | 9.3 | 2.49 | 4.2 | 3.8 |
| 2003 | 5.19 | 5.5 | 7.27 | 8.63 | 1.87 | 2.97 | 4.1 |
| 2004 | 5.5 | 5.52 | 7.76 | 8.59 | 1.84 | 2.2 | 4.42 |
| 2005 | 5.67 | 6.03 | 8.23 | 7.65 | 2.41 | 2.9 | 4.71 |
| 2006 | 5.64 | 5.98 | 7.22 | 8.65 | 2.12 | 1.91 | 3.67 |
| 2007 | 5.34 | 5.5 | 6.55 | 7.97 | 2.47 | 1.16 | 4.69 |

*Source*: Author's calculations using data from OECD.

wine exports. Falling shipping costs make markets more open and competitive with obvious benefits for consumers, but they also generate an increase in the degree of rivalry among the companies operating in the market.[8]

The wine trade's geography has undergone much more marked transformations than production. Figures 1.15a and 1.15b show the rankings of the main exporters of wine as a percentage of world production expressed in tonnes in 1961 and 2016. What is most striking is that in 1961, Algeria was the world's leading wine exporter. Viticulture had been present in the country even before French colonization (1830–1962), but it covered only two thousand hectares in 1830, climbing to five thousand in 1850. By the turn of the century wine production had grown significantly, covering 150,000 hectares in 1900, reaching 220,000 in 1928, and to 407,000 in 1951, the year of maximum expansion (di Garoglio and Desmireanu, 1961). In the meantime, Spain had taken over as the largest exporter in quantity (but not in value; this was a position held by France because of the higher prices it can command for its products—see figure 1.15d), followed by Italy and France. Four countries in the New World followed—namely, Australia, Chile, South Africa, and the United States—which won their positions thanks to very aggressive policies.

The reasons for the expansion and collapse of Algerian viticulture in less than one hundred years have been carefully investigated by Meloni and Swinnen (2014,

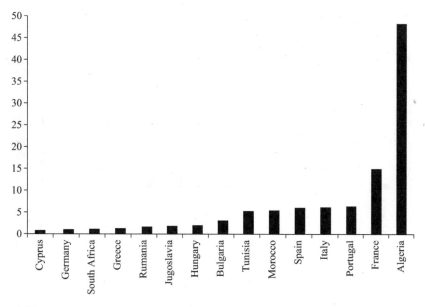

**Figure 1.15a**
Share of world export of wine (as % of total quantities), 1961.
*Source*: Author's calculations using data from FAO.

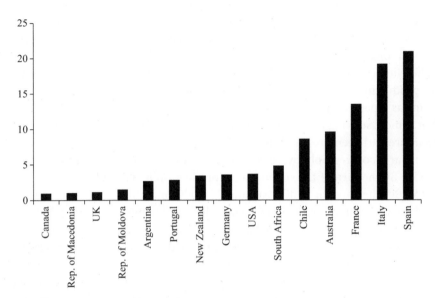

**Figure 1.15b**
Share of world export of wine (as % of total quantities), 2016.
*Source*: Author's calculations using data from FAO.

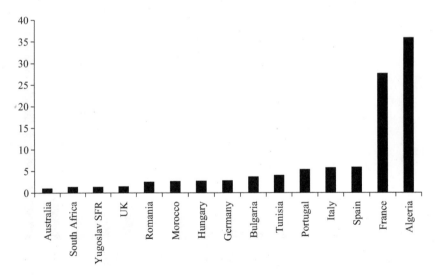

**Figure 1.15c**
Share of world export of wine (as % of total values), 1961.
*Source*: Author's calculations using data from FAO.

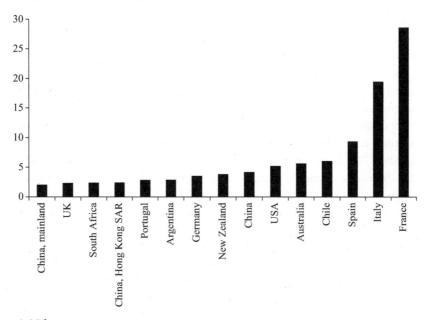

**Figure 1.15d**
Share of world export of wine (as % of total values), 2016.
*Source*: Author's calculations using data from FAO.

2018b). From 1863 onward, phylloxera gradually devastated European vineyards, leading to a drastic reduction in the quantities produced and creating persistent imbalances between supply and demand. This stimulated imports from abroad, and fraudulent activities linked to adulterated wine production became widespread. In 1880 the policy of the Franco-Algerian government to increase credit to agriculture forced many French producers on the verge of bankruptcy to emigrate to Algeria, while commercial wars with Italy and Spain between 1887 and 1892 and the absence of import duties on colonial wines quickly changed the geography of the wine supply in favor of the Maghreb territories.

The start of production in Africa had been helped by a number of technological innovations, such as cold fermentation, which allowed quality products to be obtained in areas with very hot climates (Johnson, Nye, and Franck, 2010). Ninety percent of the production, however, was destined for export to France because of the religious prohibition on alcohol consumption for Muslims. The gradual defeat of phylloxera and the consequent return of French production to precrisis levels created excess supply that led to a price collapse. The attempt by Franco-Algerian producers to market their wine in the United Kingdom as if it were French (the "Leakey case"; see Strachan, 2007 and Birebent, 2007) provoked the reaction of French wine makers, who put pressure on government authorities and succeeded in setting up

Appellations (*Appellation d'origine contrôlée*) in 1905, thus binding the name of the wine to a geographical area (*terroir*).

Between 1931 and 1935 a series of laws (*Statut Viticole*) imposed restrictions, taxes, and an obligation to uproot vines on Algerian vineyards, whose trade was later paralyzed by the outbreak of World War II. At the end of the war, Algerian production restarted, but French producers once again called for new laws with restrictive objectives, duly introduced in 1953 (*Code du Vin*). With the end of French colonization, wine production and exports collapsed in all the former French colonies, and by 2010 Algeria had slipped to thirty-sixth in the ranking of major exporters. The absence of domestic demand and the blockade of imports from France led to the return of North African production to 1880 levels.

Figure 1.16 shows the average prices of exported wine in 2016. Some of the highest prices were picked up by countries such as the United Kingdom and New Zealand where quantities are small and the average price is probably conditioned due to the wine being high-quality niche products. However, if we look at the leading world exporters, France manages to charge prices that are several times its rivals (twice the price of Italy and four times that of Spain). This difference is given by a combination of factors, ranging from higher quality to acquired reputation, advertising campaigns, and so on. However, it is hard to assess the real contribution of each variable to the determination of the export price. Figure 1.17 shows the trend of the

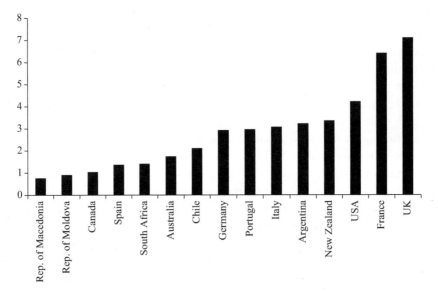

**Figure 1.16**
Average export price of wine (US$ per kilogram), 2016.
*Source*: Author's calculations using data from FAO.

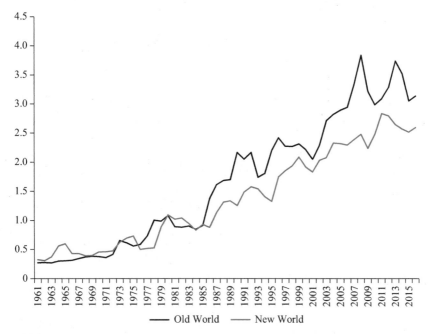

**Figure 1.17**
Average export price of wine (current US$ per kilogram), New versus Old World.
*Source*: Author's calculations using data from FAO.

average export price of wines from the Old and New World. Half a century ago the average price was almost the same in the two groups of countries while it now seems to grow at higher rates in the Old World.

## 1.2   The "Wine War"

### 1.2.1   The Wine Market Before and After the 1960s

Although wine had spread to many European colonies over the centuries, it was produced and consumed almost entirely in Europe until the 1960s. In 1961, the highest consumption levels in non-European countries were recorded in Argentina and Chile (78 and 58 liters per capita respectively)[9] because of large communities of Italian and Spanish immigrants. These levels were lower than in France and Italy (120 and 110 liters per head respectively) but similar to the 59 liters of Spain. Other New World countries had lower levels of consumption: South Africa, 16 liters; Australia, 5; New Zealand, 2.2; United States, 3.2; and some none at all—China, zero. Australia preferred beer because of its Anglo-Saxon culture and hot climate while a large number of people in the United States were teetotalers and the rest favored beer and spirits. Wine was, therefore, primarily for immigrants from Mediterranean Europe (Bartlett, 2009).

However, the situation changed radically in the following decades, and consumption grew in New World countries, with the exception of Argentina, Chile, and South Africa where levels converged as described in the previous section. In 2009, per capita consumption of wine stood at 26 liters in Australia, 1.3 in China, 8.1 in New Zealand, and 7.2 in the United States. The constant growth of domestic demand for wine played a crucial role in driving supply. Infant wine businesses in the New World were able to benefit from a number of competitive advantages—especially the widespread availability of land, low labor costs, low tax burden, and the absence of strict geographical and technical constraints such as those imposed by the appellations in European countries. In 2006 the average size of wineries in the United States was 213 hectares and in Australia, 167 hectares, as opposed to 7.4 in France and just 1.3 in Italy (Heijbroek, 2007, p. 5). New World countries, therefore, were able to benefit from large economies of scale and greater freedom to experiment with new agronomic and wine-making techniques.

### 1.2.2  Innovations in the Wine Sector

Table 1.5 shows the main innovations introduced over the centuries. Until the beginning of the twentieth century, new cultivation and production techniques spread fairly homogeneously, but over the last hundred years the Old World has been more resistant to changes than the New World. Indeed, many of the most important innovations have been explicitly forbidden by EU regulations under the pressure of the so-called "purists" who are opposed to any form of change that could "take the poetry out of wine." In contrast, the New World countries have begun to experiment and adopt new techniques—for example, drip irrigation, making it possible to plant vineyards in areas with low rainfall in Australia, and quality stabilization, a problem that has still not been resolved today in rainy and climatically unstable France. Reverse osmosis, which reduces the percentage of water in grape must before fermentation and makes colors and aromas more intense, was patented in 1992 in France but immediately forbidden, although it is used extensively outside Europe. The larger size of vineyards in the New World means many companies can make a wide-scale use of mechanical pruners and harvesters, reducing labor costs considerably.

Night harvesting, widely used throughout the world to avoid the heat and sun of the day, prevents fermentation and the acidification of grapes during transportation, but in Europe it is still limited to a small number of companies. The doubling of the density of plants per hectare that has proven to be effective in improving the quality of grapes while keeping the yields per hectare constant has been markedly slowed down by the long and complex bureaucratic process needed to change the regulations for wines with an appellation. Improvements in pruning techniques have been adopted all over the world while new fertilizing techniques have been limited in Europe by the more conservative attitude of farmers and more stringent regulations.

**Table 1.5**
Main innovations in the wine sector.

| When it was introduced | Description of innovation | Adopted in Old World | Adopted in New World | Effect on production |
|---|---|---|---|---|
| End of 18th century | Mass production of glass bottles and cork stoppers | Yes | Yes | Improvement in the preservation and storage of wine |
| Beginning of 19th century | Use of horses and introduction of rows | Yes | Yes | Improvement in production efficiency and lowering of costs |
| 1855 | Bordeaux wine classification system | Yes | Yes | Reduction of producer-buyer information asymmetries |
| 1858 | Sulfur dusting of vines | Yes | Yes | Solution of powdery mildew problem |
| 1863 | Pasteurization | Yes | Yes | No longer necessary to increase the alcohol content in wine to prevent it from turning sour during transport |
| 1880 | Introduction of American rootstocks on European vines | Yes | Yes | Solution of phylloxera problem |
| 20th century | Drip irrigation | No | Yes | Less variation in quality in different years |
| 1992 | Inverse osmosis applied to the must | No | Yes | Color and taste are made more intense |
| End of 20th century | Addition of tartaric and citric acid | Yes | Yes | Control level of acidity |
| End of 20th century | Night harvesting | Yes/no | Yes | Avoids fermentation during transportation |
| End of 20th century | Doubling density of plants per hectare, maintaining same yield | Yes/no | Yes | Improvement in the quality of grapes |
| End of 20th century | Improvements in pruning and fertilization techniques | Yes/no | Yes | Increased yield and improved aroma |
| End of 20th century | Steel tanks with built-in microchips to electronically control temperature during fermentation | Yes/no | Yes | Improvements in fermentation |
| End of 20th century | Addition of tannins | No | Yes | Control tannin levels |
| End of 20th century | Mechanical pruners and harvesters | Yes/no | Yes | Reduction in labor costs |
| End of 20th century | Use of wood shavings during fermentation | No | Yes | Gives an aroma of wood, saving on costs |
| End of 20th century | New packaging for marketing (screw caps, carton packs with soft plastic interiors, etc.) | No | Yes | Reduction of transportation costs and fewer storage problems (e.g., mold) |

The same has happened to the use of stainless steel tanks with built-in microchips to control fermentation.

Experiments have been made introducing wood shavings into fermentation tanks to give the wine the aroma of wood without the long and expensive process of aging in barrels, but the technique has been prohibited by law in the Old World. Finally, new packaging for the marketing of wine (screw caps, carton packs with soft plastic interiors, etc.) has been extensively adopted in the New World countries while in Europe it has been hampered by what is, at times, excessive attachment to tradition.

These innovations in New World countries have had two important results. The first is a drastic reduction in average production costs; in a comparison of wine producers belonging to the same quality range, costs were found to be 74 percent higher in the Languedoc region in France than in the Riverina in Australia (Heijbroek, 2007, p. 16). The second result is the increase in quality in New World countries, in spite of the criticisms about the limited variety of grapevines compared with the historical European producers and the often excessive use of wood (or its derivatives).

### 1.2.3   The "Judgment of Paris" and the Surge of New World Countries

On May 24, 1976, Steven Spurrier, an English merchant, organized a public blind wine tasting in Paris for promotional purposes linked to the bicentennial of the United States. It has since become known as the "judgment of Paris" (Colman, 2008, pp. 71–72). Eleven experts—nine French, one English, and one American—evaluated French and Californian red and white wines. Despite having the "home court advantage," the French were outplayed by the Californian wines in both categories, causing a huge stir in France. After numerous accusations of fraud, two years later a new tasting event was organized which still produced the same outcome, with Californian wines winning most of the prizes (Bartlett, 2009).

These contests acted as very effective advertising for New World wines and also convinced producers of their ability to compete globally. They encouraged and accelerated both agricultural (new vineyards) and technological (innovations applied to viticulture) investments that had particularly disruptive effects in the medium- to low-level segments of the market. These less experienced consumers had previously known only a dozen international vines and are particularly sensitive to the quality-price relationship. Large economies of scale in production and distribution have allowed New World countries to practice extremely competitive prices. Companies with thousands of hectares have begun to sell containers of decent or good quality wine very cheaply. Even in the premium sector very high levels have been reached with some American or Australian wines being knocked down in auctions at staggering prices.

While the New World was experiencing a phase of quantitative expansion and qualitative growth in the last quarter of the twentieth century, there was a drastic fall in wine consumption in Mediterranean Europe. From 1970 to 2009, per capita consumption decreased by 66 percent in France, 56 percent in Italy, 49 percent in

Portugal and 35 percent in Spain.[10] The main reasons for this are the changing preferences of the younger generation toward other alcoholic beverages (especially beer but also spirits), greater awareness of the harmful effects of alcohol abuse on health, and the stiffening of controls and sanctions to stop drunk driving. For a long time, the increase in consumption recorded in the rest of the world was not enough to offset the losses in Mediterranean countries. From the end of the 1970s to the mid-1990s, world consumption declined steadily while the imbalance between supply and demand reached a peak of nearly 65 million hectoliters, over 20 percent of production, in the five-year period 1986–1990 (see table 1.6).

This negative trend was finally reversed at the end of the 1990s, and excess production fell. Changing consumer tastes also helped to counteract the decline in wine. In Europe interest faded in low-quality table wine for everyday use while the demand for premium or super-premium wines grew as customers became better informed and more demanding. Besides, in many countries of the New World, wine is considered synonymous with elegance and distinguishes people from the "mass" of less sophisticated consumers. The demand for better quality wine went hand in hand with a rise in the average price paid, not only opening new opportunities for producers but also creating new threats. New consumers are increasingly knowledgeable, so any producer unable to offer a good product at a competitive price will be forced out of the market.

### 1.2.4 Change in Consumer Preferences

The last few decades have been characterized by profound changes not only in the geography of producers and consumers but also in the preferences of consumers. In the 1980s the United States became much more health conscious and interested in the quality and characteristics of food. This led to an increase in the demand for white

**Table 1.6**
Total world wine production and consumption (1,000 hectoliters).

| Years | Production | Consumption | Difference | Difference/ Production (in %) |
|---|---|---|---|---|
| 1971–1975 | 313,115 | 280,356 | 32,759 | 10.5 |
| 1976–1980 | 326,046 | 285,746 | 40,300 | 12.4 |
| 1981–1985 | 333,552 | 280,718 | 52,834 | 15.8 |
| 1986–1990 | 304,192 | 239,485 | 64,707 | 21.3 |
| 1991–1995 | 263,092 | 223,183 | 39,909 | 15.2 |
| 1996–2000 | 272,570 | 225,302 | 47,268 | 17.3 |
| 2001–2006 | 272,615 | 234,329 | 38,286 | 14.0 |
| 2006–2010 | 270,724 | 245,031 | 25,693 | 9.5 |
| 2011–2015 | 272,268 | 241,537 | 30,731 | 11.3 |

*Source*: Author's calculations using data from OIV.

wine and spritzers (low-alcohol drinks with white wine and tonic water). In fact, by the end of the 1980s three-quarters of the wine consumed in the United States was white, but over the next two decades, the situation changed radically once again. In 1991 an episode of the TV program *60 Minutes* drew attention to a medical study that claimed the daily consumption of red wine was one of the main explanations for the so-called "French paradox"—that is, a low incidence of heart disease in a population with a diet typically rich in fat (Colman, 2008, p. 83). As a result, within a five-year period from 1991 to 1996, the share of red wine increased from 27 percent to 43 percent.[11]

At the same time another radical change took place in the United States when the demand for sweet white wines suddenly slackened, leading to a fall in German exports of this type of wine from three million hectoliters in 1992 to two million a couple of years later. However, changes in consumer preferences do not only involve the type of wine. Americans' favorite white grape in the 1980s was Chardonnay, but ten years later consumers preferred pinot grigio and sauvignon blanc; for red wines a particular liking for Cabernet Sauvignon was followed first by a short-lived boom in merlot and later by the triumph of Pinot Noir.

Changes in consumer preferences are a serious problem for wineries since the planting of a vineyard is a costly investment and is characterized by deferred returns. For the first three years a vine does not produce fruit; from the fourth to the sixth year production is at 30 percent while from the seventh to the thirtieth year it has its highest yield; and from thirty-one years onward (generally up to forty years) there is a gradual decrease in yield per hectare, resulting in a higher quality product. Therefore, an enterprise that has made substantial investments in a vineyard, aiming at a specific vine in vogue at the time of planting, can face financial ruin if consumer tastes change suddenly.

This risk is common to all producers, although it is stronger in the New World where consumer purchases are strongly influenced by the vine. In Europe collective trademarks (appellations) play a much more important role in driving buyers' choices, thereby attenuating the changes and "anchoring" customers more to the terroir. The appellation system, on the one hand, makes the Old World consumer "loyal" to the geographical area; on the other hand, it hinders the conversion of vineyards. A producer who wants to abandon the production of an appellation wine with little demand to focus on a fashionable international vine variety has to give up the institutional recognition that they have obtained to sell a "table wine" or a "geographical indication" (see chapter 6).

### 1.2.5   Evolution of the World Wine Market

The imbalance between demand and supply mainly affected Europe because of its constantly falling consumption, and it reached a peak toward the end of the 1980s, so much so that there was talk of a "European Wine Lake." From the 1970s onward, the European Union intervened by adopting supportive policies for producers to encourage the voluntary grubbing up of vineyards and by subsidizing the crisis distillation of surpluses. As will be explained in detail in chapter 8, these measures proved costly and ineffective or even harmful since the EU purchases of bad wine perpetuated the surpluses (Thornton, 2013, p. 291). So in 2008 it was decided to change strategy and to replace support for the disposal of surpluses with the liberalization of the market.

Meanwhile, in the 1990s, the assault of new wine producers continued. In 1996 Australia launched its "Strategy 2025" initiative with "total commitment to innovation and style from vine to palate." The declared goal was to become "the world's most influential and profitable supplier of branded wines" (Winetitles Media, n.d.). Ten years later production had doubled and exports had grown five times over. Other countries of the New World also achieved notable results in both quantitative and qualitative terms so that the geopolitical balance of wine seemed destined to change quickly and definitively.

However, things did not go exactly according to plan. Market saturation and the reorganization of European producers also hit New World wine makers, though not immediately. Excess production from 2000 onward had encouraged Australia to strongly reduce its export prices between 2004 and 2006. This seriously damaged the image of Australian wine that had previously been considered "cheap and cheerful." Some observers labeled it the "coca-colarization" of wine (Aylward, 2008). The excellent value for money, at first allowing the country to triumphantly enter markets that were dominated by countries of long-standing tradition, nearly became a reputation trap. In the meantime, European producers had reorganized and could now rely on EU funds for promotion abroad in their counterattack.

Given the stagnation of domestic consumption, the export market had become increasingly important for Mediterranean Europe, and in 2010 it accounted for one-third of production.[12] The United States, with a population of around 330 million inhabitants with a very high per capita income and per capita consumption growth, became the main battlefield in the wine war. This market presents a mixture of risks and opportunities for exporting countries. On the one hand, the constraints imposed by each state on the marketing and sales of alcoholic beverages and the three-tier distribution system set up after the end of Prohibition make it traditionally complex. The three levels are the producer, the wholesaler, and retailer. In most US states, the law obliges producers to sell only to wholesalers, who then sell to retailers, and only retailers can then sell to consumers.[13] On the other hand, the younger generations have proved to be better informed and more xenophile, giving foreign producers an advantage, especially in the medium-high price range. Despite the great improvements in quality and the economic success of their domestic companies, the United States is now under pressure from rivals and is also suffering from a loss of competitiveness owing to high labor costs and the exorbitant price of land (especially in California's most famous areas), both of which are much lower in countries like Australia and, above all, Chile.

A further variable which is affecting the geography of wine is global warming (Cardebat, 2017, p. 18–20). Experts forecast that temperatures will rise by 3.6°F–7.2°F (2°C–4°C) by the end the twenty-first century. Rising temperatures allow the production of wine at higher latitudes and altitudes (e.g., in England, small mountains, etc.) which can be an advantage for some countries. However, it will be troublesome for other countries with milder climates since temperatures will become scorching and rainfall more scarce and unstable.

## 1.3    Main Differences Between the Old and New World

Table 1.7 gives a summary of the main differences between Old and New World production, distribution, and consumption of wine. Some variables favor the first group of countries, others the second.

### 1.3.1    Differences from the Point of View of Production

Production has been declining in Western Europe for decades, just as it has been rising in New World countries (except for Australia and South Africa). In the first group of countries all aspects of the market are regulated, and there is very limited freedom of entry,[14] whereas the market is free outside the old continent. When new consumption patterns move preferences toward new vines, the constraints imposed by governments or consortia may hinder the ability to adapt to market demand.

The size of companies varies, but they tend to be much larger in the New World (Cardebat, 2017, p. 31) for reasons of geography (widespread availability of land), demography (low population density), and history (fragmentation of land ownership in Europe because of inheritance and special laws). Consequently, market concentration is much higher in the New World than in the Old World (Marks, 2015, pp. 112–116). Thornton (2013, p. 289) has estimated the number of wineries and the share of the two and four largest firms for the year 2009 in the most important wine-producing countries. The difference among the two groups of countries is striking, with the four largest companies producing a much larger share: in France 15.9 percent, Italy 9.7 percent, Spain 21 percent, and Germany 3.8 percent while across the ocean in the United States it is 56 percent, Argentina 60.5 percent, Australia 62.3 percent, Chile more than 80 percent, South Africa 37.1 percent, and China 28 percent. In the United States the first twenty wineries have a combined market share of roughly 90 percent; the remaining 10 percent is left to around seven thousand firms (Thornton, 2013, p. 3). This last point has affected the control of the entire production chain, which is often impossible in Europe. As noted by Simpson (2009), wine production is dominated by small businesses and cooperatives in Europe but by major corporations in the New World. The result is a much more marked concentration of production in the second group of countries.

**Table 1.7**
Comparative analysis of the characteristics of the Old and New World.

| Variable | Old World | New World |
| --- | --- | --- |
| Production | | |
| Production (hundred liters, or hl) | Falling | Tending to increase |
| Entry barriers | Planting rights subject to EU laws | Freedom of entry into the market |
| Firm size | Generally small-to-medium size | Generally large |
| Control over production chain | Limited due to fragmentation of land ownership (cooperatives) | Large-sized firms and total control of production chain |
| Production techniques | Bound (often by law) to traditions | Innovative |
| Product differentiation | Hundreds of native vines and appellations | Few international vines that are planted all over the world |
| Production costs | High | Lower (on average) |
| Wine flavor | The naturalness and authentic flavor of the vine are preferred | Very fruity and strong aroma of wood from the use of chippings |
| Distribution | | |
| Quality signaling | Principally based on the classification system and collective brands | Based on firm brand |
| Power of distributors | Lower | It depends. For example, in the US it is very high |
| Consumption | | |
| Domestic consumption (hl) | Generally falling | Generally increasing |
| Purchasing power of domestic consumers | Falling strongly | Different trends, but not negative as in the Old World |

In 2012 only two European wineries (both French) were among the top ten in the world per surface area, with the first three cultivating more than ten thousand hectares each (Mediobanca, 2014, p. 8). This state of fragmentation in Europe is favored by the political influence of numerous producers, both independent and cooperative members, who want to maintain the status quo by hindering the aggregation and consolidation that has taken place in other continents (Simpson, 2009).

In the Old World the small size of many plots of land has encouraged the creation of wine-making cooperatives to which members contribute their grapes. In 2000, this type of company had 49 percent of the market share in Portugal, 52 percent in France, 55 percent in Italy, and 70 percent in Spain (Anderson, Norman, and Wittwer, 2004, p. 18). There is, however, a real risk of opportunistic behavior (free riding) since the price per kilogram of grapes is generally fixed to a large extent, reducing the incentive to make major investments to pursue qualitative improvements.

From the point of view of production techniques, producers in the New World are much freer to experiment and innovate than in the Old World where they are limited by the regulations of the appellations. Europe can, however, count on its unique heritage of vines as an important element of strength while the production of new competitors is limited to a few international grape varieties. Table 1.8 shows the number of indigenous vines used in the production of wine by country,[15] clearly showing the wealth of countries like Italy, Portugal, and France. Data for New World countries is not available, but apart from some cultivar of American grapes in the United States and Canada, creoles in South America, Pinotage in South Africa, and some new hybrids in Australia, no "native" vines from *vitis vinifera* exist in these areas of the world. The huge variety present in the Old World can intimidate the newcomer and does not mean that all varieties are of high quality, but it can become a key advantage in winning over the most experienced clientele, especially considering the converging trends in the use of vines of recent years.

As shown by Anderson (2013), globalization has heightened the concentration of varieties. Whereas in 2000 half of the new vines planted in the world belonged to twenty-one different species, in 2010 that figure stood at fifteen. Concentration is greatest in the New World countries where half of the newly planted vines belong

**Table 1.8**
Number of native vines by country.

| Country | Table grapes | Wine grapes | Including hybrids | Including foreign varieties | Including native varieties | Total |
|---|---|---|---|---|---|---|
| Italy | 87 | 421 | 48 | 32 | *341* | 508 |
| Portugal | 117 | 339 | 34 | 37 | *268* | 456 |
| France | 48 | 248 | 53 | 8 | *187* | 296 |
| Greece | 0 | 197 | 1 | 37 | *159* | 197 |
| Spain | 71 | 158 | 10 | 29 | *119* | 229 |
| Croatia | 44 | 163 | 18 | 53 | *92* | 207 |
| Germany | 6 | 142 | 95 | 7 | *40* | 148 |
| Hungary | 27 | 88 | 49 | 21 | *18* | 115 |

*Source*: Schneider (2011).

to only seven species. In recent years, the trend toward French varieties has intensified; from 2000 to 2010 the total surface area in the world covered by these clones increased from 26 to 36 percent (from 20 to 27 percent in the Old World and from 53 to 67 percent in the New World). In Australia the percentage of French vines has gone from just over 30 percent in 1975 to almost 90 percent in 2010.

All this makes it increasingly difficult for competitors to distinguish themselves from the others, especially in New World countries—many of which have no native plants—while France sees its vines planted all over the world. Although identical clones can produce very different wines depending on the characteristics of the soil and the agronomic and wine-making techniques adopted, it is becoming increasingly difficult to horizontally differentiate products in a market that is globalized and characterized by a smaller variety of vines. As a result, companies are forced to undertake vertical differentiation (linked to quality), with the arduous task of achieving an excellent sensory profile. On the other hand, there are those who claim that imitation by competitors is the best form of flattery a producer can receive. By using French vines as a benchmark, followers implicitly and tacitly recognize the superiority of the leader's viticulture that influences the opinion of experts and enthusiasts.

Another useful indicator is the varietal similarity index that intuitively resembles a correlation coefficient and provides a measure of how similar the portfolio of varieties planted in one country is compared with that of the rest of the world.[16] This index tends toward one when the mix of varieties planted in a country is identical to that of the rest of the world and, on the contrary, is equal to zero when there is no overlap. Table 1.9 shows the data for some countries for 2000 and 2010. From the data analysis it emerges that this index has increased in many countries in both the Old World and the New World, a further demonstration of the homogenization in progress. For the same reason, it is no surprise that France has the highest value of all, at 0.72 in 2010, and it is growing strongly compared with the previous decade. Figures 1.18a and 1.18b, taken from Anderson (2013), show the rankings of the first thirty red and white grapes sorted by world surface percentage; the share of the first five to six vines changed significantly between 2000 and 2010 and saw the advance of both red and white French grapes.

As a consequence of the larger size of companies and the freedom to adopt technological innovations, production costs are more contained in the New World. The wines from this part of the world are sweeter and fruitier with a more marked hint of wood than European ones, mainly thanks to the use of technology.

In a competitive market, characterized by a large number of companies and product homogeneity, rivalry is based exclusively on price. If all companies have the same technology, then larger companies will have lower average costs and therefore can set lower prices in the presence of increasing economies of scale. In this globalized

**Table 1.9**
Varietal similarity index.

| Country | 2000 | 2010 |
| --- | --- | --- |
| Argentina | 0.30 | 0.38 |
| Australia | 0.45 | 0.62 |
| Austria | 0.12 | 0.15 |
| Chile | 0.46 | 0.60 |
| China | n.a. | 0.47 |
| France | 0.57 | 0.72 |
| Germany | 0.36 | 0.26 |
| Greece | 0.19 | 0.21 |
| Italy | 0.36 | 0.44 |
| New Zealand | 0.34 | 0.30 |
| Portugal | 0.46 | 0.29 |
| South Africa | 0.29 | 0.50 |
| Spain | 0.69 | 0.62 |
| United States | 0.41 | 0.65 |
| Uruguay | 0.21 | 0.23 |

*Source*: Data come from tables 54 and 55 in Anderson (2013), pp. 530–542.

context, European businesses, smaller in size and with strong regulatory restrictions on the use of technologies, would automatically be losers. For this reason, the New World—having neither native vines nor the appeal of history—focuses on the improvement of globalized products and price cutting.

Europe, in contrast, tries to exploit diversity, hence its native varieties, and to protect them through the appellation system that binds production to precise geographical boundaries.[17] The European Union has succeeded in preventing the production of wines with names similar to "Champagne" or "Chianti" outside Europe through commercial agreements. The idea is to create a market similar to monopolistic competition in which various companies are present (in this case, business groups) that produce differentiated goods. Each manufacturer (or group) can specialize in a certain segment differentiating itself from competitors horizontally (by type of product) or vertically (by quality and, in time, their individual or collective reputation).[18] (See Cardebat [2017, pp. 20–23] for a description of wine segmentation and Marks [2015, pp. 134–135] for an explanation of the way producers develop monopoly power through supply restrictions and the establishment of wine appellations.)

### 1.3.2    Differences in Distribution

The marketing strategies of non-European companies are mainly based on the indication of the vine and the promotion of the corporate brand while in the old continent the large number and small size of businesses have stimulated the creation of collective trademarks (appellations) and favored the classification of quality based on a hierarchy established by public authorities (e.g., in Italy, the classifications DOCG—Controlled and Guaranteed Designation of Origin; DOC—Controlled Designation of Origin; IGT—Indicazione Geografica Tipica; and "table wine" or Vino da Tavola—VdT). In the New World, wineries generally control the whole value chain. By relying on large distributors wineries can increase their bargaining power and, therefore, their profit margin. In some countries like the United States, however, distributors have become so large and powerful (Thornton, 2013, p. 3) that their profit margins erode those of alcohol producers. Further, distributors often

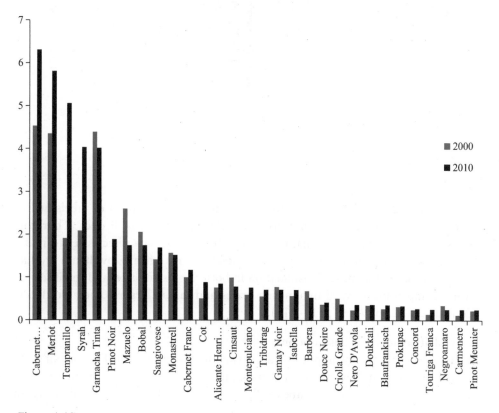

**Figure 1.18a**
Share of the world surface area cultivated with the top 30 red vines.
*Source*: Anderson (2013), p. 20, figure 13.

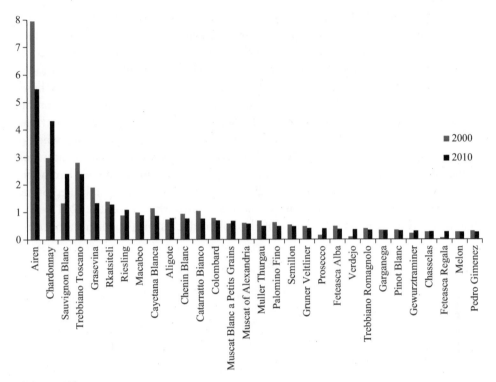

**Figure 1.18b**
Share of the world surface area cultivated with the top 30 white vines.
*Source*: Anderson (2013), p. 20, figure 14.

favor large wine producers that can offer a diversified portfolio of labels at low prices with regular deliveries and provide little support to small ones (Thornton, 2013, p. 175). Given that in many US states the law forbids direct sales to retailers and consumers, this can be a serious obstacle to small firm growth and product differentiation and quality. Consolidation in both the distribution and retail sectors pushes competition toward minimizing prices and favors larger-volume producers (Colman, 2008, pp. 91, 97, 114–116).

### 1.3.3   Differences in Consumption

As discussed above, consumption is decreasing in Mediterranean Europe but is on the increase in most New World countries. Since purchasing power influences consumption in both quantity and, above all, in value, careful attention should be given to trends in this variable in the various geographical areas. The economic crisis that started in 2008 has led to a contraction in growth all over the world, but Europe, and especially the Mediterranean, has suffered most from the collapse of gross domestic product and rising unemployment.

## 1.4   Challenges for the Wine Market

Over the next decades the world wine market will have to face two big challenges. The first one concerns the development of the Chinese market, and the second, climate change.

A new giant is entering the wine market—China (see Cardebat, 2017, pp. 16, 81). With a population of almost 1.5 billion people and a steadily growing economy, the total consumption of wine in China has tripled in less than twenty years, reaching 19.1 million hectoliters in 2016 (source: OIV). Given the size of its population compared with Western countries, per capita consumption (among people aged 15 years and older) is still very low—just 1.7 liters as opposed to approximately 51.2 liters in France and 43.6 in Italy in the same year. The opportunities for market growth are, therefore, enormous. As shown by Masset et al. (2016), Cardebat et al. (2017), and Cardebat and Jiao (2018), the Chinese market is already an important driver of prices for the finest international wines. China, however, is not content with just importing products from the West and has invested heavily with joint venture agreements and by hiring European oenologists, and this has led to a rapid increase in both the quantity and the quality of its own wine.

There are, therefore, two rival sides in the market. Producers of the old continent hope the decline in domestic consumption may be counterbalanced by increasing exports through intense reorganization and support from restyled EU policies while producers in the New World, located in countries with rising consumption, enjoy a number of competitive advantages ranging from lower production costs to less stringent regulations. Considering this premise, and in view of the entry of new giants into the market, one cannot rule out a further heightening of global competition in the future.

As to the second issue, global warming is producing dramatic changes in climate conditions all over the world with respect to temperature, rainfall, and the frequency and intensity of extreme events such as hurricanes and floods. In their extensive study Hannah et al. (2013) show that the area suitable for viticulture could decrease by 25 to 73 percent in major wine-producing regions by 2050. Viticulture will move toward cooler climates at higher latitudes and altitudes. In their attempt to preserve the quantity and quality of grapes, many producers could be forced to increase the water usage for irrigation and other technologies which might produce negative environmental effects. Some (warm) countries will suffer from these changes while other (cold) ones will benefit (Ashenfelter and Storchmann, 2016; Leeuwen and Darriet, 2016).

# 2

# Consumption, Quality, and Prices

Life is too short to drink bad wine.
—A phrase attributed to Johann Wolfgang von Goethe (1749–1832)

This chapter will first discuss the variables affecting the consumption and abuse of alcohol and secondly those that affect the quality and price of wine. Before reviewing the literature, a premise is necessary. Most consumers drink alcohol moderately, and this produces positive effects on people's physical and mental health. Alcohol abuse, however, generates a series of negative consequences for the consumers themselves and for others (see chapter 7), but fortunately this involves only a minority of individuals. It is important, therefore, to understand what influences people's behavior by distinguishing between moderate consumption and abuse, the latter being further differentiated into consumption exceeding the World Health Organization's (WHO's) recommended number of alcohol units, binge drinking, and alcoholism.

## 2.1 The Consumption of Alcoholic Beverages

The variables that determine the consumption of alcohol have been classified in four groups (ICAP, 2009):

1. genetic predisposition;
2. individual characteristics;
3. social and economic factors; and
4. environmental factors.

The following discussion starts from the specific and moves to the general, instead of the other way around. Genetic predisposition is unique and subjective for every individual as are a good part of an individual's characteristics. On the other hand, social, economic, and environmental factors influence entire categories of people (if not peoples). The decision to proceed from the specific to the general is based on the consideration that the variables conditioning the consumption of an individual tend to be constant in time, whereas the general can change substantially.

### 2.1.1   Genetic Predisposition

Genetics influences the consumption and, above all, the abuse of alcohol and drugs. There is a strong element of hereditariness in alcohol dependence. A study conducted on a sample of people adopted at a young age showed that 18.2 percent of those with biological parents with alcohol problems developed forms of dependence, against 6 percent of those who had normal parents (Hawkins, Catalano, and Miller, 1992). As dopamine is the main neurotransmitter influencing the pleasure experienced during the consumption of alcohol, genetic differences that can affect the functioning of dopamine pathways are considered important determinants of alcoholism.

Gene polymorphism, in which genetic variations occur with an incidence of at least 1 percent in the population, is believed to influence the consumption of and dependence on alcohol. Choi et al. (2005) analyzed polymorphism in a sample of 352 individuals, 106 of them with alcohol dependence and 246 without. They found genetic abnormalities only among those who showed alcohol dependence, and when they analyzed subgroups, they identified a gene that would appear to be most responsible for this risk. Other genes have been identified by Edenberg and Foroud (2006) in a study of families with more than one member suffering from alcoholism. Although an increasing amount of medical research is being carried out on this subject, much remains to be discovered. However, 50–60 percent of the risk of developing life-threatening behaviors depends on the hereditary genetic predisposition and the remaining proportion on other environmental factors (Le Strat et al., 2008).

### 2.1.2   Individual Characteristics

*Gender*
The consumption of alcoholic beverages differs considerably between men and women. In a large-scale study done between 1997 and 2007, questionnaires were submitted to a sample of citizens in thirty-five countries. Wilsnack et al. (2009) found a prevalence of regular drinkers among men and teetotalers among women. The lower levels of consumption recorded for women are due to both social and physiological factors. In many cultures it is not socially acceptable for a woman to consume alcohol or, in any case, to drink heavily.[1] The task of child rearing is also predominantly or exclusively entrusted to women and requires clear-headedness and self-control, and so motherhood greatly reduces alcohol consumption (Little et al., 2009). Further, the female body is composed of a higher percentage of lipids and a lower quantity of water than the male body, so the same amount of alcohol per kilogram of weight produces a greater amount of alcohol per liter of blood (Ramchandani, Bosron, and Li, 2001).[2]

The harmful effects of alcohol on the body are also inversely proportional to an individual's weight. This has two important consequences (Ely et al., 1999). First, the negative consequences of alcohol on health and self-control are greater for women than for men, even with the same level of consumption, because women have a lower body weight. Second, women tend to consume less alcohol since a smaller amount of alcohol has the same negative effect on their nervous system.

## Age

A vast amount of literature has shown that age influences both total consumption and patterns of consumption significantly. Consumption levels and the frequency of binge drinking are inversely correlated with age (Wilsnack et al., 2009), with notable differences between countries. After administering questionnaires to a sample of 13,553 people living in twenty urban areas in the United States, Johnson et al. (1998) found higher levels and frequency of consumption among men, young people, and whites. The peak is reached around the age of twenty-one for both men and women; after that age, total consumption tends to decline while the frequency tends to remain stable. When people enter adulthood, they must take on a range of responsibilities, both in the family (e.g., married life, relationships, care of children and elderly parents) and at work, and these responsibilities are often incompatible with irresponsible behavior (Little et al., 2009).

Aristei, Perali, and Pieroni (2008) studied expenditure on the purchase of alcoholic beverages in Italy using Italian National Institute of Statistics (ISTAT) data for the period 1997–2002. According to their purely descriptive analysis, total consumption reaches a peak between the ages of forty and fifty. The authors separated the effect of age at the time of the interview from that of the birth year (cohort) and the general trend in the country (time trend). In line with the literature, the study confirmed the negative correlation between the interviewee's age and the level of consumption. Moreover, net of other sociodemographic variables and the age of the respondent, the generations born around the mid-twentieth century present greater participation and consumption levels than the later ones. In recent decades, radical changes in lifestyles and the rhythms and types of work and greater awareness about the harm caused by bad habits have led to a sharp decrease in the purchase of alcoholic beverages.

For the elderly, the consumption of alcoholic beverages, especially in European Mediterranean countries, follows a traditional pattern as it takes place daily at mealtimes and rarely with more than six glasses. Despite this, those most likely to have nonmoderate daily consumption levels are often the elderly, especially among males.[3] This is probably due to a lack of knowledge about the correct amount of alcohol that should be consumed. WHO, in fact, recommends that, during meal consumption, men should not exceed two to three alcohol units per day; women should not exceed one to two; and elderly people should not exceed one. It is most likely that the elderly will maintain the habits they have acquired during their lifetime, unaware of the increased health risks in advancing age (ISTAT, 2013, p. 9).

Age also strongly influences the frequency of episodes of alcohol *abuse* (ICAP, 2009), which is significantly higher in adolescents and twenty-year-olds. These excesses are particularly dangerous for young people who are less able to metabolize alcohol properly and are inexperienced in managing states of drunkenness. Alcohol abuse is the cause of serious accidents in the home and on the roads and damages health in the long run. In adulthood, alcohol abuse decreases, consumption becomes more regular, and the reasons for consumption change. In the past, wine and beer were considered more than just a drink. They were part of the meal—rich in carbohydrates, sugars, vitamins, proteins, and minerals.[4] Today, however, the reasons for drinking alcoholic beverages have very little to do with enriching our diet; such reasons include the following items (Agrawal et al., 2008):[5]

1. To socialize: drinking to facilitate interaction with other people or to celebrate an event or a person
2. To feel good: people drink so that alcohol can act on their central nervous system and alter their state of mind and perception of reality
3. To console oneself and to reduce stress: some people "drink to forget" (e.g., to forget personal or economic problems)
4. To conform: people drink because others do

The reasons for drinking vary from one person to another, but what is of greater interest is that they tend to change with age (Peterson and Hektner, 2008). Adults usually drink in moderation and mainly to socialize, or when they are going through a difficult time, they may exceed consumption to combat stress (the first and third reasons). In contrast, consumption among adolescents is usually concentrated during the weekends, with frequent episodes of alcohol abuse for the deliberate intention of losing control and imitating peer behavior (the second and fourth reasons).

Contrary to what is often believed, the age at which a person begins to consume alcohol does not in itself affect the likelihood of having problems with alcoholism. It is the *abuse*, not the mere consumption, of alcoholic beverages at a young age that produces harmful effects (Bonomo et al., 2004). In any case, people who start drinking heavily at an early age often have a series of deviant behaviors that characterizes them as being more at risk (Clark and Bukstein, 1998). It is therefore very difficult to attribute any responsibility to the age at which alcohol is approached, especially as the risk of dependence decreases when it is the family that introduces children to the alcohol (Warner and White, 2003), as is the case for most people (Coleman and Cater, 2003).

*Personality traits*

Consumption and alcohol abuse are also correlated with personality traits, such as risk aversion, impulsiveness and strong emotion seeking. Risk aversion has been widely studied by both psychologists and economists, but the definitions for classifying an individual as having a propensity or an aversion to risk differ in the two disciplines. Psychologist C.R. Cloninger (1987), for example, developed a tridimensional personality questionnaire to assess three elements—namely, harm avoidance, novelty seeking, and reward dependence. These three features are considered by psychologists as the signs of a personality that loves, or at least does not fear, risk.

Economists, on the other hand, evaluate risk aversion based on preferences for risk-return combinations expected in a number of possible investments. Although they rest on different logical bases, the definitions provided by psychologists and economists lead to rather similar classifications of individuals. Moreover, various studies in both disciplines have found a strong negative correlation between risk aversion and alcohol consumption.[6] The negative correlation between risk aversion and the consumption of alcohol is generally accepted, although in Howard, Kivlahan, and Walker (1997), only the link between seeking new stimuli and alcohol consumption is positive while the other two components of the Cloninger test have only a marginal role. As for economists, Dave and Saffer (2008) included risk aversion in the demand function and, using two different American databases (Panel Study of Income Dynamics [PSID] and Health and Retirement Study [HRS], both from the University of Michigan), found a 6 to 8 percent prevalence of alcohol consumption in less cautious people.

Consumption and alcohol abuse levels are also 50 percent higher among people with mental health problems (Cleary et al., 2009), while the abuse of psychoactive substances among the very young affects individuals with psychological problems, such as anxiety, depression, and low self-esteem, in 60 percent of cases (ICAP, 2009).[7] Although it is sometimes difficult to clearly identify the direction of causality and isolate the effect of one variable net of other confounding elements from a methodological point of view, genetic variables and individual personality characteristics obviously play a key role in determining consumption habits of alcoholic beverages.

### 2.1.3 Social and Economic Factors

Levels and patterns of alcohol consumption are influenced by social environment, socioeconomic status, and the absolute and relative prices of beverages.

The social environment—formed by family, friends, and work contacts—affects the development of drinking patterns over time. The influence of the family will last into adulthood, but it is stronger in adolescence (Halebsky, 1987) and can be positive when parents' consumption is moderate or negative in the opposite case. People with strong ties to their family are less influenced by the behavior of their peers, which is a protective factor when parents consume alcohol moderately. Nash, McQueen, and Bray (2005) administered questionnaires to 2,573 American high school students and showed that disapproval of alcohol abuse by parents reduces the influence of friends and is a strong deterrent to excess. Conversely, individuals whose parents have alcohol problems tend to assimilate their behaviors and say that they feel unloved and socially excluded and are left to themselves (Burke, Schmied, and Montrose, 2006).

Older siblings (McGue, Sharma, and Benson, 1996) as well as friends (Valliant, 1995; Ali and Dwyer, 2010) are also very important in forming the consumption habits of young people.[8] Wrong friendships can lead to increased alcohol consumption with a fall in school and university performance (Kremer and Levy, 2008). Personal relationships are another important determinant; alcohol consumption is, in fact, more moderate among people in a stable relationship (Prescott and Kendler, 2001) but tends to grow when a relationship breaks down.

Socioeconomic status indicators generally focus on (1) education, (2) income, and (3) unemployment. These variables are correlated with each other, though each one centers on different aspects. Education and income, in fact, influence access to intangible and material assets respectively while employment reflects prestige and the power associated with a certain type of job (Van Oers et al., 1999).

From a theoretical point of view there are valid reasons to believe that these three variables can be correlated both positively and negatively with consumption and alcohol abuse. Education can be negatively correlated with consumption and, above all, with the abuse of alcoholic beverages because it improves risk perception and awareness of its negative effects on health.[9] However, most educated people have a greater sense of self-control and often work in environments where the consumption of alcohol is not only tolerated but also expected (Huerta and Borgonovi, 2010).[10]

As for the question of income, the economic models based on the maximization of consumer welfare include a budget constraint so that an increase in spending capacity should also increase, inter alia, the consumption of alcoholic beverages. This point becomes even more relevant if greater economic means correspond to a more intense social life. However, the opportunity cost of reducing time spent at work to consume alcohol (the cost of leisure time) and the potential damage to one's reputation as a result of a state of intoxication are greater for people with a high income. For obvious reasons this problem is felt more by those who have a job as alcohol consumption decreases job performance, thus increasing the risk of dismissal, and during an economic crisis when the probability of finding a new job decreases. Yet people on a low-income and the unemployed "have little to lose" when they drink heavily for consolation and for the release of tension, using alcohol as "self-medication" (see Hill and Angel, 2005). The effect of income on alcohol consumption is, therefore, uncertain.

Similarly, the economic and psychological costs of unemployment affect both the unemployed and the employed.[11] Those who have lost their job suffer a drastic drop in income while the community as a whole has to bear an increase in tax burden to cope with increased social spending (e.g., unemployment benefits). Adverse economic conditions should encourage the purchase of necessary goods rather than alcoholic beverages (Ruhm, 1995), especially in those countries where health care is largely private and health insurance is suspended in the event of dismissal.

However, psychological costs connected to the state of anguish for current and future economic situations concern, to a greater extent but not exclusively, those who have lost their job and are responsible for the upkeep of a family. As alcohol can be consumed to alleviate stress and for comfort, it is reasonable to expect an increase in consumption among the unemployed. This may also be true, though to a lesser extent, for those in employment, especially in times of economic crisis. We can, therefore, find the opposite effect of a fall in income. Further, unemployment is like a virus; those who suffer most are the direct victims, but as the disease spreads, there is a greater risk of contagion. Individuals whose jobs are in jeopardy may be encouraged to increase consumption because of stress or conversely, given the increasing difficulties in finding a new job, reduce consumption and increase effort at work (Catalano et al., 1993). Lastly, unemployment increases the free time that can be dedicated to enjoyable pastimes during which alcohol is usually consumed (television, parties with friends, etc.), even though there are fewer economic resources and opportunities to meet people (Devalos, Fang, and French, 2012), so the net effect is ambiguous.

Economic theory, therefore, has worked out valid arguments for both a positive and a negative effect of socioeconomic status on the consumption and abuse of alcohol. The question at this point becomes merely empirical. From an econometric point of view, however, it is very difficult to measure the contribution of socioeconomic status to consumption and alcohol abuse since the variables considered—income, education, and employment—are strongly correlated with each other, generating collinearity that may affect the results.[12] Another problem, which is always present in econometric analyses, is the risk of omitting some important regressors (omitted variable bias). This is not a minor point in models of alcohol consumption that are applied to existing databases as it can lead to a distorted estimate of the parameters.[13] A third issue is reverse causality; for example, if we agree that there is a link between unemployment and alcohol consumption, which one influences which? The loss of a job could induce an individual to seek consolation in alcohol, but it is also true that alcohol abuse negatively affects productivity and increases the risk of dismissal. Hence, it is often rather difficult to identify and measure the direction of causality that can be two-way. The differences that have emerged in various studies may, therefore, depend on the quality of the databases and the econometric techniques and instrumental variables used as well as the peculiarities of some countries or historical periods.

### *Empirical studies on the effect of education*
In some studies the level of education increases alcohol consumption and the frequency of abuse, while in others it reduces them. Examples of the first kind are the Los Angeles County Department of Public Health (2001) that interviewed a sample of 8,354 adults in the county in 1999 and Schoenborn and Adams (2010) using National Health Interview Survey (NHIS) data on a sample of American families in the period 2005–2007. They both recorded with purely descriptive evidence greater consumption by people with higher educational qualifications. Similarly, the ISTAT (2013) annual survey showed that in Italy graduates presented higher levels of alcohol consumption

and drank less on a daily basis but more frequently outside of meals, which is dangerous for health. Comparable results can also be found in econometric studies: Dawson et al. (1995), Moore et al. (2005), and Dave and Saffer (2008) (all using American data) and Strand and Steiro (2003) (using Norwegian data) found a greater percentage of drinkers among highly educated people. Huerta and Borgonovi (2010) applied a two-stage model to longitudinal British data for a sample of people born in 1970 (in the British Cohort Study) and found a positive relationship between school results on the one hand and consumption and alcohol problems on the other.

The list of studies that have shown the negative relationship between education and alcohol consumption is, however, equally long. Crum, Helzer, and Anthony (1993) used data on a sample of households between 1980 and 1984 and found a higher risk of developing disorders related to alcohol abuse in individuals who had dropped out of school or university compared with those who had a degree. Droomers et al. (1999) (using Dutch data for 1991) and Casswell, Pledger, and Hooper (2003) (using longitudinal data on American individuals aged eighteen, twenty-one, and twenty-six) found higher levels of consumption among people with lower education. Parry et al. (2005) using 1998 South African data detected a strong association between symptoms of alcohol problems and lack of schooling. Karlamangla et al. (2006); Aristei, Perali, and Pieroni (2008); Kestilä et al. (2008); Lee et al. (2009); and Schnohr et al. (2009) came to the same conclusions using American and European databases. The effect of education on consumption and alcohol abuse is not, therefore, unequivocal.

*Empirical studies on the effect of income*

The elasticity of alcohol consumption to income is important for both firms (in a monopolistic or competitive market) and legislators. In fact, demand which is inelastic to changes in income reacts weakly to economic cycles. This is good during economic downturns. However, in the long term consumption patterns are going to be flatter because demand is not very responsive to an increase in gross domestic product (GDP) and buyers' purchasing power.

The study of the effects of income on consumption and alcohol abuse has also produced conflicting results. When the consumption of alcohol was considered without distinguishing the types of drinks, positive correlations were reported by Gottlieb and Baker (1986), Dawson et al. (1995), Moore et al. (2005), and Kerr et al. (2009) (using American data); Pietilä, Rantakallio, and Läärä (1995) (using Finnish data); Strand and Steiro (2003) (using Norwegian data); and Aristei, Perali, and Pieroni (2008) (using Italian data). Opposite results were obtained by Karlamangla et al. (2006) (using American data), Batty et al. (2008) (using Scottish data), Caldwell et al. (2008) (using British data), and Kestilä et al. (2008) (using Finnish data).

However, there is a much larger number of analyses examining the three main types of alcoholic beverages—beer, wine, and spirits—separately. These results are less controversial and the effect of income almost always appears to be positive. Fogarty (2010a) reviewed 141 studies that analyzed the elasticity of consumption with respect to income and to price. Elasticity is defined as the ratio between percentage variations of two variables—in our case the percentage variation in consumption divided by the percentage variation in income or the price of the drink. If elasticity to income is positive, the good is considered normal as the increase in economic means corresponds to an increase in the demand for the good; whereas if it is negative, it is called an inferior good because cheaper, low-quality goods can be replaced with other more expensive ones.[14] When elasticity is positive and between zero and one, the good is considered necessary since the reduction in consumption is less than proportional to the reduction in income ("inelastic" demand), whereas if it is higher than one it is considered a luxury good (demand adapts more easily to economic circumstances).

Fogarty (2010a) reported summary statistics and distributions of the elasticity of demand with respect to income and the price of the three alcoholic beverages—beer, wine, and spirits—found in the 141 studies reviewed. A series of interesting points emerged from this analysis. First, the distribution of the elasticities in the six histograms shows a certain dispersion; if the data, countries, years, and econometric methodologies change, different results are obtained.[15] Second, despite this variability in the estimates, very few studies (less than 8 percent for wine, beer, and spirits) showed income elasticity to be negative (or an inferior good), in line with what economic theory and common sense says. Third, income elasticity is on average higher for spirits (1.15) and lower for beer (0.64), with wine in an intermediate position (1.10). The demand for beer is more rigid than wine, and the demand for wine is in turn more rigid than for spirits. Beer is a necessary good while wine and spirits are luxury commodities. The reason for this is probably the average price, which is lower for beer than wine, whereas spirits are the most expensive.

### Empirical studies on the effect of unemployment and economic cycles

The effect of unemployment on consumption and alcohol abuse is uncertain. Ruhm (1995) used data on American states and found that alcohol consumption is procyclical; when unemployment falls, liquor consumption increases, together with road accidents. Similar results were obtained by Freeman (1999) with an analysis of logarithmic differences. The problem of these two databases is that they did not include individual characteristics that may have important implications for the study in question, which is why Ettner (1997) used longitudinal NHIS data from 1988. With the use of instrumental variables, he showed that unemployment increases alcohol consumption but reduces the symptoms of addiction. Dee (2001), using longitudinal BRFSS data for the period 1984–1995, found the opposite. In periods of economic crisis alcohol consumption decreases, but the cases of binge drinking increase by 1.5 percent in the presence of a 1 percent increase in unemployment.

Ruhm and Black (2002) extended Dee's (2001) study using very similar data (the US Centers for Disease Control and Prevention's Behavioral Risk Factor Surveillance System surveys [BRFSS], 1987–1999). The authors confirmed the pro-cyclicality of alcohol consumption and showed that changes are driven by the behavior of heavy drinkers (those who drink more than one hundred alcohol units a month). However, two other studies found a countercyclical relationship. Devalos, Fang, and French (2012), with the US National Epidemiologic Survey on Alcohol and Related Conditions data for the period 2001–2005 and fixed effects panel models, found a positive relationship between the unemployment rate on the one hand and abuse, drunk driving, and alcohol dependence on the other. Mossakowski (2008), with data from the US National Longitudinal Survey of Youth from 1979 to 1992, showed that the duration of poverty and involuntary unemployment affect alcohol abuse in people between the ages of twenty-seven and thirty-five years. Lastly, Charles and DeCicca (2008), with NHIS data for the period 1997–2001, found the effects of unemployment on consumption and heavy drinking are not statistically significant. Similar conclusions were also reached by Jiménez-Martín, Labeaga, and Vilaplana Prieto (2006) who used BRFSS data for the period 1987–2003 and a similar methodology to Dee (2001). The problem of reverse causality was addressed by Mullahy and Sindelar (1996) who applied instrumental variables to US data from the 1988 NHIS Alcohol Supplement and showed how alcohol abuse has a negative impact on employment.

In summary, most of these studies found a pro-cyclical effect of income on alcohol consumption, while the effects of unemployment are more controversial. At the aggregate level Helble and Sato (2011), using data on fifty-nine countries in the period between 1961 and 2004, analyzed the relationship between per capita change in GDP and per capita variation in alcohol consumption. Overall, the authors found a robust pro-cyclical relationship; taking into consideration income effect, unemployment, and all the other mechanisms illustrated above, the positive effects prevail over the negative when the economy grows, and vice versa.

There is, however, a general consensus on the deleterious effect of economic and social marginalization on consumption habits. Numerous studies in Europe and in the United States have shown how the rates of alcohol dependence are much higher among the homeless and the destitute than in the rest of the population (FEANTSA, 2009; Fazel et al., 2008), although there is some variation between countries due to demographic differences, unequal access to health systems, and sampling quality. Alcoholism rates among the homeless in urban areas stand at 72.7 percent in Munich, 62.9 percent in Los Angeles, 46 percent in Melbourne, and 24.9 percent in Paris while in Brazil over 70 percent of street children consume large quantities of alcohol (ICAP, 2009). In the study of a random sample of 621 Swedish citizens born in 1914 in Malmö, Hanson (1994) found a strong correlation between various indices of social isolation and alcohol consumption/abuse. Although the study failed to identify the causal direction, the author supported the need to promote policies that encourage social inclusion in order to reduce alcohol consumption and protect public health.

## Empirical studies on the effect of price

After this review of the socioeconomic variables, the discussion now moves on to deal briefly with the relationship between prices and consumption. Unlike income, it is normal to expect an inverse relationship between the two variables (negative elasticity). Price elasticity may be lower, higher, or equal to −1. In the first case, the variation in demand is more than proportionate to that in price (of the opposite sign) and demand is said to be elastic, whereas it is inelastic or rigid in the second case and unitary in the third.

The elasticity of demand is important for both a monopolist entrepreneur and the government (see Thornton, 2013, pp. 225–235; Marks, 2015, pp. 143–144; Towse, 2010, pp. 146–147). In fact, with monopoly the demand curve of the firm and that of the market coincide. Therefore, it is essential to know how consumers will react to a price increase to maximize profits. If the demand is inelastic, the firm can increase the price without losing many customers, and the overall impact on revenues will be positive. If the marginal revenue is greater than the marginal cost, the best strategy is to raise the price.

The wine market is, however, very competitive. Consumers can easily find valid alternatives; therefore, wineries are not free to raise their prices whenever they want. The elasticity of demand is, instead, much more relevant for legislators. To implement effective public policies to discourage alcohol abuse and its negative consequences on risky and criminal behaviors, they have to know to what extent consumers will react to tax and price increases. If the demand is inelastic, a larger tax increase is necessary to produce significant results.

The three main factors that affect the price elasticity of demand are

1. the degree of substitutability with other goods (the greater the substitutability, the greater the elasticity);
2. the time horizon considered (demand is generally more inelastic in the short than in the long run since it is more difficult to change and adapt habits to new prices); and
3. the type of product, given that necessity goods have more inelastic demand and luxury goods more elastic.

Fogarty (2010a) shows that only a few studies (far less than 5 percent and close to zero) report positive price elasticity of a good. As in the case of income elasticity, price elasticity is on average higher for spirits (−0.73) and lower for beer (−0.45), with wine in an intermediate position (−0.65). The consumption of beer is less elastic than wine, which in turn is less elastic than spirits.[16] The lower income and price elasticity of beer is attributable to price and is much lower than that of other alcoholic drinks. In other words, faced with a fall in income or an increase in price, it is much more likely that the consumption of the cheaper drinks will remain unchanged (Helble and Sato, 2011). Another possible explanation is that, as a consequence of a decrease in income, part of the consumption of alcoholic beverages (mostly spirits) in bars and restaurants is replaced by wine and beer at home (Ruhm, 1995).

Most scientific studies have focused on estimating the price elasticity of alcoholic beverages at an aggregate level without making a distinction among low, medium, or high levels of consumption. Manning, Blumberg, and Moulton (1995) used 1983 data from the NHIS to check whether a price increase has the same impact on alcohol consumption in all three categories. The results demonstrated that price elasticity is much lower among people with low and high levels of consumption, probably for very opposite reasons. For light drinkers the impact of a price increase is irrelevant in absolute terms, whereas heavy drinkers often develop forms of addiction that make it difficult to reduce consumption. This has important implications since price elasticity can seriously undermine the effectiveness of a policy aimed at reducing alcohol abuse based on an increase in taxation and, consequently, prices.

The consumption of a specific alcoholic beverage, however, is also affected by changes in the prices of other drinks since they can represent valid alternatives to the preferred drink if it becomes too expensive (for a simple constrained maximum utility model with a Cobb-Douglas function and two goods—wine and beer—see box 2.1). If the cross elasticity of a good is positive at a certain price of another good, then the two products are called "substitutes" (when the price of good X increases, the consumption of good Y increases). If, instead, it is negative, they are said to be "complements."

Empirical studies have shown limited substitutability for alcoholic beverages. Clements and Johnson (1983) (using Australian data for the period 1955–1956 to 1976–1977); Nelson (1997) (using quarterly US data from 1974 to 1990); and Angulo, Gil, and Gracia (2001) (using Spanish data on domestic consumption) found positive cross elasticity for wine, beer, and spirits, even if the results of the first study were not always statistically significant. In contrast, Australian data for the period from 1975–1979 to 1988–1989 in Chang and Bettington (2001) indicated that wine and beer are complements while in Ornstein and Levy (1983) and Wohlgenant (2009) the cross elasticities were not statistically different from zero. The evidence, therefore, seems to point toward a rather limited substitutability of alcoholic beverages while price elasticity, as shown above, is generally significant and inverse to the starting price.[17]

### 2.1.4  Environmental Factors

Alcohol consumption is significantly influenced by the environment in which a person grows up and lives. The dominant culture, understood as the social acceptance of consumption and tolerance toward alcohol abuse, influences people's behavior and reflects on the regulations that govern the marketing and supply of alcoholic

**Box 2.1**

A simple model of constrained utility maximization applied to the consumption of alcoholic beverages.

To show the typically expected relationship between the price of alcoholic beverages and consumption, let us imagine that we want to maximize a Cobb-Douglas utility function $U(w,b) = w^\alpha \cdot b^\beta$ in which $\alpha + \beta = 1$ (i.e., $\beta = 1 - \alpha$) and that consumer satisfaction depends on the quantity of wine $w$ and beer $b$ consumed (one could, alternatively, consider the categories of wine and other alcoholic beverages) and on preferences for the first ($\alpha$) and second $(1 - \alpha)$ drink. This must be done within a budget $y$ so that the value of the wine and beer purchased ($p_w \cdot w + p_b \cdot b$), in which $p_w$ and $p_b$ are the prices of the two goods, must not exceed a preestablished amount of money. If it is assumed that this sum is a constant share of income, it would follow that as income increases, the consumption of both drinks will increase according to the preferences expressed for each of them.

In formal terms the problem of maximization looks like this:

$$Max_{w,b} U(w,b),$$

so that:

$$p_w w + p_b b = y,$$

with $w$ and $b \geq 0$, $p_w$ and $p_b > 0$ and $0 < \alpha < 1$. The Lagrangian is given by:

$$L = w^\alpha b^{1-\alpha} - \lambda(p_w w + p_b b - y),$$

with $\lambda$, which is the Lagrange multiplier. First-order conditions are obtained deriving the Lagrangian with respect to the two goods $w$ and $b$:

$$\frac{\partial L}{\partial w} = 0 \Rightarrow \alpha w^{\alpha-1} b^{1-\alpha} = \lambda p_w \Rightarrow \lambda = \frac{\alpha \left(\frac{b}{w}\right)^{1-\alpha}}{p_w} \tag{1}$$

$$\frac{\partial L}{\partial b} = 0 \Rightarrow w^\alpha (1-\alpha) b^{1-\alpha-1} = \lambda p_b \Rightarrow \lambda = \frac{(1-\alpha)\left(\frac{w}{b}\right)^\alpha}{p_b} \tag{2}$$

Then putting the two equations in the system we obtain

$$\frac{\alpha \left(\frac{b}{w}\right)^{1-\alpha}}{p_w} = \frac{(1-\alpha)\left(\frac{w}{b}\right)^\alpha}{p_b}.$$

The equation indicates that equilibrium occurs when the relationship between marginal utility and price is the same for wine and beer. If these two relationships were different, the consumer could increment his utility by increasing the consumption of one good

*(continued)*

**Box 2.1** (continued)

and reducing that of the other, thus achieving the optimum. With some adjustments the following relationship between beer and wine consumption is reached:

$$\frac{b}{w} = \frac{1-\alpha}{\alpha} \frac{p_w}{p_b}.$$

The relationship between beer and wine consumption depends on the preferences of each consumer toward the first $(1-\alpha)$ or the second $(\alpha)$ beverage and their relative prices. If the preference for one or the other product changes in the course of their life, the consumer will modify the share of the consumption of the two goods to maximize their utility. Similarly, the relationship between the consumption of beer and wine is a direct function of the price of wine and an inverse function of the price of beer: if the price of beer increases, more wine and less beer will be consumed, and if the price of wine increases, the opposite will happen. This is valid, however, only under the conditions imposed in the model, such as the utility function of the Cobb-Douglas type; while it does not apply to perfect substitutes and perfect complements, if income were to increase, the consumption of both drinks would increase.

beverages. There are, however, marked differences between countries in this respect. European Mediterranean countries are characterized by high alcohol consumption generally distributed throughout the week and at mealtimes and by a firm condemnation of abuse (Naboum-Grappe, 1995). Nordic countries, on the contrary, are less permissive toward the levels of total consumption, which is discouraged with high taxes and restrictive regulations, while the intake of alcohol takes place mostly at the weekend and outside of mealtimes. This leads to frequent cases of loss of control and excesses which are, nevertheless, socially tolerated (Heath, 1995).

The culture of a country is formed in the course of centuries and is the result of a series of elements such as climate, foreign domination and religion. In most countries, however, women drink less than men even before motherhood, since consumption and above all alcohol abuse are socially less acceptable for women than men (Cottino, 1995).

Finally, alcohol consumption is also influenced by catastrophic events and wars that cause serious psychological damage to the populations involved. About 4–5 percent of survivors of natural disasters tend to develop post-traumatic stress disorder (PTSD), as for example in the case of Hurricane Katrina in New Orleans where there was a very high incidence of PTSD among survivors (Coker et al., 2006). Vetter et al. (2008) studied the impact of the 2004 tsunami in Asia on Swiss tourists who survived the disaster and found an increase in symptoms of depression and the use of drugs, cannabis, and alcohol.

Vlahov et al. (2002), Stein et al. (2004), and Schiff (2006) all focused on terrorist attacks. Vlahov et al. (2002) analyzed data on telephone interviews made with New Yorkers five to eight weeks after the attack on the World Trade Center. Of the 988 people considered, 9.7 percent reported an increase in cigarette consumption, 24.6 percent reported an increase in alcohol consumption, and 3.2 percent in marijuana. Similar results were found by Stein et al. (2004) in a sample of Americans residing outside the city of New York and showed how anxiety spread simply through a passive involvement with the media during the dramatic events. Schiff (2006) examined the effect of prolonged exposure to terrorism in six hundred adolescent Jews, religious or otherwise, living in Jerusalem. Exposure to the risk of attack—particularly intense between September 2000 and August 2005 with 889 episodes, 1,064 Israeli victims, and 7,441 wounded—caused an increase in post-traumatic stress symptoms, depression, and alcohol consumption, though they were restrained by their religious faith.

## 2.2 The Quality of Wine

The previous section discussed the main micro- and macroeconomic variables that influence the choice and level of alcohol consumption. When a decision has been made, the individual generally sets an amount to spend and looks for the best quality based on the information in their possession. Therefore, consumers position themselves in a certain segment of the market at the time of purchase depending on how much they want to spend. Product differentiation can be horizontal or vertical. In the first case we refer to the type (white, red, rosé, sparkling wine, liqueur, etc.) while in the second case we refer to the quality of the drink. Two types of viticulture and enology have always coexisted: an ordinary level with wines for the masses ("jug" or "basic wine") and a top-quality level with fine wines ("premium wine"), historically intended for nobles and clergymen and then later intended for the upper middle class (Mariani, Boccia, and Napoletano, 2006).

In the past, the production of top-quality wines was very limited, but now it has grown together with purchasing power so that a classification based on a mere dichotomous distinction of basic premium wines is no longer sufficient. Over the years various classifications based on the price have been made that implicitly assumes a strong positive correlation between the price and quality of the product. Rabobank, for example, considers "basic" wines to be priced at $5; "premium" wines range from $5 to $7.99; "super premium" wines range from $8 to $13.99; "ultra-premium" wines from $14 to $49.99; and "icon" wines are over $50 (see Schirmer, 2012).[18]

Some wineries specialize in just one segment—for example, mass wines or those of the highest level—but more and more companies differentiate production both horizontally and vertically to attract new buyers, to diversify risk, and (partially) to help guard against the consequences of changes in tastes and consumption. Although the current exploitation of vineyards has shortened their duration (AIS, 2005a, p. 12), the life cycle of a vine can last seventy years. Given the amount of investment needed to plant a vineyard, having to reconvert the entire production halfway or one-third of the way through its life cycle can jeopardize the financial stability of a company.

Although quality and price are positively correlated, investing in quality is not necessarily profitable. As reported by Rust, Zahorik, and Keiningham (1995),

1. quality is an investment;
2. economic efforts to pursue quality need to be calculated;
3. there is a risk of making too many investments in quality; and
4. not all investments in quality are the same.

Before deciding whether to aim for excellence, a company should evaluate carefully the costs and benefits of this type of investment and calculate the net present value. Unfortunately, apart from Castriota (2018), there is little or no scientific literature on the profitability of quality investments in the wine sector (see chapter 3). In any case, since quality is one of the key elements in the selection of wines, it is important to understand the variables that influence it. Studies on the subject mainly rely on the opinions of experts in the wine guides of famous tasters such as Robert Parker, Hugh Johnson, and Luigi Veronelli. Further, studies on the determinants of wine quality have to solve a number of methodological issues that are discussed in box 2.2.

The variables affecting wine quality can be grouped into three main categories: (1) the terroir, (2) agronomic and wine-making techniques, and (3) company characteristics.

### 2.2.1   Terroir

In the Old World the deep-rooted conviction that "quality was linked to *terroir*, the almost mystical combination of soil, aspect, microclimate, rainfall and cultivation that the French passionately believed gave the wine from each region—and indeed, each vineyard—its unique character" (Bartlett, 2009) still holds true.[19]

As any wine maker knows, soil plays a fundamental role because its chemical composition influences the sensorial characteristics of wine. The same vine planted in two different parts of the world can, in fact, give completely different results even with the same climate and production techniques. The influence of the composition and chemical-physical-microbiological structure of the soil as well as the genetics of the rootstock have been widely studied by agronomists and wine makers.[20]

Another fundamental element of the terroir is the climate, which can influence the quality of wine temporarily or structurally. In fact, some areas have a very stable climate while others generally guarantee the production of quality wines even though uncertain atmospheric conditions, especially during the harvest period, mean the

**Box 2.2**
Methodological issues in empirical studies on wine quality.

When reviewing the literature on the determinants of wine quality six points need to be clarified. First of all, the samples that are analyzed in the wine guides are not representative of the real world of wine: the ordinary wines sold in supermarkets and bulk wines tend to be excluded. There is, therefore, a strong imbalance in favor of medium-high range products purchased mainly in the Horeca. However, there is no reason, except for a few exceptions, why the variables influencing the superior quality wines should not influence also those in the lower ranges in a similar way.

A second point concerns the reliability of the ratings given in guides whose aim is to reduce information asymmetries (Cardebat, 2017, pp. 53–55). Hodgson (2009) analyzed over four thousand wines entered in thirteen US wine competitions and showed that 84 percent of wines that received an award in one competition were not awarded in another. Morrot, Brochet, and Dubordieu (2001) made an experiment with fifty-four experts; they tasted a real red and a real white wine. A few days later they tasted the same wines again, but some of the white wines were colored red with a neutral-tasting food colorant. Results show that the experts described the red wines in a similar way even though some red wines were actually white. Opinions can also be influenced by personal tastes and preferences. Using data from the Guida dei Vini di Veronelli from 2004 to 2009, Castriota, Delmastro, and Curzi (2013) showed that the evaluators' opinions are conditioned by two types of subjective distortions—namely, generosity and preferences for some characteristics of the product.[a] Nevertheless, sommeliers' evaluations follow the rules that are well established at an international level and in time have proven their validity. Castriota and Delmastro (2012), for example, studied what determines the reputation of Italian wineries by using the ratings awarded by Hugh Johnson's international guide and L'Espresso's national guide and found the results were extremely similar since there was a very strong correlation (0.62) between the two values.

Third, as written in bold by Robert Parker in his website when explaining his rating system: "Scores, however, do not reveal the important facts about a wine. The written commentary that accompanies the ratings is a better source of information regarding the wine's style and personality, its relative quality vis-à-vis its peers, and its value and aging potential than any score could ever indicate" (Parker, n.d.). The scientific literature on the usefulness of tasting notes is reviewed by Storchmann (2012, p. 25) who showed that the ability among nonexperts to identify wines after reading the notes is random.

Fourth, the methodology of tastings can affect the results. As pointed out by Colman (2008, pp. 120–121) with respect to Parker's ratings, tasting tens or even hundreds of wines within a few hours favors big, concentrated wines which can "shout louder" than their rivals. These types of wine—sometimes named "fruit bombs"—perform well when tasted in isolation but are not necessarily the best choice for a dinner. Further, some observers are skeptic about Robert Parker's ability to evaluate thousands of wines each year and wonder whether the 0–100 scores he assigns are expressed in relative (within vintage and region) or absolute terms. Next, experiments have shown that the sound and lighting conditions during wine tastings can strongly influence the description and the scores awarded (Goode, 2016, p. 66).

*(continued)*

**Box 2.2** (continued)

Fifth, experts do not necessarily experience wine like untrained drinkers. Castriota-Scanderbeg et al. (2005) made an experiment with seven professional sommeliers and seven untrained people matched by gender and age. They made them taste three wines and a glucose solution as a control while having their brain scanned with neuroimaging technologies. Results show that during the tasting different areas of the brain of sommeliers and of nonexperts are activated. This study suggests that learning changes the way of thinking and of tasting wine.

A last point concerns the use of sensory evaluations (e.g., persistence of taste or level of acidity) as determinants of quality ratings in econometric regressions: the relationship between the tasters' ratings and sensory variables is tautological (Delmastro, 2007) and of no interest from an economic point of view. For this reason, the discussion in section 2.2 focuses mainly on inherent variables such as (1) the terroir, (2) agronomic and wine-making techniques, and (3) company characteristics.[b]

*Notes*: [a]The wines were reviewed by two tasters, the first of whom systematically awarded—all other things being equal—more generous votes than the second taster, showing a preference for red wines, sweet wines, and those from the northeast of Italy. This is particularly important because the random assignment to one or another taster can influence the rating given to the wine in a statistically significant way, with consequences for sales and the price that can be applied.

[b]For an econometric analysis of the ratings on the quality of wine that includes sensory variables, see Combris, Lecocq, and Visser (1997).

results can vary from one year to the next. Yet other areas are not very suitable and have no hope of excelling. Corsi and Ashenfelter (2000) used data on a sample of Barolo and Barbaresco wines and concluded that only rainfall in the months of August and September exerts a significant (and negative) influence on the quality while other variables are irrelevant. Grifoni et al. (2006), using data on six prestigious appellations of central and northern Italy, showed that the best wines are obtained in the years with scarce rainfall and high temperatures. Jones and Storchmann (2001) reached the same results with data on twenty-one French Crus Classés châteaux. Even within the same area, there are plots of land with exposures and altitudes that guarantee optimal sun, temperature, humidity, and ventilation.

Gergaud and Ginsburgh (2008) showed the role played by exposure with French data: land facing the south and east is protected from winds that come from the west and is more exposed to the sun that dries the grapes, reducing the risk of mold. Given the importance of the sun in determining the quality of wine and considering the sudden climatic changes of the last decades, some cold and rainy regions in northern Europe could benefit considerably from a rise in temperature and a reduction in rainfall. Ashenfelter and Storchmann (2010) applied a hedonic model of sunlight to the Moselle area in Germany and came to the conclusion that a rise of three degrees in the temperature would double wine prices and an increase of one degree would lead to an increase of 20 percent.[21]

### 2.2.2 Agronomic and Wine-Making Techniques

Investments in vineyards and cellars are expensive, but they generate substantial improvements in the quality of the wine produced. There is a long list of potentially important techniques and innovations, but literature has reviewed only a part of them. In a study made with Italian data from the Istituto Agrario di San Michele all'Adige (now known as the Edmund Mach Foundation), Zago (2009) demonstrated the relevance of agronomic variables, such as the number of vines per hectare, the number of bunches per vine, the depth of the roots, and yield per hectare on Chardonnay and Merlot wines produced from 1994 to 1996. In particular, there emerged a trade-off between quantities and quality: as the wine yield per hectare increased, the wine quality decreased unrelentingly.

Wine-making techniques are equally important. Delmastro (2007), using Italian data from Piedmont, proved the contribution of variables such as the length of the aging process, the use of barrels or barriques (225-liter),[22] an increase in the alcohol content ("superior" wine), and an increase in production standards ("reserve" wine). Alston et al. (2015) analyzed the alcohol content of more than one hundred thousand wines from eleven countries from 1992 to 2009 and showed that it has increased from 12.7 to 13.7 percent. However, the cause of this rise in wine alcohol was not global warming but rather the rational decision of wine makers to use over-ripe grapes to produce more mature, flavored, and intense wines (Thornton, 2013, pp. 120–121). Investments in technology, when they do not distort the product,[23] can be expected to lead to qualitative improvements; what remains to be seen is whether the costs exceed the revenues. In other words, it is not clear whether the companies that produce quality wines, sustaining large investments in the long run, are more or less profitable than the average company.

But how important is the terroir and how important is technology? Gergaud and Ginsburgh (2008) analyzed the determinants of the quality of Bordeaux wines using data for the period 1980–1992. Quality is strongly influenced by both the terroir (e.g., the chemical and organic characteristics of the soil and the exposure of vineyards) and technology (from the initial choice of the vine to the final bottling stage).[24] The results of the study showed, in fact, that technology matters far more than terroir, whose role is negligible. In Ginsburgh, Monzak, and Monzak (2013), using French data from the Médoc, once again the importance of technology proved to be far superior to land.[25]

These studies consider relatively small areas, but if the results were to be confirmed by new surveys with data on other countries and years, this would strengthen the conviction of many operators (including the famous wine maker Michel Rolland) that good wine can be produced anywhere in the world; all that would be necessary are the right techniques.[26] In turn, this leads to important implications for industrial policy: if the terroir matters little and adopting technology is expensive, then the Old World is at a disadvantage compared with new competitors. European Community regulations, in fact, strongly limit or even prohibit the use of many innovations while the small size of companies often prevents big investments.

Lastly, any discussion of the role played by wine-making techniques has to consider the wine classification system, which will be discussed in more detail in chapter 6. In Italy, like in the other EU countries, there are five hierarchically ordered levels: at the bottom we find the ex–table wines (Vino da Tavola, or VdT); then the varietal wines, the typical geographic indication (Indicazione Geografica Tipica, or IGT); Controlled Designation of Origin (Denominazione di Origine Controllata, or DOC); and finally, the Controlled and Guaranteed Designation of Origin (Denominazione di Origine Controllata e Garantita, or DOCG). The production of IGT, DOC, and DOCG wines is possible only within clearly defined geographical borders and requires compliance with a set of rules established by law concerning production techniques, such as the vines to be used, the maximum yield per hectare, and minimum aging. This is to guarantee the particularity of the wine—linked to the tradition and characteristics of the terroir—and a minimum level of quality. The closer to the top of the pyramid (DOCG), the more stringent the rules become and the higher the quality expected.

A positive correlation between DOCG and wine quality was found by Delmastro (2007), although there is a certain variability in the ratings given to wines that have different appellations but are of the same level (Barolo and Barbaresco stand out among the Piedmontese DOCG) or that belong to the same designation (with some particularly distinguished and historic vineyards—for example, crus). Corrado and Oderici (2008), using 1997–2006 data from the Guida dei Vini di Veronelli, documented that a DOC or DOCG designation has become progressively less important for determining the quality of the wine while the role of some wine-making practices has increased. This is explained partly by the fact that the sample analyzed by a guide is, as mentioned previously, biased in favor of high-end wines. Many companies that aim for excellence, in fact, have started to produce table wines or IGT to avoid the strict rules of the law and to experiment with new vines and agronomic and wine-making techniques. The companies that can afford to produce successful and expensive wines theoretically classified as "low end" are those that over the years have built a solid business reputation and therefore do not have to resort to the collective brand.

It is a completely different story for table wines sold through large-scale retail trade channels. In this case it is not a question of excellence but of cheap wines. Therefore, there is a stronger correlation between quality and belonging to a particular quality segment in wines purchased in large-scale retail trade channels than in the hotel and catering industry (or Horeca in Europe, as in hotel, restaurant, and catering). In other words, table wines and IGTs sold in wine shops are of a much higher quality than products sold in supermarkets, and sometimes they can reach very high levels.

### 2.2.3  Business Characteristics

This group of variables includes the age, size, and ownership structure of wineries. The age of the company can have a positive effect on quality if it is related to the age of the vines because older vines—those over thirty or forty years old—decrease in productivity and increase in quality. In addition, agronomists and wine makers experiment with techniques and products, and they learn from their mistakes ("learning by doing").

Company size, measured by the number of bottles produced or hectares owned or cultivated, can exert a positive effect on the average quality of wine, given that large companies have greater financial resources and can adopt large-scale technological innovations to cut costs or increase quality.[27] However, the opposite is also true: there can be a negative relationship between company size and average quality because of the growing difficulty of placing high-end products on the market at prices that are beyond the reach of most people. In other words, it is one thing to be able to place 250,000 bottles of precious wine per year on the market and another to sell 250 million bottles. Therefore, while a small cellar can focus exclusively on excellence, large companies must necessarily diversify by focusing largely, if not totally, on consumer goods. The empirical result depends on the sample considered; it can change radically depending on if all kinds of companies are reviewed by guides or if only the high-end ones are reviewed. Unfortunately, the existing studies consider mainly or exclusively this last segment of the market and do not come to conclusive results. Frick (2004) found company size had a positive effect on the average quality of wine while Delmastro (2007) did not record any statistically significant effect.

As for the ownership structure, companies can be privately owned or state-owned companies, cooperatives, or foundations or be part of a group of companies or conglomerate. In private companies the control of the entire production chain and, therefore, of quality lies in the hands of the owner, who can pursue a certain quality according to the chosen market segment and degree of vertical integration. In fact, wine firms can produce (1) both the grapes and the wine, (2) only the wine, with grapes purchased from suppliers, or (3) neither of them, in which case these "bottlers" sell the wine produced by other wineries with their own label. Usually, the higher the degree of vertical integration, the higher the quality (see chapter 4). As mentioned above, some wineries belong to conglomerates with multibillion revenues and diversified portfolios of products ranging from food to alcoholic beverages and so on. Belonging to a conglomerate can have a negative impact on quality, especially if wine is the core business, in which case selling millions of bottles to the mass market requires competitive prices.[28]

An important aspect that can affect the quality of wine is the separation of owner-ship and management since the owner has every interest in managing the company as best as they can, but they do not necessarily have the ability (or, at least, may not be the most suitable person to do so). In state-owned companies the control of the production chain is in the hands of managers, but often there are no incentives to pursue results in qualitative and quantitative terms as in private companies.

Cooperatives only have full control of production in the winery during vinif-ication since the decision about techniques and machinery for wine making is a prerogative of the management. However, the actual production of grapes—the upstream phase—is decentralized and delegated to individual landowners who have to comply with a set of instructions and rules issued by the cooperatives. There is a strong incentive here to behave in an opportunistic way since the cost of growing the grapes falls entirely on the individual while the gains from a superior quality of grape are divided equally between all the members, regardless of the merits of each one (Pennerstorfer and Weiss, 2013).

It is reasonable to assume that this incentive becomes stronger as the number of mem-bers increases, and this also makes peer control more difficult. In the absence of effective control mechanisms or economic incentives, the cooperative becomes a place to unload the grapes produced at the lowest possible cost. The situation is even worse if the coop-eratives attract grape producers who are less motivated or have poor land so that they cannot produce good wine on their own or sell their grapes to a private winery.

Cooperatives, on the other hand, are nonprofit companies. When members deliver their grapes to the winery, they receive not only payment for the raw material but also any profit from the production and sale of the wine, which is then distributed proportionally among members. Well-managed cooperatives that make a capital gain and adopt a differentiated payment system according to the quality of the raw material can attract small owners with the best land and contrast or even reverse unfavorable selection.

Foundations are legal entities created through irrevocable donations; they do not have an owner and are managed by a committee that establishes the internal regu-lations in accordance with the goals set by the donor. Foundations are nonprofit organizations that may define social objectives in their statute and are exempt from the strict controls to which private companies are subject. The effect of this type of company on the quality of wine could be negative as managers are not rewarded on the basis of business results or positive as they can have a longer time horizon and greater patience in investment strategies. Lastly, companies that are part of a group can exploit economies of scale and adopt technologies, distribution, and marketing.

From an empirical point of view Frick (2004), using panel data on over 3,200 bottles produced by 305 German wineries in the period 1996–1999, demonstrated how com-panies run by external managers produce better wines. The opposite, however, is true for private companies managed by their owners, state-owned companies, cooperatives, and foundations. Delmastro (2007), with data on 2,046 wines from 414 wineries in Piedmont, confirmed the results of Frick (2004) for private companies and cooperatives. It demonstrated that the average quality of wines increases when the enterprise is administered by the owner who carries out the function of wine maker. When the management, therefore, is entrusted to an expert, be this an external manager or the wine-maker owner, the company benefits, but the opposite happens when it is administered by the owner who does not have specific skills. This occurs frequently in some countries in the Old World where the small size of companies and the family corporate culture perpetuate the "do-it-yourself" approach. Similar conclusions were reached in a qualitative analysis by Mediobanca (2014, table 17); this analysis showed that private companies tended toward "great wines" (with a price above €25) and wines with an appellation. The same study (table 18) also revealed a greater propensity for these companies to sell directly and through Horeca (hotel, restaurant, and catering industries) and wine shops while cooperatives rely heavily on large retailers.

The question of the supply of grapes produced by third parties or on rented land also deserves consideration. Many wineries buy a certain percentage of raw material from external producers and therefore do not have any control over the growing of the grapes. In this case, however, the temptation of the farmer to adopt opportunistic behavior could be neutralized by incentive mechanisms that are more effective than those in cooperatives. Removing a member from a cooperative can prove to be much more complicated than changing a supplier.

Relations between grape producers and wineries can—but do not necessarily have to—be formalized with contracts, which generally provide for a series of instruments to guarantee a minimum level of quality, such as the definition and monitoring of agronomic practices and a system of rewards and penalties that vary according to the characteristics of the goods produced.[29] Goodhue et al. (2003) analyzed the type of sales agreements used by Californian companies distinguishing between written contracts, verbal agreements, and the absence of any formal contract. The study showed that larger companies and companies that produce premium grapes are more likely to protect themselves through the signing of formal contracts. Given the high perishability of the raw material, in fact, wine makers can try to increase their own profit margin to the detriment of agricultural firms by forcing them to accept significant price reductions given the short time in which they can sell their goods.

Zylbergsztain and Miele (2005) analyzed 139 Brazilian farmers and found a greater stability in contracts among producers of quality grapes, given the larger potential damage from a missed or delayed sale of the raw material. Indeed, large companies find it more economical to sign formal contracts. As transaction costs are generally fixed, they decrease in unitary terms as production and the price of grapes rise. Long-term relationships can increase trust between parties, making the use of formal contracts superfluous, and they are often replaced by oral agreements (Allen and Lueck, 2002).[30] The empirical evidence from Californian companies provided by Goodhue et al. (2003), however, contrasts with this principle that is well established in agricultural economics. Mutual knowledge and trust built in the course of time do not seem to replace written contracts.

Fernández-Olmos, Rosell-Martínez, and Espitia-Escuer (2009) analyzed the factors that influence the decision of Rioja wineries to either produce their own grapes or buy them from external producers and found a positive correlation between wine quality and vertical integration. These results confirm the importance of control over the production chain. The size of the company, however, is negatively correlated with vertical integration. Given the very high cost of land in the most famous and prestigious wine-growing areas, the investment required to buy tens, hundreds, or thousands of acres often becomes unsustainable. Further, it is difficult to find land for sale in the most prized areas, a problem that is exacerbated in Europe by EU regulations on appellations and on planting rights that hinder the growth of companies even where there are still plots available.

The question of renting land is more complex. In this case the winery directly controls the cultivation of the vine, but if the lease contract does not have an adequate time horizon, it may be discouraged from making agronomic investments. Unfortunately, there is not sufficient documentation on this last point. Malorgio, Hertzberg, and Grazia (2008) did demonstrate with Italian data that wineries that do not own land produce a much greater share of table wines than those who do (68 percent and 19 percent respectively) while the share of cooperatives is in an intermediate position. This study, therefore, clearly proves the importance of vertical integration.

## 2.3   The Price of Wine

After reviewing the determinants of the quality of wine we will now look at how prices are established. Empirical models, commonly called "hedonic regressions," link the price of an asset (in the case of wine that is usually a 75 centiliter [a little over 25 ounces] bottle) with a number of characteristics of the product, the production structure, and the market. Hedonic models were first proposed by Court (1939) and later perfected by Griliches (1961) and Rosen (1974). The methodological issues arising in studies on the determinants of wine price are discussed in box 2.3.

Empirical studies[31] have shown that wine prices are determined essentially through five channels: quality, consumer preferences, production costs, scarcity, and reputation (see table 2.1). The factors that affect the price of wine are

- product quality,
- type of wine,
- vine,
- vintage,

**Box 2.3**
Methodological issues in empirical studies on wine price.

It is of utmost importance to specify the model used to study the determinants of wine prices. The most commonly adopted are the linear, log-linear, log-log, and the Box-Cox (1964) transformation.

In the first, all the dependent and independent variables are used without being transformed, which means they can be easily interpreted from an economic point of view. Sometimes, however, the relationship between dependent and independent variables is nonlinear. In this case, if the variables are transformed appropriately, it is possible to return to the linear relations. The model can be estimated with the ordinary least squares method, though the interpretation of the coefficients changes from one model to another. The variables can be modified in many ways, but the most interesting for the wine sector are those that use logarithmic and Box-Cox transformations. In the log-linear model, only the dependent variable is transformed into a logarithm; in the lin-logarithmic model, only the independent variables get converted; and in the log-log model both dependent and independent variables are modified. With the Box-Cox methodology, a variable is transformed by means of an iterative procedure to normalize the original data:[a] the new variable becomes $X_\lambda = (X^\lambda - 1)/\lambda$. In the wine sector the log-linear model is the most frequently used, even if there are applications that make use of the Box-Cox transformation (see the study by Nerlove, 1995, with data on Swedish consumers). The choice of the functional form, therefore, is fundamental because it can lead to biased conclusions if incorrect.

The choice of the sample is equally important. Costanigro, McCluskey, and Mittelhammer (2007)—using data for 13,024 wines from California and the state of Washington reviewed by the Wine Spectator guide between 1991 and 2000—showed how wine is a strongly differentiated product even within the white and red categories. Hedonic model estimations by product categories and price ranges strongly improve the ability to explain data variability and produce more accurate results. On the other hand, it is wrong to hypothesize that a certain variable has the same effect on the price of the goods considered as a whole.

Another relevant methodological aspect concerns the inclusion or otherwise of quality in the set of regressors. Indeed, some studies consider price as a proxy for wine quality (see, for example, Ginsburgh et al., 2013), arguing that the two variables are strongly correlated. In the long run a good cannot be systematically priced more than it is worth. Prices and quality are not, however, the same thing. In the first place, although many consumers are influenced by the ratings expressed in wine guides, their own opinions may differ greatly because of their different knowledge and experience. Using 6,175 observations from seventeen blind tastings organized in the United States between 2007 and 2008, Goldstein et al. (2008) found that only experts prefer the most expensive wines while the correlation between price and quality is fairly negative for nonexperts. Secondly, several studies have shown that quality is only one of the various elements affecting the determination of wine prices and that there are many others that explain a relevant part of the variance of regression. In some studies, the quality was even of little or no importance at all. Combris, Lecocq, and Visser (1997), for example, studied the price determinants of a sample of Bordeaux wines and found that sensory characteristics influence the quality

*(continued)*

**Box 2.3** (continued)

expressed by judges but not the price, which instead reflects the objective characteristics shown on the label (e.g., type of wine, vintage, classification system, designation, etc.).

The correlation between sensory quality and price is not, therefore, necessarily strong. Various elements, such as production costs, past quality/reputation, and marketing campaigns, can all influence the willingness of buyers to pay (demand side) and the prices charged by producers (supply side). It is therefore appropriate to analyze the price determinants net of quality. While the identification of the variables that affect the quality reflected in the opinions of experts and consumers is more relevant for agronomists and wine makers, what is most interesting from an economic point of view are the variables that increase or decrease the price of products of equal quality.

Methodological aspects are, therefore, of fundamental importance because they strongly influence the results of the econometric investigations.

*Note*: [a]The Ramsey Regression Equation Specification Error Test (RESET) test, where the null hypothesis is that the best specification uses all the variables expressed (see Oczkowski, 2001), can be used to choose the most suitable functional form from the various possible transformations.

**Table 2.1**
Determinants of wine price.

| Variable | Channel | | | | |
| --- | --- | --- | --- | --- | --- |
| | Quality | Consumer preferences | Production costs | Scarcity | Reputation |
| Quality of product | X | | X | X | X |
| Type of wine | X | X | X | | |
| Vine | X | X | X | | |
| Year | X | | | X | X |
| Aging | X | | X | | |
| Aging potential | X | | X | X | |
| Technology | X | | X | | |
| Famous external oenologist | X | | | | X |
| Firm reputation | X | | X | | X |
| Collective reputation (belonging to an appellation) | X | | X | | X |
| Institutional reputation (public classification system) | X | | X | | X |
| Biological/biodynamic production | X | X | X | X | |
| Firm size | X | | X | | |

- aging,
- aging potential,
- technology,
- the hiring of a famous external oenologist,
- expectations about quality,
- belonging to an appellation/geographic area,
- official classification system,
- organic/biodynamic production, and
- firm size.[32]

Product quality has been widely discussed and, apart from the findings of the study by Combris, Lecocq, and Visser (1997), has been shown to influence both the willingness of consumers to pay (see Bombrun and Sumner, 2003; San Martín, Brümmer, and Troncoso, 2008; Crozet, Head, and Mayer, 2012) and production costs. Paroissien and Visser (2018) showed that producers of medaled wines can increase their price by 13 percent. Further, as quality improves, wine production falls, which leads to rationing that, in turn, raises the price. The production of quality wines is the first step toward building a solid business reputation.

The type of wine (white, red, sparkling wine, etc.) and grapes (Pinot Noir, Sangiovese, etc.) reflect the quality, the preferences of the consumers, and production costs (see Bombrun and Sumner, 2003; Costanigro, McCluskey, and Mittelhammer, 2007; San Martín, Brümmer, and Troncoso, 2008). Red wines, sparkling wines (above all the metodo Classico or Champenoise method), and straw wines (passiti) cost more than the others on average. The first two, in fact, require more complex technologies and aging while straw wines have a lower yield. All these factors affect production costs.

Consumer preferences for product types can also differ from one country to another, depending on the climate, and they may change in time according to trends. The same holds for vines, which have different yields and production costs and can fall in or out of favor with consumers. Cuellar, Karnowsky, and Acosta (2009), for example, examined the effect of the film *Sideways* on the American consumption of wine. The film, shown in theaters from October 2004 to May 2005, received five Academy Awards and grossed $100 million at the box office, with $70 million of that in the United States alone. In a memorable scene from the film the protagonist belittles and refuses to drink merlot but exalts Pinot Noir. The authors of the study showed that the film had positive effects on the sales of Pinot Noir, whereas sales of merlot had slowed down, though not as much as expected.

Vintage influences quality since the best years are produced after summers with little rainfall (Jones and Storchmann, 2001; Ashenfelter, 2008). Moreover, low rainfall negatively affects the quantities produced so that scarcity rations the supply and increases the average production cost, which, in turn, drives prices upward. Finally, some vintages enjoy a "reputational reward" that increases the willingness of consumers to pay beyond what is justified by the volume and quality levels achieved (Oczkowski, 1994; Combris, Lecocq, and Visser, 1997; Costanigro, McCluskey, and Mittelhammer, 2007).

It is well known that aging improves the sensory characteristics of wine, especially red and some sparkling wine, but it is expensive (purchase and maintenance of barrels, storage of wine, deferred gains), so corporate decisions must be based on careful cost-benefit analysis. Dimson, Rousseau, and Spaenjers (2015) found that the aging of young fine wines has a positive effect on the return of the financial investment. As noticed by Cardebat (2017, pp. 12–13), we have to distinguish the enological from the rarity value of aging. In fact, aging increases the quality of certain types of wine up to a certain number of years, but then it decreases while over time the number of available bottles decreases. Therefore, in the first years the higher value from aging is due to better quality, but after the peak it is due to rarity.

Aging *potential* is another element that can positively affect the price of wine (Jones and Storchmann, 2001), but it involves just the niche of high-end products that lend themselves to long aging. The use of technology in a broad sense, such as the use of wooden barrels mentioned above, includes all those agronomic and enological techniques that affect not only quality but also costs of production. Ginsburgh et al. (2013) found that technology and climate change can explain more than two-thirds of price variance. Once again, companies need to make a careful evaluation of the real benefits of expensive investments, bearing in mind the market and segment of interest.

It is difficult and expensive to build up a company's reputation, which may be understood as the buyers' expectations about the average quality of the current product based on the quality provided in the past. It requires significant investments both in production and marketing policies, but in time it can repay the expense and efforts of producers. A study by Ali and Nauges (2007) about Bordeaux wines showed that individual reputation influences the prices of en primeur wines more than variations in the short-term ratings expressed by critics.[33] To increase their reputation and charge higher prices, some firms hire famous external oenologists as consultants. Using data on Californian red and white wines, Roberts, Khaire, and Rider (2011) compared prices before and after hiring the new oenologist. Since the old wine depended on the previous oenologist, the positive price difference found is due solely to the reputation of the new oenologist.

Membership of an appellation (e.g., Aglianico del Vulture) or a certain segment of the classifications established by authorities (e.g., DOCG) influences quality since the minimum standards (e.g., maximum yield per hectare, alcohol content, etc.) differ significantly, which will as a consequence affect the willingness of buyers to pay.[34] Numerous studies have found positive effects for appellations (Oczkowski, 1994; Combris, Lecocq, and Visser, 1997; Costanigro, McCluskey, and Mittelhammer, 2007; San

Martín, Brümmer, and Troncoso, 2008; Cross, Plantinga, and Stavins, 2011) and for official classifications (Combris, Lecocq, and Visser, 1997; Corsi and Strøm, 2013).[35]

The production of organic or biodynamic wines[36] increases the willingness to pay of those consumers who are more sensitive to quality issues (Barber, Taylor, and Strick, 2009), intended above all as product wholesomeness and respect for the environment (Mollá-Bauza et al., 2005).[37] Positive price differentials in favor of organic wines were discovered by Corsi and Strøm (2013) with questionnaires administered to 171 Piedmont wineries. Schmit, Rickard, and Taber (2013) measured the willingness to pay for wines produced with environmentally friendly techniques by using experiments. The authors confirmed the idea that promoting these techniques leads to an increase in demand and consequently in price increases but only if the sensory characteristics meet the expectations of consumers.

Kallas, Serra, and Gil (2010) analyzed what determines the decision to adopt organic techniques of production by working with data on a sample of Catalan wineries. Older producers, those who are mainly driven by short-term economic motivations and those that run large companies, are all less likely to adopt organic practices. As highlighted by Vastola and Tanyeri-Abur (2009), the price strategies of organic producers must take into account the fact that these practices involve lower soil yields and high labor and certification costs that increase total unit costs by at least 30 percent. Again, costs and benefits must be weighed up, bearing in mind that the return on investment is realized in the long run by building a reputation as a fair and sustainable producer.

Finally, company size affects the ability to make large-scale investments to cut average production costs, which will have some repercussions on consumer prices (Oczkowski, 1994; Corsi and Strøm, 2013). This price containment effect is amplified if larger companies place large quantities of wine in the medium-low range of the market.

Companies have an interest in communicating all the characteristics of the wine that reflect its quality to the consumer. Reputation and signals are particularly relevant because wine quality is discovered only at the time of consumption. A buyer can rely on an abundance of information to find the best product, but it is expensive to obtain and process. This is why consumers rely on signals such as price, label information, and expert opinions. A good part of economic theory considers price to reflect the market structure, but price can also be used as a marketing tool as, for example, when it influences consumers' perceptions of quality. This has been shown in experiments where people were told the retail prices of the wines they were going to taste (Goode, 2016, pp. 79–80). Participants tasted the same wines more than once but, even if the wine was the same, the price displayed was changed. Results show that there is a strong correlation between the declared price and the subjective evaluation. In their evaluation process people are influenced by the price which is perceived as a signal of quality.

In a study on the use of signals by over six thousand European consumers, Gergaud and Livat (2007) showed that price is used as a quality signal, especially by nonexperts.

**Table 2.A.1**
Quality determinants of wine, Veronelli Guide 2004–2009.

| Variable | Description | (1) | (2) | (3) | (4) |
|---|---|---|---|---|---|
| New entrant | DV = 1 if the wine has been reviewed for the first time | -0.889*** (0.0247) | -0.788*** (0.0402) | -0.753*** (0.0404) | -0.755*** (0.0404) |
| Red | DV = 1 if the wine is red | 0.837*** (0.0223) | 0.396*** (0.037) | 0.444*** (0.0376) | 0.378*** (0.0416) |
| Rosé | DV = 1 if the wine is rosé | -0.523*** (0.0717) | -0.337*** (0.0988) | -0.305*** (0.101) | -0.322*** (0.113) |
| Sweet | DV = 1 if the wine is sweet | 0.406*** (0.0491) | 0.370*** (0.0753) | 0.462*** (0.0749) | 0.217*** (0.0809) |
| Age | Age of wine in years | 3.228*** (0.094) | 3.667*** (0.146) | 3.545*** (0.147) | 3.545*** (0.147) |
| (ln) Bottles | Ln of number of bottles produced by the firm | -0.264*** (0.00952) | -0.125*** (0.0152) | -0.139*** (0.0154) | -0.144*** (0.0155) |
| (ln) Hectares | Ln of number of hectares in vineyard | 0.169*** (0.00968) | 0.124*** (0.0153) | 0.131*** (0.0154) | 0.134*** (0.0155) |
| Cooperative | DV = 1 if the wine is produced by a cooperative | 0.291 (0.198) | 0.0228 (0.27) | -0.0222 (0.269) | -0.0561 (0.269) |
| (ln) Hectares * Cooperative | Slope DV | -0.137*** (0.0341) | -0.101** (0.0464) | -0.0915** (0.0465) | -0.0863* (0.0463) |
| Barrels | DV = 1 if the wine is aged in barrels | | 0.529*** (0.0385) | 0.553*** (0.0391) | 0.556*** (0.0392) |

| | | (1) | (2) | (3) | (4) |
|---|---|---|---|---|---|
| Barriques | DV = 1 if the wine is aged in barriques | | 1.388*** | 1.369*** | 1.373*** |
| | | | (0.0361) | (0.0363) | (0.0364) |
| IGT | DV = 1 if the wine is IGT | 0.138* | 0.0782 | 0.0788 | 0.0914 |
| | | (0.0714) | (0.117) | (0.118) | (0.119) |
| DOC | DV = 1 if the wine is DOC | −0.349*** | −0.216* | −0.263** | −0.255** |
| | | (0.0684) | (0.113) | (0.114) | (0.114) |
| DOCG | DV = 1 if the wine is DOCG | 0.218*** | 0.051 | 0.00578 | 0.0206 |
| | | (0.0734) | (0.12) | (0.121) | (0.121) |
| N | | 47,227 | 21,103 | 21,103 | 21,103 |
| Pseudo R2 | | 0.119 | 0.174 | 0.203 | 0.206 |

*Note:* Results come from ordered logit regressions with robust standard errors (in brackets). Regressions are run with the ologit Stata command, which estimates the proportional odds/parallel lines model. DV (dummy variable) is a binary (1/0) variable. Regressors with *** are significant at 1 percent level, with ** at 5 percent, and with * at 10 percent. All the specifications include region, year, and denomination (IGT, DOC, and DOCG) dummy variables. Results come from table 2 of Castriota, Delmastro, and Curzi (2013).

**Table 2.A.2**
Price determinants of wine, Veronelli Guide 2004–2009.

| Regressors | Description | (1) | (2) | (3) | (4) |
|---|---|---|---|---|---|
| Quality | Number of stars | 1.384*** | 1.232*** | 1.238*** | 1.242*** |
| | | (0.0142) | (0.0228) | (0.0235) | (0.0236) |
| New entrant | DV = 1 if the wine has been reviewed for the first time | −0.412*** | −0.258*** | −0.259*** | −0.261*** |
| | | (0.0225) | (0.0359) | (0.0359) | (0.0359) |
| Red | DV = 1 if the wine is red | 0.695*** | 0.0456 | 0.0442 | 0.00612 |
| | | (0.0202) | (0.0331) | (0.0332) | (0.0361) |
| Rosé | DV = 1 if the wine is rosé | −0.745*** | −0.636*** | −0.636*** | −0.657*** |
| | | (0.0695) | (0.0965) | (0.0965) | (0.106) |
| Sweet | DV = 1 if the wine is sweet | 0.922*** | 0.777*** | 0.774*** | 0.899*** |
| | | (0.0495) | (0.0676) | (0.0676) | (0.0717) |
| Age | Age of wine in years | 1.681*** | 1.817*** | 1.816*** | 1.806*** |
| | | (0.0724) | (0.115) | (0.115) | (0.115) |
| (ln) Bottles | Ln of number of bottles produced by the firm | −0.527*** | −0.412*** | −0.411*** | −0.413*** |
| | | (0.00946) | (0.0144) | (0.0144) | (0.0144) |
| (ln) Hectares | Ln of number of hectares in vineyard | 0.276*** | 0.207*** | 0.207*** | 0.211*** |
| | | (0.00892) | (0.0135) | (0.0135) | (0.0135) |
| Cooperative | DV = 1 if the wine is produced by a cooperative | 1.102*** | 0.34 | 0.339 | 0.317 |
| | | (0.173) | (0.234) | (0.234) | (0.233) |

| | Slope DV | | | |
|---|---|---|---|---|
| (ln) Hectares * Cooperative | −0.346*** | −0.197*** | −0.197*** | −0.195*** |
| | (0.0307) | (0.0413) | (0.0413) | (0.0412) |
| Barrels DV = 1 if the wine is aged in barrels | | 0.840*** | 0.839*** | 0.841*** |
| | | (0.0359) | (0.0359) | (0.0358) |
| Barriques DV = 1 if the wine is aged in barriques | | 1.416*** | 1.415*** | 1.410*** |
| | | (0.033) | (0.033) | (0.0331) |
| IGT DV = 1 if the wine is IGT | −0.583*** | −0.977*** | −0.976*** | −0.988*** |
| | (0.0715) | (0.11) | (0.11) | (0.109) |
| DOC DV = 1 if the wine is DOC | −0.852*** | −1.066*** | −1.065*** | −1.076*** |
| | (0.0692) | (0.106) | (0.106) | (0.106) |
| DOCG DV = 1 if the wine is DOCG | 0.288*** | −0.0522 | −0.0512 | −0.0666 |
| | (0.0743) | (0.114) | (0.114) | (0.113) |
| N | 47,052 | 21,099 | 21,099 | 21,099 |
| Pseudo R2 | 0.1847 | 0.1991 | 0.1991 | 0.2 |

*Notes*: Results come from ordered logit regressions with robust standard errors (in brackets). Regressions are run with the ologit Stata command, which estimates the proportional odds/parallel lines model. DV (dummy variable) is a binary (1/0) variable. Regressors with *** are significant at 1 percent level, with ** at 5 percent, and with * at 10 percent. All the specifications include region, year, and denomination (IGT, DOC, and DOCG) dummy variables.

Almenberg and Dreber (2011), with an experiment on 135 people conducted in Boston in 2008–2009, demonstrated that the price of wine strongly influences the judgment of quality, especially if this information precedes tasting, which happens whenever it is not a repeat purchase. Heffetz and Shayo (2009), however, found that high prices used as a quality signal work only in laboratory experiments and not in everyday life. The use of high prices seems, therefore, to influence the willingness to pay when declared in experiments or possibly when consumers have limited knowledge.

The opinions of experts and wine guides are another signal used by a large number of enthusiasts; proof of this lies in the proliferation of guides and websites evaluating bottles of wine all over the world. Ali, Lecocq, and Visser (2008) showed that gurus like Robert Parker are able to influence the price of en primeur wines from Bordeaux. His ratings are generally published in spring of each year before prices are established. In 2003, however, the ratings were published in autumn, after pricing. This "natural experiment" allowed them to isolate the impact of the expert's opinion on the price of wine. Dubois and Nauges (2010) came to similar conclusions and, using panel data on 108 châteaux of the Bordeaux region, distinguished the effect of the experts' ratings from the unobservable quality of the product. Using Swedish sales data, Friberg and Grönqvist (2012) found that favorable expert reviews increase prices by around 6 percent for more than twenty weeks, whereas negative ones do not have any effect.

While signals are important, the information on the label is strategic. Lecocq et al. (2005) demonstrated with experiments that, if participants first read the labels in wine auctions, then sensorial information becomes irrelevant, whereas if they taste the wine first, then the information on the labels increases the willingness to pay. San Martín, Brümmer, and Troncoso (2008) found, using data on Argentine wines sold in the United States and judged by Wine Spectator, that labeling practices are more influential than the opinions of experts. This is probably due to the fact that more people read the label than consult a guide before proceeding with a purchase.

In conclusion, the price of wine is influenced by a set of variables. Some of these concern the quality of the wine and therefore mainly interest agronomists and wine makers while others (choice of the type of wine that attracts demand, information on the label, etc.) are strictly the responsibility of those who manage the company. However, as shown by Jaeger and Storchmann (2011), there is a certain degree of dispersion in wine prices that is higher for expensive bottles that are purchased infrequently.

## Appendix 2.1

Tables 2.A.1 and 2.A.2 show the results of regression analyses with robust standard errors in which the dependent variables are, respectively, the quality and the price of wine. The database, used also in Castriota, Delmastro, and Curzi (2013) and Castriota (2018), contains information on approximately fifty thousand wines produced by more than four thousand companies and reviewed by the Veronelli Guide from 2004 to 2009. For each wine tasted the guide reports the year and the region of production; if it has been judged for the first time; the type (white/rosé/red, sweet/dry); if it has been aged using wooden barrels or barriques; and the size and legal nature of the company. The information on the use of barrels and barriques is not so complete so that their use as regressors reduces the number of observations available. The analyses of table 2.A.1 and 2.A.2 show that the variables influencing prices are the same as those that determine the quality, net of use of quality as a regressor (see table 2 of Castriota, Delmastro, and Curzi, 2013).

# 3

## Competition and Firm Profitability

I have no friends and no enemies—only competitors.
—A phrase attributed to Aristotle Socrates Onassis (1906–1975)

The structure of the market inevitably determines, together with other variables, the level of competition between companies which, in turn, influences their profitability. The first section will discuss monopolistic competition in the wine sector and explain why and how firms in the Old and New World try to differentiate from their competitors to avoid price wars and falling profit margins (for a theoretical review of the main market forms of interest for the wine sector, see appendix 3.1). In the second section we apply, point by point, Porter's five forces model to the wine sector. The aim is to identify the forces that operate most in favor of or against profitability in the sector. Finally, the third section analyses the profitability of the wine sector in light of Italian data on company balance sheets and international literature.

### 3.1 Monopolistic Competition and How to Differentiate from Competitors

The wine sector is characterized by very strong product differentiation, both horizontal and vertical,[1] and the presence of thousands of producers scattered over five continents. There are also very strong information asymmetries between producers and consumers, making signals such as price, reputation, ratings in wine guides, and advertising important determinants of consumer choices. The market, therefore, can rightly be defined as monopolistic competition. Each producer chooses the type of wine to put on the market (generally more than one variety) and the quality to be achieved, which may vary to reach different types of consumers and diversify risk. Each bottle of wine is unique in both its objective and subjective (as perceived by the buyer) characteristics. The entrepreneur has to discern the preferences of one or more niches of the market so that he can become a monopolist and make abnormal profits (Thornton, 2013, pp. 3–4). In contrast, a company that produces an undifferentiated good is automatically placed in a market of perfect competition in which price pressure will make profits disappear. (See Aylward's [2008] concerns about the "coca-colarization" of Australian wines by large corporations.)

In the long run, however, even in monopolistic competition profits will disappear as new businesses enter the market and offer other products that are imperfect substitutes, unless the third condition necessary for monopolistic competition—freedom of entry and exit—is violated. In the Old World, in fact, the European Union has imposed a ban on the planting of new vineyards to rebalance demand and supply while there is a scarcity of available land in the best areas of the New World, limiting the entry of new firms. The long-term profitability of firms depends on their ability to differentiate products in terms of their real, but also perceived, characteristics and quality. For this reason, the role of appellations seems to be crucial, since they were established in Europe with two main objectives:

1. To create unique and inimitable products: a wine with an appellation can only be produced within specific geographic boundaries established by law. The enhancement of territorial uniqueness protected by specific legal rules creates an inescapable barrier to entry that makes a group, and not the individual producer, the monopolist of a market niche. Verdicchio di Matelica and Taurasi, for example, can only be produced in the municipalities and in the provinces authorized by the Italian state. Anyone marketing wines with the same name that is produced outside the authorized area would be accused of infringing the rules. Each appellation can be produced by a group of companies operating within specific geographic boundaries; there may be many companies in the groups or just a few. The French appellations, La Romanée and Château-Grillet, which cultivate a total of 0.84 and 3.8 hectares respectively, are examples of pure monopoly given that only one manufacturer is authorized to produce the AOC-branded wine ("AOC" meaning *Appellation d'Origine Contrôlée*, or Controlled Designation of Origin—the equivalent of the Italian DOC);

2. To enhance the reputation of a group of businesses (i.e., collective brands; see chapter 6) by setting common rules and minimum quality standards. In a world characterized by information asymmetries, a consumer relies on signals such as price, wine guide ratings, the institutional classification system of products, and appellations. When this generates higher expectations in consumers than can actually be matched by the real quality, it creates a surcharge that is reflected in persisting profit margins.

The production of an undifferentiated wine, on the other hand, exposes a company to fierce competition based on price.[2] In this case unit costs must be contained through an efficient production structure, and the company or collective brand has to be promoted through advertising campaigns to retain the loyalty of customers and convince them that the product has better characteristics and qualities than its rivals, even if this is not exactly true. In Italy the ex–table wines, now called simply "wines," are an example: it is forbidden by law to label the vine, the vintage, the area of production, and production standards for this range of product, except for varietal wines (see chapter 6). The only aspects that can be exploited are packaging and advertising campaigns.

To avoid this challenging situation the Old and the New World have adopted different strategies. The European Union has established wine appellations (Colman, 2008, p. 45) to prevent the use of geographic names like Champagne and Barolo by producers outside their borders. To exploit the collective brand, wines have to be produced according to strict rules that discipline every aspect, from the grapes used to the yields per hectare and so on. In addition to this public solution aimed at generating monopoly, a private solution has been provided by some producers whose wineries are not located within the borders of famous wine appellations. In this case wine makers are rediscovering and promoting local grape varieties that have been forgotten over the last decades or centuries (e.g., Bellone and Nerello Mascalese in Italy).

In the New World the strategy is similar but also different. Many countries have established "wine areas" to create monopoly power. However, in this case producers can freely choose what (white/red/sparling/sweet, grape variety, etc.), how much (yields per hectare, how many hectares, etc.), and how to produce (agronomic and enological techniques). In this way wine areas end up being simple borders. The majority of producers opt for the most famous international (often French) grape varieties and deliver very flavorful, intense, and approachable wines (Marks, 2015, p. 193). These products are easy to understand but also difficult to distinguish from those of other countries or continents. Since there are no native *vitis vinifera* grape varieties in the New World, some producers are trying to differentiate from competitors not by rediscovering abandoned vines but rather by planting new grapes created as hybrids in US universities and research departments (McKee, 2016), as happened when Abraham Perold at the University of Stellenbosch in South Africa in 1925 mixed Pinot Noir and Cinsaut and created the successful vine Pinotage. It is difficult to predict whether these varieties will be successful at a national and international level.

## 3.2   Analysis of Competition in the Wine Market: Porter's Five Forces in the Wine Sector

The models presented in the previous section and in appendix 3.1 describe the functioning of the main market forms but are necessarily subject to simplifications that are often reductive or even unrealistic. The main conclusion is that competition erodes long-term profits: companies have to differentiate in some way from their competitors and remain monopolist in their niche market. To better understand what factors affect the level of competition and consequently a firm's profitability, Michael Porter's five forces model (1979) in figure 3.1 will be applied to the wine

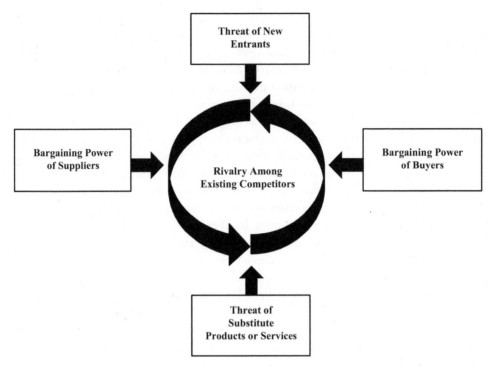

**Figure 3.1**
Porter's five forces model.

sector. The model was presented in an article in the *Harvard Business Review* when the author was a young associate professor. It triggered a revolution in the field of strategic business analysis, and the model was later integrated with reflections and insights to clarify some points and updated in 2008 to take account of the development of new sectors using high technology. Porter (2008) will therefore be the reference used in what follows.

The five forces model is applied to the wine sector (table 3.1), highlighting wherever possible the differences between the Old and New Worlds.

### 3.2.1    Five Forces Analysis

(A) *Threat of new entrants into the industry*
This is a concrete and ever-present risk, as shown by the large-scale entry of New World (and recently Chinese) firms in a market that was dominated by Europe until the 1980s.

**Table 3.1**
Porter's five forces in the wine sector.

| Force | Presence |
|---|---|
| **A. Threat of new entrants** | Present |
| *Barriers to entry* | Present, but moderate |
| 1. Economies of scale on the supply side | Present, especially in mass-produced goods, but not so much as to discourage the entry of new firms in the market |
| 2. Benefits of scale on the demand side | Absent; there is no benefit from the consumption of the same wine by other buyers |
| 3. Cost of change for the customer | Absent |
| 4. Capital requirements | Variable but not impossible |
| 5. Advantages of existing firms independent of their size | High; there is a scarcity of available land and exorbitant prices per hectare. Other advantages are mostly in the New World (brand, control over value chain) |
| 6. Unequal access to distribution channels | Present but not relevant |
| 7. Restrictive government policies | Strict in the EU; variable in other countries |
| *Expected retaliation* | Minimal. There are a very large number of producers already in the market with generally small market shares. |
| **B. Power of suppliers** | Limited. Most producers own or rent land. In Europe many small grape producers are members of cooperatives while others are weak links in the chain. |
| **C. Power of buyers** | Variable. Distribution is more concentrated in the New World, and price elasticity is higher for lower quality wines and among low-income consumers. |
| **D. Threat of substitutes** | Low in static terms (low cross elasticity of goods), and variable in dynamic terms (growing share of beer in Mediterranean Europe, growing share of wine in other countries) |
| **E. Rivalry among existing competitors** | High |
| *Intensity of competition* | High |
| Number of producers | High |
| Growth of sector | Persistent imbalance between demand and supply that is leading to a process of concentration and an increase in economies of scale |
| Barriers to exit | Average; part of the investments can be recovered. Further, the EU's funding for grubbing up vineyards have lowered barriers |
| Commitment to the business | High; strong noneconomic motivation (intrinsic motivation) of many producers |

*(continued)*

**Table 3.1 (continued)**

| Force | Presence |
|---|---|
| *Competition front* | Average |
| Product differentiation | High |
| Incidence of fixed costs | High; land and machinery tie up capital |
| Need for large size | Moderate because not essential |
| Perishability of product | Limited (white and sparkling wine) or minimal (red and fortified wine) |

| Factor | Presence |
|---|---|
| Growth rate of sector | Low. Imbalance between demand and supply (though falling) is leading to a process of concentration to increase economies of scale. |
| Technology and innovation | No product innovations, and innovations in production are slow. |
| Government | In the New World state intervention is minimal or aimed at increasing production. In the EU there are restrictive policies and subsidies. In both, there are campaigns against drunk driving and excessive consumption of alcohol. |
| Complementary products and services | Absent |

Barriers to entry. They are present but moderate.

1. Economies of scale on the supply side: as in almost all sectors, there are economies of scale on the supply side that allow the reduction of average production costs. Therefore, whoever enters the market with small plots of land and limited production is at a competitive disadvantage compared with the large companies already present (Thornton, 2013, p. 4). It has been estimated that in the United States the investment to build a wine company with good—but not full—economies of scale (around five hundred thousand cases) amounts to about $35 million (Thornton, 2013, p. 176). This, however, is especially true for lower-level wines that are marketed through large-scale distribution: a low sale price means unit costs have to be contained and production made in large quantities. But, the medium-high segment can reach satisfactory profit margins, or at least a balanced budget, even with smaller quantities. An estimate of the costs to build a two-thousand-case wine facility is about $600,000 (Thornton, 2013, p. 177).
2. Benefits of scale on the demand side: there are none since there is no benefit from the consumption of the same wine by other buyers. Some benefit could result

from imitation and adjustment to the behavior of others when a particular product or type of wine becomes fashionable, but this circumstance does not seem very likely in a differentiated market like wine.

3. Cost of change for the customer: no cost for consumers and almost none for traders who only have to choose a new producer and sign a new supply contract.

4. Capital requirements: variable, but not unbearable. Starting the production of wine involves buying vine cuttings,[3] land (if it is not rented or grapes are not bought from third parties), machinery, and barrels; constructing buildings for processing and storage; hiring employees (administrative and technical staff); bearing advertising costs (especially in the New World where the company brand counts more than geography); and looking for buyers. The financial commitment of a company is made even more burdensome by the fact that, as reported in chapter 1, a vine does not produce fruit in the first three years; from the fourth to the sixth year production stands at 30 percent; and from the thirty-first year onward (usually until the fortieth) the yield per hectare decreases. The eventual aging of the wine in barrels for one or more years raises costs, postponing revenues even further. The latter is not only deferred over time but is also random: many of the variables that affect the quality of wine, such as climate and soil quality, are beyond the control of the producer and may vary over time. Nevertheless, the minimum commitment is in the order of millions, not billions, of euros, so it is not enough to discourage the entry of new producers into the market (as happens, for example, in the pharmaceutical industry).

5. Advantages of existing companies independently of their size: high. For some time now there has been a shortage of land in the most prestigious areas (e.g., Champagne), which—when available—is sold at exorbitant prices.[4] In the European Union this difficulty is amplified by the legislation on planting rights.[5] In the New World other advantages—such as the company brand and control of the value chain—are more relevant. The experience accumulated over time by agronomists and wine makers, on the other hand, plays the same important role everywhere. Further, it takes time for a firm to build a reputation among consumers; therefore new entrants have to sustain high marketing and promotion costs before earning a positive reputation price premium (Thornton, 2013, p. 177).

6. Unequal access to distribution channels: present, especially for small wineries. It is true that distributors are always looking for new products that can satisfy the curiosity or the needs of a heterogeneous and constantly evolving public. However, small producers in the Old World are facing declining domestic demand and have to increasingly rely on foreign markets to survive; this can be difficult if a firm does not have skilled export managers and large, diversified portfolios of products. The problem can be even more severe in the United States where the three-tier system prevents direct-to-consumer sales and shipments in many states and counties (see chapter 8).

Here producers must sell to distributors, who sell to retailers, who finally sell to buyers. The problem is that, to maximize profits, distributors often privilege the large producers who can offer all the products they need at low prices. Small producers have a limited number of bottles and labels and higher production costs. Relying on many small producers can be better from the point of view of consumers who can enjoy a greater variety of products, but it is inefficient for distributors.

7. Restrictive government policies: very strict regarding production in the European Union because of planting rights; variable in the New World. After the repeal of Prohibition, every state and county in the United States was allowed to regulate the production, distribution, and sale of alcoholic beverages. Nowadays a significant number of counties are still "dry."[6] In Australia the authorities openly support the growth of wine production with long-term planning (see the Strategy 2025 program). But as far as consumption is concerned, in both countries these groups have a variety of instruments used with greater or lesser intensity to contain alcohol abuse and alcoholism (see chapter 7).

Expected retaliation. The risk that companies in the market react in a vigorous way to the entry of new companies is minimal in Europe and limited in the New World. In the wine sector there is, in fact, a very large number of producers, each of whom has a fairly limited market share, especially in the Old World.[7] Given the large number involved, incumbents rarely respond strongly to the entry of a new competitor since they are usually unaware of it. On an aggregate level, however, there may be collective reactions, planned or otherwise, immediate or delayed, that lead to a reduction in price.

(B) *Power of suppliers*
The power of suppliers is limited for the following reasons.

- The markets for the supply of cuttings, fertilizers, and other chemical products; machinery; and labor (administrative staff, agronomists, oenologists) are competitive enough to guarantee that none of these suppliers may compromise the profitability of wineries.
- Most of the wineries, especially the high-end ones, own or rent land. Those who have plots of land that are too small to produce wine in an economically sustainable way (e.g., have less than five hectares) may decide to rent land, sell grapes to a private winery, or take them to a cooperative. In Europe, where there is a deeply rooted cooperative tradition dating back to the second half of the nineteenth century, the latter option is often preferred because it adds an additional link in the value chain. The cooperative, in fact, pays members not only the price of the grapes (raw material) but also a share of the profits from their transformation into wine. Those who, on the contrary, decide to sell only the raw material have to

accept the lower remuneration which simple grape growers are entitled to: in this case the compensation is likely to be low, especially if it has not been previously agreed on with buyers in formal supply contracts.

- Small producers of wine grapes depend heavily on wineries. In fact, wine grapes can be used profitably only to produce wine.[8] Therefore, suppliers cannot sell their goods to the best bidder in other sectors of the food industry.
- The cost (if any) of changing supplier is very low. There are no learning costs for using grapes from different producers, whereas costs connected with the use of new technologies are bearable.
- Wine producers can threaten suppliers to proceed with upstream vertical integration if there is land available or if regulations allow the planting of new vines.

Two factors can partially mitigate this imbalance of bargaining power that is heavily weighted in favor of wine grape buyers.

- Grapes are a highly differentiated product. This can increase suppliers' contracting power but only for high quality grapes.
- There is no substitute for grapes in wine production.

## (C) *Power of buyers*

The power of buyers is variable, depending on the context. In the wine sector, it is reduced by the following circumstances.

- Consumers cannot buy directly from producers. Direct purchase is allowed in most countries, but in others, like the United States, it can be prohibited, and consequently, organized distribution is more concentrated.
- The products are highly differentiated both horizontally and vertically, making it more difficult for clients to exert downward pressure on prices.
- Buyers, either consumers or intermediaries, cannot threaten to proceed with upstream vertical integration.

Contrarily, the power of buyers is partially increased by the fact that the cost to change suppliers is zero for final consumers and low for intermediaries. Moreover, in some countries—as for example Scandinavia and Canada—the purchase and distribution of alcoholic beverages lie entirely in the hands of state companies that consequently have great bargaining power.[9] Although the final purchase is made by thousands of individuals in public-owned stores, intermediation by a single large entity acting as a cooperative of consumption reduces the power of sellers.

Price sensitivity also varies as wine generally forms a very small part of the family budget, making the consumer less responsive to price. However, the economic situation of buyers differs greatly depending on the country and time and therefore will influence their willingness to pay and their attention to price. In general, however, less wealthy customers are, for reasons of necessity, more aggressive and pay more attention to price.

## (D) *Threat of substitutes*
The threat of replacing wine with other drinks is quite variable. On the one hand, it is difficult to imagine radical product innovations that can bring new alcoholic drinks onto the market that are so successful that those already present lose significant shares. On the other hand, considering the existing drinks, we have to distinguish the behavior of the individual in the short term from that of a community in the long term. As we saw in chapter 2, there is limited cross elasticity of alcoholic beverages to changes in the price of others: consumer preferences are given, and people often prefer to reduce quality rather than change the type of product, limiting expenditure but maintaining the same quantity.

As shown in chapter 1, however, there are clear changes in the habits and preferences of peoples at a collective level, with a process of convergence that sees a fall in the share of wine in European Mediterranean and Latin American countries out of the total amount of alcohol consumed but that is an increase in other countries. The threat of substitution is, therefore, very serious for some countries (Cardebat, 2017, pp. 57–60) while it works in a favorable sense for others. The risk of a change in consumer preferences for the type of wine (e.g., white, red, sparkling wine) or vine (e.g., Nebbiolo, Chardonnay, Tempranillo) is, instead, a serious problem for all producers as these choices involve decade-long investments. On an aggregate level it is a "zero-sum game" while the effects can be quite dramatic at an individual level.

## (E) *Rivalry among existing competitors*
There is a high level of rivalry between companies operating in the wine sector.

- Although there are considerable differences in the size of companies, the large number present in the market and the process of internationalization make competition quite strong. As reported by Anderson, Norman, and Wittwer (2004), however, concentration in the wine sector is much lower than in other sectors. Rabobank data from the end of the 1990s showed that the world market share held by the top three companies was 6 percent in the wine sector, compared with 35 percent for beer, 42 percent for spirits, and 78 percent for nonalcoholic beverages.
- World consumption shows very low growth rates and marked differences among geographic areas, with some in expansion and others in serious difficulty.
- As in all sectors of the economy there are barriers to exit (e.g., grubbing up vineyards),[10] even if part of the investments (e.g., land and machinery) can be recovered in the case of closure.
- Many producers are driven by strong noneconomic motivations. A significant number of producers, especially small- and medium-sized ones, do not pursue

maximization of profits (profit maximizers) as a priority objective but rather pursue utility (utility maximizers) that tends to coincide with the prestige of the company and the quality of its wines (Scott Morton and Podolny, 2002). These entrepreneurs are willing to give up part of the return on invested capital (ROIC) to excel in their business and they tend to specialize in the production of high-end wines. This increases pressure on other profit maximizers in the higher segments.

Fortunately, competition among producers is not based solely on price, and this reduces price pressures since:

- the wine sector is strongly differentiated, both horizontally and vertically;
- some consumers are enthusiasts and are willing to dig into their pockets to satisfy their palates; and
- the product can be stored for several years.

In summary, the level of competition in the wine sector seems to be quite high due to the risk of new companies entering the market, the substitution of wine with beer (for Mediterranean European countries), and the degree of rivalry among existing competitors.

### 3.2.2    The Factors, Not Forces, that Influence Competition

Four factors—which Porter recommends should not be confused with forces—can further affect competition: growth rate of the sector, technology and information, government, and complementary products and services. The first three seem to play against the Old World only, while the fourth can benefit both.

- At the aggregate level, consumption is almost static, with a contraction in Mediterranean European countries and growth in others. There is a persistent imbalance between supply and demand, which decreased for a while but encourages a process of business concentration through mergers and acquisitions (especially in the New World) that aim to increase economies of scale.
- There are no product innovations, and process innovations are rather slow.
- Government policies differ greatly from one country to another. In the New World, state intervention is not aimed at decreasing production and in some cases (as in Australia) is even aimed at increasing. In the European Union, in contrast, there is a set of restrictive measures and subsidies whose effects have been frequently criticized. In compliance with the provisions of the World Health Organization, all countries have adopted policies to combat drunk driving and alcohol abuse that have contributed to the reduction of per capita consumption.
- There are products and services complementary to wine production—and above all, tourism—which can promote virtuous competition focused on the quest for quality.

As highlighted by Porter (2008), a complete and rigorous analysis of the five forces would require data collection and scientific studies on each of the points listed—a book in itself. In any case, there are no homogeneous data on many aspects discussed in this chapter.

### 3.3   Analysis of Profitability in the Wine Sector

After discussing the main market forms, the forces that influence competition, and consequently, the profitability of companies, we can now turn to an analysis of profitability in the wine sector. Is making wine profitable? And if so, how profitable? To answer these questions, we need to analyze the balance sheets of wineries. The most appropriate measure for this task is ROIC,[11] given by the ratio between operating income and net operating invested capital. It enables us to assess whether and to what extent management is able to remunerate all the capital invested—namely equity and credit capital—in the running of the company (Porter, 2008).

Below are the results of some analyses made on Italy's Aida (financial analysis of companies) data from 2015 for private companies only.[12] Figure 3.2 shows the estimate of the (nonparametric) Kernel distribution of the ROIC and that of a normal variable. This plot clearly shows how the ROIC does not have a normal distribution, since it is characterized by a very pronounced peak near the median value and has fat tails. On the one hand, ROIC is on average low (1.73 percent), with most companies

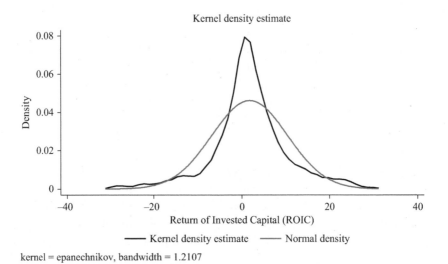

kernel = epanechnikov, bandwidth = 1.2107

**Figure 3.2**
Distribution of ROIC of Italian wineries, 2015.
*Source*: Author's calculations using data from Bureau van Dijk's Aida database.

in the sector having weakly positive or weakly negative values. On the other hand, however, there is a certain dispersion of data with a significant number of companies that make extreme returns on capital in both directions.

In other words, the very strong dispersion present in the graph shows that this sector is not only on average very competitive and not very profitable but also strongly diversified with profitability between a minimum of –30 percent and a maximum of 30 percent. For every company that records very heavy losses, there is another that makes huge profits. Figure 3.3 shows the cumulative distribution of ROIC: about 20 percent of private companies are running at a loss while three quarters have ROIC between –5 percent and 5 percent. It should be underlined that these data concern Italy, a country with high levels of tax evasion and avoidance, and could therefore underestimate the real profitability of wineries. Further, profitability in other countries could be higher because of the larger size of businesses. On the other hand, it should be kept in mind that those companies that went bankrupt are not in the database anymore; therefore the average profitability could be overestimated.

Intangible expenses deserve a separate in-depth analysis. They are generally classified into four types: human, intellectual, organizational, and relational capital. Over the years the cost of training or hiring qualified staff, developing or buying patents and software, or establishing solid relationships with customers has grown

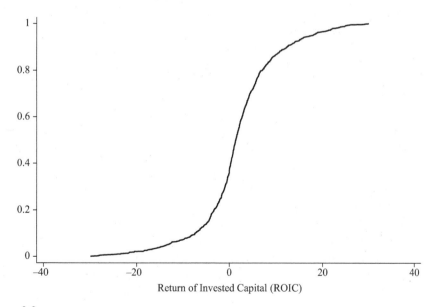

**Figure 3.3**
Cumulative distribution of ROIC of Italian wineries, 2015.
*Source*: Author's calculation using data from Bureau van Dijk's Aida database.

constantly. According to advocates of the resource-based view, intangible resources are the key to maintaining sustainability in the long run (Itami, 1987)[13] since they cannot be easily copied and acquired by competitors (Nelson, 1991). From an empirical point of view Villalonga (2004) demonstrated with US panel data that intangible expenditure plays an important role in supporting the competitive advantage of companies, which is defined as profit persistence. For the agricultural and food sectors, however, there is a negative relationship between intangible expenditure and sustainability. Using data from five European countries from 1993 to 2004, Casta, Ramond, and Escaffre (2008) came to the opposite conclusions to Villalonga (2004): intangible expenditure does not imply any improved competitive edge. The evidence, therefore, shows rather conflicting results.

The extent and type of investments required and the impact they have on company profitability vary greatly from one sector to another. While advertising costs are huge in areas such as drinks and cosmetic products, in others, such as pharmaceuticals, the most onerous item is the multiyear investments in research and development (Megna and Mueller, 1991). The sector in which a firm operates will affect the impact each of these items produces on profitability, as shown by Villalonga's (2004) results.

In light of all these considerations, an analysis must obviously refer to specific sectors. As far as wine is concerned, there is often the widespread belief that the sector is characterized by traditional production techniques and, above all in the Old World, by the prevalence of small- and medium-sized businesses that make intangible expenses superfluous or nonproductive. Actually, fierce competition from New World countries and the important and costly innovations in production introduced in recent decades have clearly shown the need to make investments in the fields of agronomy, wine making, organization, advertising, reputation, and so on (Zahra, 1999; Berthomeau, 2001).

This belief is reinforced by the conclusions reached by Amadieu and Viviani (2011) who, in a longitudinal study conducted on a sample of 196 French wineries (101 cooperatives and ninety-five private companies), empirically demonstrated that intangible expenditure, consisting mainly of advertising and promotion since research and development costs are small, increase expected profits and reduce variance (and therefore risk). In a sector where large investments in fixed assets—primarily in land and buildings—are necessary, the amount of intangible expenditure was understandably lower than for tangible expenses. The authors of the study argued, however, that to effectively improve the strategic positioning of French wineries on international markets, investments in intangible capital need to be massive. For this strategy to be successful there must be greater concentration, so companies should be helped to grow sufficiently to bear these burdensome investments. In addition, intense and fruitful cooperation among producers should be encouraged to promote prestigious collective branding.

The literature review in the chapters 2 and 6 has shown that excellence, as measured by the quality of wines and the reputation of wineries, is an important determinant of wine prices. For this reason, there is a strong belief that firms should achieve excellence to increase prices and sales. However, whether increasing quality and building a famous firm or collective brand also increase firm profitability is not clear. In fact, when firms increase quality, the costs often increase at an exponential rate and can exceed the additional revenues. Further, even if profits are higher, profitability might be lower if—as in the wine sector—huge investments are required (e.g., owning land, equipment, consultants, etc.). The overall impact might also depend on the sector of activity since excellence implies a number of fixed costs which can be covered only with large quantities.

Business scalability can, therefore, be the key to making a good investment. In the mobile phone industry, there is a huge initial investment to design a new product, but afterward the variable production costs are minimal, and it is possible to produce millions of devices. In the wine sector, instead, scalability is limited by the availability of land and by the EU laws on planting rights while variable costs are significant. A similar line of reasoning holds for vertical integration. In some sectors like wine, the direct control of the whole value chain is necessary to increase quality while in others like electronics or automobiles where hundreds or thousands of components are necessary it is better to rely on specialized suppliers.

However, the in-house production of grapes requires large investments, and it is not clear whether being a simple bottler—a company which buys cheap wines from other firms and resells them—is less profitable (Thornton, 2013, p. 169). Thornton (p. 153) provides the example of Castle Rock Winery, estimated to be the twenty-sixth largest winery in the United States and having no vineyards at all. The company has long-term agreements with a number of independent producers in California, Oregon, and the state of Washington to deliver wine according to detailed instructions and supervised procedures and then to sell its wine through a network of distributors in forty-eight states and over the internet.

Castriota (2018) used Italian data from the Veronelli wine guides of 2004–2009 on more than fifty thousand bottles and confirmed the results (obtained in the literature) that quality and vertical integration are important drivers of price; better wines as well as those produced by private rather than cooperative firms are more expensive. However, in a second analysis the author collected balance sheet data from Aida on a sample of around seventeen hundred Italian firms over the period 2006–2015. The database is enriched with data coming from telephone surveys and wine guides on the type of activity carried out and firm and collective reputation. The results show that firm reputation is positively influenced by vertical integration, firms producing both grapes and wine having a better firm reputation. However, using a number of different econometric methodologies it turns out that neither firm nor collective reputation are significant drivers of ROIC; the ROIC's main determinant is firm size due to economies of scale and enhanced export capabilities. In sectors with limited scalability, firms should carefully consider their investment strategies to avoid overinvestment in quality and reputation.

Net of reputation effects, wineries producing grapes and wines are more profitable, but the highest performance belongs to bottlers that sell large volumes of cheap products and have no land and little invested capital. Three more advantages have to be taken into account. First, bottlers can diversify their portfolio horizontally across regions to satisfy clients' needs. Second, the costs of market entry and exit are almost null; bottlers can change suppliers if tastes change without need to uproot vineyards. Third, bottlers are not subject to any production constraint and can grow limitlessly.

### Appendix 3.1: Main Market Forms of Interest for the Wine Market

In the course of time, economists have developed four fundamental models of market structure: perfect competition, oligopoly, monopoly, and monopolistic competition. This taxonomy is based on the number of companies present in the market and the differentiation of the product (figure 3.A.1). If there is only one producer we speak of a monopoly while if there are a few companies the market is an oligopoly. Firms can differentiate their production horizontally (type of good) or vertically (quality) if they want to extract a high share of consumer surplus and maximize their profits. In the presence of many companies we can have perfect competition if the good is undifferentiated or monopolistic competition if the products are characterized by vertical and/or horizontal differentiation. The three forms that are conceptually the

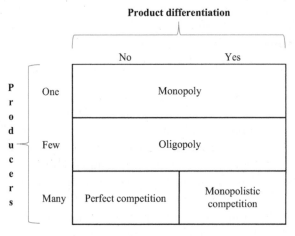

**Figure 3.A.1**
Type of market structure.

most interesting for the wine production market are perfect competition, mono-poly with homogeneous goods, and monopolistic competition (oligopoly is not rel-evant).[14] Monopsony—where only one buyer exists—is relevant for the market of grape suppliers and will be briefly discussed at the end of the appendix.

### Perfect Competition

Five conditions must be met for a market to be defined as perfect competition.

1. The product should be a standardized or undifferentiated product. Neither hori-zontal nor vertical differentiation is considered.
2. There should exist a large number of firms and consumers who are not able to influence the market with their individual behavior and are, therefore, price tak-ers, not price makers. Consumers are usually price takers while companies can often influence the market price, as in the case of monopolies and oligopolies.
3. There should be perfect information. Firms and consumers know all about the prices, features, and quality of the products. Producers know the prices and fea-tures of all the production factors like labor, tools, machinery, and so on.
4. All companies should have access to the same technology.
5. There should be freedom of entry and exit from the market in the long term at no cost, except for capital investment.

If these conditions are met, and a traditional cost function as in figure 3.A.2 is hypothesized where marginal cost (MC) and average total cost (ATC) first decreases and then grows, the firm will maximize profits in the short term by choosing to produce the quantity at which marginal revenue (MR) is equal to marginal cost (point B). Since no firm has market power, MR is equal to price and is graphically

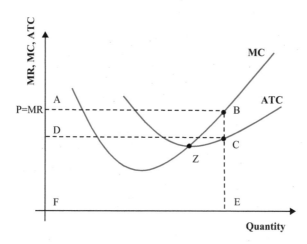

**Figure 3.A.2**
Short-term costs and revenues in perfect competition.

represented by a horizontal line. In other words, the individual demand curve is horizontal, and no matter what quantity is produced by the firm, price does not change. Profits are made up of the rectangle A-B-C-D, the result of the difference between revenues (A-B-E-F) and costs (C-D-E-F).[15]

In the long run, however, firms are free to enter and exit the market. The presence of profits encourages new firms to enter the market, leading to an increase in supply and a consequent decrease in market price (figure 3.A.3), in turn lowering the marginal price-revenue line (figure 3.A.4). At an aggregate level the quantities produced increase

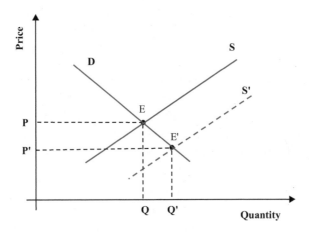

**Figure 3.A.3**
Long-term supply curve in perfect competition.

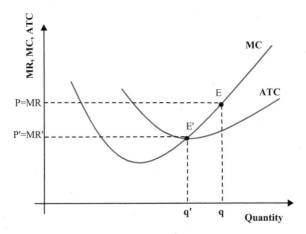

**Figure 3.A.4**
Long-term costs and revenues in perfect competition.

due to the entry of new companies in the market, even if at an individual level firms reduce the volume to keep marginal costs equal to revenues. New firms stop entering the market when the price reaches the minimum value of average costs: at this point profits are zero, and no one enters the market. If someone decided to enter, price would fall below average costs, and in the long run, someone would leave the market.

Perfect competition is considered by classical microeconomics as an optimal mechanism for the efficient allocation of resources because:

- every production factor is remunerated on the basis of its marginal productivity;
- the producer maximizes their profits, even if they are zero in the long run; therefore, there are no extra profits and the entrepreneur is remunerated only for their own work; and
- the consumer pays the lowest possible price that covers minimum average production costs.

### Monopoly

In the basic model with homogeneous goods a market is said to be a monopoly if there is only one supplier of a good that does not have valid substitutes. Monopolistic firms have market power, so they can raise the price above the competitive level to maximize profits. The possibility of achieving profits should entice new companies to enter the market, such as in the model of perfect competition. The question therefore arises: what prevents companies from entering a potentially lucrative market? The answer can be found in four types of entry barriers.

1. Control over a market of scarce resources or production factors: an example is when Cecil Rhodes took control of almost all the largest diamond mines through a series of acquisitions at the end of the nineteenth century.
2. Economies of scale: if unit costs are always falling (the first section of the curve shown in figure 3.A.2) and large investments are required, then larger companies are more profitable and tend to exclude small ones that are at a competitive disadvantage in the market. This type of market supported by economies of scale is called a "natural monopoly."
3. Technological superiority: a company that continuously innovates and always guarantees process or product innovations to reduce costs or win over customers/buyers can create or maintain a monopoly situation. In this case, however, the barrier is necessary in the short run since competitors will not just stand by and watch, so the advantage can fade quickly. For example, Nokia, the leader of the mobile phone market in the early 2000s, ended up on the verge of bankruptcy with the advent of Apple-branded smartphones.
4. Legal barriers: a monopoly can be either public or private. In the latter case, it is established by the state through patents and copyright. Patents give owners an exclusive

twenty-year right to the exploitation of an invention to repay the investments incurred, and copyright gives the exclusive original right to the diffusion and exploitation of a work for seventy years. When a patent or copyright expires, anyone can exploit the invention or the work, and a monopoly suddenly switches to competition.

A firm's demand is, for the sake of simplicity, represented by a straight line rather than a curve so that marginal revenue is a line that is also straight but with a double slope (figure 3.A.5). While marginal revenue is constant and equal to price in perfect competition, in a monopoly the firm is the price maker: market price decreases as the quantity produced by the monopolist increases. The increase in volume has two opposite effects on the firm's revenue: a quantity effect that leads to increased revenue and a price effect that tends to reduce it. Overall, total revenue increases up to a peak point. Once a certain amount has been reached, it begins to decrease. Maximization of monopoly profits follows the same rules as competition and needs marginal revenue to be equal to marginal costs (point Z), even if the former is decreasing and no longer constant. The optimal level of production will be lower and the price higher than in perfect competition. The monopolist's profits correspond to the rectangle A-B-C-D, given by the difference between revenues (A-B-E-F) and costs (C-D-E-F). If the barriers to entry are not lowered in the course of time, the monopolist will make positive profits both in the short and in the long period, in contrast with competition. This happens to the detriment of the consumer who sees his own surplus decrease. This redistribution process, however, is not a zero-sum game since it generates a deadweight loss of wealth for society, attributable to the mutually beneficial transactions that did not take place because of the monopolist's behavior.

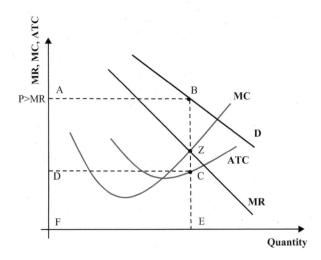

**Figure 3.A.5**
Short-term costs and revenues in monopoly.

Figure 3.A.6a shows the demand curve (which traces the decreasing marginal utility of the good) and supply curve (which reflects the increasing marginal costs of production) in competition. The equilibrium between demand and supply is at point B. The area in triangle A-B-C is consumer surplus or the positive difference between the price that an individual is willing to pay for a specific good or service and its market price, while the area in triangle B-C-D is producer surplus or the positive difference between the price paid and what the producer would have been willing to accept. Conversely, figure 3.A.6b shows a monopoly's demand, supply, and marginal revenue curves. In this case the equilibrium between supply and demand is at

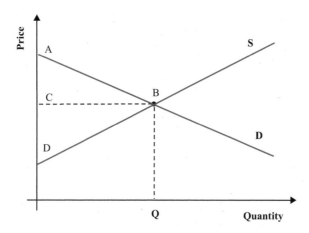

**Figure 3.A.6a**
Consumer and producer surplus in perfect competition.

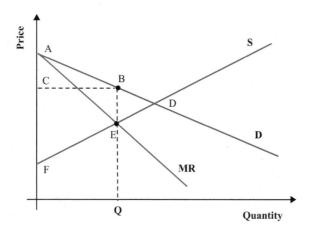

**Figure 3.A.6b**
Consumer and producer surplus in monopoly.

E instead of D, with consumer surplus reduced to the triangle A-B-C to the advantage of the increased producer B-C-E-F surplus and with a deadweight loss given by B-D-E. In a monopoly the total surplus is less than in a competitive company. Some mutually beneficial exchanges do not happen.

Producers and consumers find themselves in two opposing positions; producers are interested in maintaining a status quo that assures them market power and positive profits also in the long run while consumers want to maximize their well-being by buying goods and services at the lowest price possible. The state, which pursues the interest of society as a whole, may adopt legislative measures to prevent ex ante or ex post monopoly situations by imposing public ownership or by regulating the market. In the first case, the market remains a monopoly, but control by a public authority should guarantee the pursuit of consumer interests by setting fair prices and adequate quality standards. In the second case, however, the market can establish prices or maximum market shares. If the authorities decide to intervene when a monopoly already exists, the monopoly can be broken by requiring the company to split into two or to sell a part of the company to a competitor. It is always an open question whether regulatory (state) intervention is advisable, with a large number of economists convinced that the cure is more harmful than the disease. Public decisions are, in fact, subject to political influences that vary according to election results and generate opportunities for corruption.

**Monopolistic Competition**
Monopolistic competition has to satisfy the following conditions.

1. A differentiated product: each company produces a good or service that buyers consider to be different from that offered by competitors. As a result, every seller has some market power, even though it is less than in a monopoly, because the products are imperfect substitutes. For example, there are a number of restaurants on a street in the center of town, each with its own characteristics that distinguish it from the others, but they are still fundamentally businesses that supply food and beverages.
2. A large number of firms: from this point of view it looks more like perfect competition than a monopoly or oligopoly.
3. Freedom of entry and exit from the market in the long run: as in competition, there are no economic or legal obstacles in the long run impeding the entry or exit of companies, which will depend on the opportunities for profit.

Monopolistic competition is therefore different from other forms of the market. It is different from competition as companies have some market power and information is imperfect; different from a monopoly because of product differentiation and competition between companies; and different from oligopoly because freedom of entry and exit prevents collusion among competing companies. As implicit agreements cannot be made with rival companies to reduce competitive pressure, product differentiation becomes of fundamental importance and the only way for firms to acquire some market power. There are three forms of product differentiation.

* Horizontal differentiation is based on style or type. The producer tries to carve out a niche or market share by discerning a group of consumers' preferences, which depend on a number of sociodemographic and cultural factors. Willingness to pay varies from one individual to another and is increased by product differentiation; the same individual readily accepts to pay a higher price if the product reflects his best preferences or meets his needs. Goods are substitutes but are imperfect. Sometimes, however, the differences are less marked than consumers think and are the result of appropriate marketing campaigns.

* Vertical differentiation is based on quality. Here again, consumers have different preferences, needs, and willingness to pay. Some manufacturers specialize in the supply of low-level products sold at low prices; others in high-end products sold at high prices to wealthy clients.

* Geographic differentiation is based on location. When goods or services are of the same quality or type, often what counts is the position of a business. Consumers who are pushed for time generally make purchases near their home or workplace. This is even truer for small purchases: few people will travel long distances daily to have a coffee at a bar, but their range of action widens in direct proportion to a chef's reputation when eating out for dinner. With varying degrees of incisiveness, however, location is important for the vast majority of businesses.

The differentiation of a product, whatever form it takes, is an advantage for both the consumer, who has a much wider choice from which to find the good or service that best meets their needs, and for the producer, who can carve out a niche and provide for customers with greater economic means. Differentiation can, therefore, increase consumption both in volume and, above all, in value. Given, however, that the products are imperfect substitutes and the market is limited, the entry of new companies reduces the opportunities for sales even if they produce slightly different goods and services. If a new restaurant opens in a downtown street or a new petrol pump opens on a highway, sales can be expected to fall in the other businesses. In other words, if the cake remains the same size and the number of table mates grows, the size of the slices will decrease.

In the short term, profit maximization works as in a monopoly (figure 3.A.5): the firm has an individual demand curve that is negatively inclined and a marginal revenue curve with a slope that is twice that of demand. The optimal quantity is near the point at which marginal revenue is equal to marginal cost. A market price that is higher than the average cost of production guarantees extra profits.

In the long run, however, this form of market looks more like perfect competition: in the presence of profits and freedom of entry new businesses will enter the market, leading to an increase in the supply of competitor products—imperfect substitutes (figure 3.A.3). This, in turn, has repercussions on the individual demand curve since customers now want to pay less for the same amount of goods or services (figure 3.A.7a). When a new restaurant opens in a shopping mall, the market power of sellers decreases since the average number of customers decreases. New businesses stop entering when the individual demand curve is tangent to the average total cost at the optimum point (point B, where cost and marginal revenue are equal). Here firms' profits are zero because the price is equal to ATC (figure 3.A.7b). The long-run equilibrium of monopolistic competition is therefore characterized by zero profits:

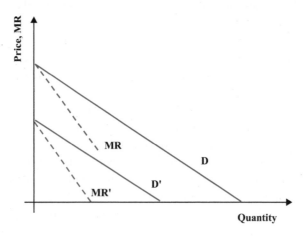

**Figure 3.A.7a**
Long-term demand and marginal revenues in monopolistic competition.

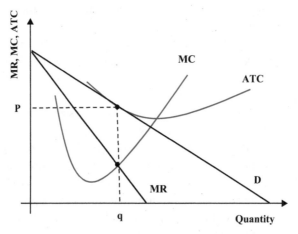

**Figure 3.A.7b**
Long-term costs and revenues in monopolistic competition.

these producers are monopolists without monopoly profits. Monopolistic competition in the long run is very similar to perfect competition, with an important difference in the prices charged and the quantities produced. As can be seen from the graphs, the optimal point is not at the minimum value of the average total cost but rather at a point on the demand curve to its left. As in a pure monopoly, consumer welfare is lower because they pay higher prices and consume fewer goods and services. Since the price is higher than marginal cost, some mutually beneficial transactions do not take place. However, it is not clear whether this circumstance is a source of inefficiency since consumers benefit from the horizontal and vertical differentiation of products.

## Monopsony

A market where there is only one buyer and several sellers is called a monopsony. While it is difficult to find an example which holds at the global level, there exist several local markets where one buyer has substantial purchasing power or even a local monopoly in purchasing raw materials, labor, or final goods. A typical example is depressed areas where one large firm controls the entire labor market. Large firms which process agricultural goods are another example: in the wine sector some large wine makers have supply contracts with most producers in their valley, county, or region. In the United States, due to the three-tier system, in many states producers must sell to distributors who sell to retailers. Over the last decades, however, the market of alcohol distributors has become more concentrated, with a handful of companies controlling a large share of the market, especially at a local level.

Figure 3.A.8 represents a monopsonistic market with one buyer only. While in competition the price (of the grapes supplied, labor or whatever other good or

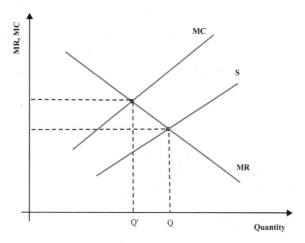

**Figure 3.A.8**
Costs and revenues in monopsony.

service) is determined by the demand and supply curves, in monopsony the buyer is a monopolist. Whenever they decide to increase the purchases, they cannot increase the price only for the last unit and have to increase it for all the previous units as well. Therefore, the marginal cost curve lies above the supply curve and determines an equilibrium with both quantities and prices that are lower than those of perfect competition. This generates a deadweight loss with some transactions which do not occur.

In addition to this problem, the monopsonistic buyer is very powerful because they can threaten the supplier by not buying anything if the price does not fall below a certain level. This is a problem for the seller because they will not be able to find another buyer, and it is particularly problematic if the good is perishable.

# 4

## Types of Companies

The best fertilizer for the vineyard is the owner's shadow.
—Proverb

As in every sector of the economy, very different types of companies coexist in the wine-producing sector in terms of both objectives and legal forms. The goals of a firm affect its organization and the average quality of its products, so the first section of this chapter will discuss the motivations that drive people and institutions in their work and, consequently, identify the objectives of a company depending on its type. In the second section there is a description of the main business types, this time classified by the degree of vertical integration[1] and their organizational and qualitative differences. In the third section, the cooperative model is dealt with in greater detail, showing its strengths and weaknesses. Finally, the last section tackles the question of industrial districts and their contribution to local economies.

### 4.1 Types of Worker Motivation and Company Objectives

Labor economists have long analyzed the motivations that drive people in their work. There are two types: extrinsic and intrinsic motivation. The first consists in economic aspects, such as salary, productivity bonuses, and benefits (car, computer, mobile phone, rent paid by the company, salary supplement for expats, etc.). Traditional economic theory holds that the supply curve of labor is upward sloping since more people are willing to work as wages increase. The second type can be egoistic (self-regarding) or altruistic (other-regarding) (Ben-Ner and Putterman, 1998). Selfish motivation concerns professional growth and recognition of one's own abilities, interest in the work, and the pleasantness of the working environment while altruistic motivation pushes individuals to work to achieve social goals (e.g., helping people in need or defense of the environment).

This distinction is of fundamental importance for the study of the behavior of both workers and companies. In the first case, the neoclassical theory of the labor market states that equilibrium is reached when supply equals demand, with the first sloping upward and the second downward. In practice we have seen that people endowed with strong intrinsic motivations are willing to accept lower wages, "donating" their labor,[2] so to speak, or in the extreme case of volunteering, people "work for nothing" (Freeman, 1997). They are also more productive (Becchetti, Castriota, and Tortia, 2013).

There are organizations such as social enterprises that can attract a high number of workers despite low salaries and flattened career paths. Using data from over 2,066 employees in a sample of 228 public, private, and social enterprises operating in the social services sector, Borzaga and Tortia (2006) showed that people employed in the later (cooperatives) declare greater satisfaction and loyalty to their institution. Becchetti, Castriota, and Depedri (2014), using data on Italian social enterprises, found a large number of people had decided to leave their jobs in the for-profit sector to migrate to the nonprofit sector, even if this resulted in a reduction in salary. Although their economic situations had deteriorated, many declared greater job satisfaction as a result of flexible working hours, better relations with colleagues and superiors, and the type of work that is more in line with their studies. The literature has therefore shown that the variables that motivate people can differ substantially from one person to another and that there is a mix of incentives to motivate workers that go well beyond the mere economic aspect.

If we move from workers to consider organizations, traditional economic models start from the assumption that companies pursue profit maximization. This hypothesis is reductive and appears to be motivated by the need to simplify theoretical-mathematical models (Scitovszky, 1943).[3] For many economists, the dogma of maximizing profits appears justified from an evolutionary point of view in the case of private firms (which are the majority) (Scott Morton and Podolny, 2002): any company that does not pursue this objective will disappear in the long run as it involves a very strong incentive to minimize costs.

The motivation that drives owners/founders can, however, differ significantly. Schumpeter (1911) claims that entrepreneurs create new businesses for a variety of reasons that go beyond mere gain. First of all, there is the desire to create at least a kingdom, if not a dynasty. Then there is the desire for conquest, the impulse to fight to prove their superiority over competitors, and the yearning for victory—not for the fruits of success, but for success itself.[4] According to Knight (1921), the "prestige of the entrepreneur" and the "satisfaction of being one's own boss" must be taken into account when explaining an entrepreneur's decision-making process.[5]

Scott Morton and Podolny (2002) defined companies as *profit maximizers* if they pursue the goal of maximizing profits and *utility maximizers* if they (primarily) pursue noneconomic objectives (e.g., product quality or social well-being). Reaching this second type of objective may lead to an increase in production costs and, as a consequence, a reduction in profits giving rise to a trade-off (Thornton, 2013, p. 169). Therefore, the owner/founder of a company (table 4.1) may also be driven by extrinsic

**Table 4.1**
Owner/founder motivation and company objectives.

| Motivation | Objective |
| --- | --- |
| Extrinsic | Profit maximization; and perhaps quality |
| Intrinsic: egoistic (self-regarding) | Quality and reputation |
| Intrinsic: altruistic (other-regarding) | Development and employment |

motivations (profit maximizer) or by intrinsic motivations (utility maximizer) of the egoistic type (quality of the wine and reputation of the company) or altruistic type (development and creating jobs). These objectives are not necessarily alternatives and can sometimes be pursued in parallel, as for example when an increase in quality in the long run is reflected in an increase in corporate profitability.[6]

Table 4.2 presents a classification of the main corporate types, with a description of the objectives of each one to give a better understanding of the differences between the various types of companies. Private companies reflect the neoclassical model more faithfully: the prime objective of most entrepreneurs is maximizing profit while social purposes are generally alien to the owner function. The pursuit of quality and corporate reputation is important for both profit maximizers if it increases long-term profitability and for utility-maximizer producers willing to give up part of their profitability to achieve noneconomic objectives.

Scott Morton and Podolny (2002) analyzed the intrinsic and extrinsic motivation of wine producers who were classified as profit or utility maximizers. The empirical results, obtained with surveys submitted to 184 Californian companies, show a strong variability in motivation, confirming the hypothesis that for all enterprises the sole objective of maximizing profits is not only reductive but also misleading. Further, entrepreneurs who are driven by intrinsic motivation (utility maximizers) on average produce wines of superior quality and are rarely ranked at the bottom. To reach the prefixed goals—whether they be profit, quality, and so on—the production inputs have to be chosen and combined appropriately. In a study that classified agricultural businesses in the state of Utah as utility or profit maximizers, Singell and Thornton (1997) found that utility maximizers have more capital and more family members working in the firm and fewer employees compared with profit maximizers. Dunkelberg et al. (2013) reached the same conclusions by analyzing the allocation choices of about three thousand new American companies created between 1984 and 1985.

The conglomerate is another type of private firm which is very profit-oriented. Some—as, for example, E. & J. Gallo, which owns other wine brands as well—have wine as their core business, in which case they have huge economies of scale and

**Table 4.2**
Legal form of companies and corporate objectives.

| Legal form | Main objective | Profits/members' income | Quality and reputation | Development and employment |
|---|---|---|---|---|
| Private companies | Profits and quality | Yes. The main objective of entrepreneurs, especially utility maximizers, is to make profit. | Yes. Quality and reputation are important objectives, especially for utility maximizers, and may be a priority. | No. Apart from a few exceptions, entrepreneurs do not follow social objectives. |
| Cooperatives | Members' income | Yes. The main objective is to maximize price paid to members who contribute grapes. | No. There are strong incentives for opportunistic behavior and little sense of loyalty to the company brand. Excellent results can be achieved only in regions/cooperatives with high social capital, strict rules, and effective control mechanisms. | Yes, through the income earned from the grapes delivered to the cooperative. |
| Social enterprises | Employment of disadvantaged people | No. There are no owners or profit-making objectives. The business aims to cover costs and to offer employment to disadvantaged people. | The aim is to sell products, overcoming buyers' mistrust. | Yes. The main objective is to find employment for disadvantaged people. |
| Foundations | Objectives other than economic gain | No. There are no owners or members, and they do not work for gain. The business aims to cover the necessary costs to reach its statutory objectives. | They exist only if they are expressly written in the statute. | Yes. The founder decides the objectives that generally concern economic development and employment. |
| State-owned companies | Development and employment | Not a priority. The business aims to cover the necessary costs to reach its statutory objectives, usually to generate development and create employment. | They exist only if they are expressly written in the statute. | Yes. The objectives are decided by public authorities and generally concern economic development and employment. |

large portfolios of products, increasing their bargaining power with distributors. Others instead belong to huge groups with well-differentiated activities, ranging from alcoholic beverages (including beer and spirits, like the British firm Diageo or the US company Constellation Brands) to fashion and luxury (like the French LVMH). These firms can affect competition by using profits from other sectors to lower their wine prices and force other firms out of the market (Thornton, 2013, p. 178).

The main goal of production cooperatives is to generate income for their members through payments in exchange for raw material and, in this way, create economic development and employment. They were set up to take over a further link in the value chain in wine making and to find a way around the critical moment of the harvest when the risk of grapes overripening dramatically shortens the period for negotiation. Wine production through a cooperative vastly extends this period because the finished product, unlike the raw material, can be kept for a long time. However, product quality, especially a good sensory profile,[7] is more difficult to achieve, given the strong incentive for opportunistic behavior and members' weak sense of identification with the company brand. Further, quality is relatively undifferentiated for many agricultural products, holding little fascination for consumers, and therefore members do not feel obliged to make any extra effort to improve the result (as, for example, in cereal production).

The wine sector, from this point of view, is more fortunate since it is not a question of the simple agronomic production of fruit: the transformation of grapes into wine involves an additional step that enhances differentiation, translating into a heightened identification of the cooperative member with the final product. The nature of wine as a product can also contribute favorably to the cause through its charm, history, and ties with the terroir. However, the prestige derived from the production of quality wine is shared with hundreds or thousands of other members, reducing the incentive to commit resources to obtain high-quality raw material. Excellent results can be achieved only in regions/cooperatives with high social capital, strict rules, and effective control mechanisms.[8]

Social enterprises can perform activities to offer socio-health and education services (classified in Italy as "type A," the most common) or to integrate disadvantaged people, such as the physically and mentally disabled, ex-prisoners, former drug addicts, and others into the labor market (classified in Italy as "type B") (Becchetti and Castriota, 2011). As there are no owners or profit-making objectives, social enterprises aim to cover costs and to offer services to the community and jobs to disadvantaged people. The purpose of achieving high-quality standards, if present, serves to allay buyers' suspicion of goods and services made by disadvantaged people.

Foundations are nonprofit legal entities created through an irrevocable donation and have neither owners nor members. They are managed by a committee that defines their goals (e.g., social or linked to a certain economic sector) and operating methods. On the one hand, managers are not fully remunerated on the basis of the results of the foundation, which should negatively affect the quality and efficiency of management, but on the other hand these business types are characterized by a lower intertemporal discount rate (they are more "patient"), which can be a strong point in long-term investment planning (Frick, 2004).

Finally, in state-owned companies, profit is a secondary and possible goal while business activities are aimed at covering the costs necessary to achieve statutory goals that generally refer to development and employment. If the company operates in a competitive market, quality will aim at guaranteeing economic sustainability. If it operates in a state monopoly, quality is only necessary if it can compromise consumers' and users' consensus toward the political class, which is responsible for the appointment of managers.

In the wine sector, social cooperatives, foundations, and state-owned companies are rare while production cooperatives constitute a real economic power, especially in Europe.

## 4.2   Vertical Integration and Quality

A second criterion for the classification of companies in the wine-making sector is the level of vertical integration since this influences the quality of the product and, in turn, the reputation of the company (figure 4.1). In fact, to achieve qualitative excellence, the entire production chain from the vineyard to the cellar has to be closely controlled. Economides (1999) demonstrated with a theoretical model that the quality offered by only one

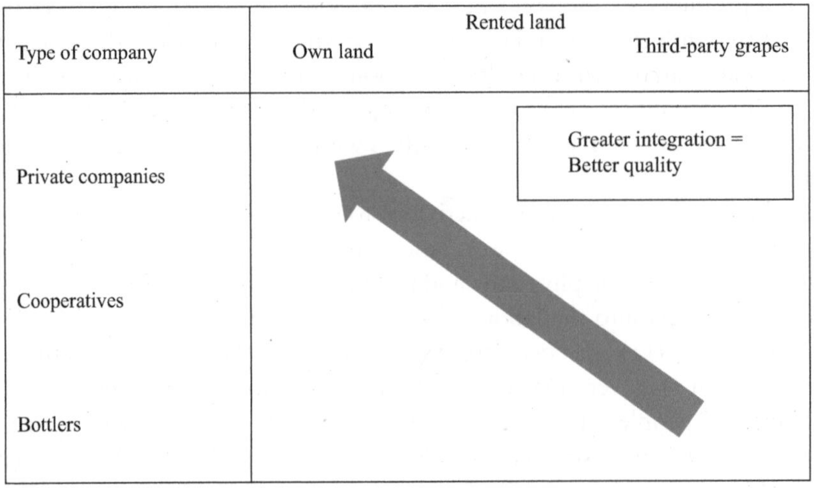

**Figure 4.1**
Type of company and level of vertical integration.

vertically integrated monopolist is higher than with a group of monopolists linked to each other in a direct vertical relationship. When there is an increase in the number of links in the production chain that are not subject to the direct control of the producer, it is reasonable to expect a decrease in the quality of the final good. From this point of view, there are three macro-groups of companies with a decreasing level of vertical integration and expected quality: private companies, cooperatives, and simple bottlers.

### 4.2.1 Private Companies

They vinify grapes that can come from land they own or rent or be purchased from third parties. The grapes coming from land that has been directly cultivated by the company, whether owned or rented, should be of the same quality, unless the rental agreement is short term, which could discourage expensive and irreversible investments. If grapes are purchased from third parties, suppliers could in principle guarantee higher quality and/or lower prices as they specialize in the production of only one good. These benefits, however, are frequently offset by incentives to opportunistic behavior because any increase in the price of wine sold only benefits the wine maker. The information asymmetry between supplier and buyer generates classical moral hazard (Pauly, 1968; Arrow, 1968). To solve this problem, the buyer can try to influence the behavior of the supplier and therefore the quality of the raw material in four ways (Hueth et al., 1999; Olmos, 2008), although these solutions are unlikely to lead to qualitative results on a par with those of land managed directly by the owner.

- Measure the quality of the product at the time of delivery to decide whether to accept the goods and how much to pay for them. An objective evaluation of the quality of grapes is, however, rather expensive and difficult to accomplish in a short time and on a large scale (Oczkowski, 2001). The most frequently used techniques, involving reduced costs and time, are based primarily on the analysis of the sugar content and color but lead to results that are at best satisfactory and certainly not perfect (Fraser, 2003).
- Periodically and directly monitor the supplier through field inspections. This activity does not necessarily have to be conceived as "quality control" and can serve to share information, experiences, and professionalism. An optimal number of inspections needs to be fixed to maximize the economic return because visits involve a cost in terms of specialized staff, and if they are too frequent they lose their value.
- Impose rules and production standards on vines and planted clones, agronomic techniques, frequency, and timing of pruning, irrigation, and so on.
- Make the supplier responsible for the price paid by binding it to the price obtained by the producer for the sale of the wine.

What matters most, however, in addition to capital and raw materials, is labor—that is, the managerial activity carried out by the owner or the external manager. Before

Berle and Means (1932), who were the first to document the rise of large corporations, economic theory did not make a clear and explicit distinction between ownership and management. In the following decades the literature made in-depth analyses of the contribution of these variables to company performance. In the wine industry private companies can be managed by the owner or an external manager. The decision to delegate control of the company to somebody outside the family unit can be taken for a number of reasons. First, an external manager may have more skills and experience and contribute to the winery's economic success. Delegation is more frequent as the size and complexity of the company grows (Colombo and Delmastro, 2004). Second, while it is reasonable to expect the company founder to be actively involved, later generations often cultivate and pursue interests other than the management of the family business. However, there is little literature on the relationship between the age of the company and management by external managers.

What is of greater interest here is to understand the impact of management on the quality of wine. Is the quality superior, net of all other variables, in the wineries managed by the owner or in those managed by an external manager? In an analysis of the Piedmontese wine market, Delmastro (2007) found that cellars managed by the owners in person got better qualitative ratings on average in the three main national guides (I Vini di Veronelli, the Slow Food Wines of Italy/Gambero Rosso, and the Italian Sommelier Association's Duemilavini), confirming the importance of many producers' noneconomic motivations found by Scott Morton and Podolny (2002).[9] The manager, on the other hand, benefits only in part from the prestige enjoyed by the winery. They may be more interested in maximizing their pay, especially if the variable is linked to short-term profit, while investments in quality have a very long-term horizon. The possible differences between the objectives pursued by the owner (principal) and the manager (agent) represent a typical case of moral hazard (Pauly, 1968; Arrow, 1968; Zeckhauser, 1970), to which we can add the problem of adverse selection (Akerlof, 1970) caused by the recruitment of qualified personnel in a context with information asymmetries about the skills and talent of candidates.

Dilger (2004) and Frick (2004), however, come to opposite conclusions, both from a theoretical and an empirical point of view. Dilger hypothesizes that (1) an external manager has a greater interest in favoring quality over quantity because the costs connected to the improvement of agronomic and wine-making techniques fall on the owner, (2) an external manager has more skills compared with the owner, and (3) the owner evaluates the manager's skills not on the basis of the quantities produced or the profits made but rather the quality of the wine. Although the third of these hypotheses appears questionable, the author finds, using data relating to over 77,000 German wines produced by 309 companies in the period 1996–1999, that the bottles produced by wineries overseen by external managers on average receive better ratings in the *Der Feinschmecker* ("The Gourmet") guide.

Frick (2004) empirically investigated two antithetical hypotheses. First, companies managed by the owner should be more efficient and produce better wines because, all things being equal, it is difficult and expensive to monitor the work of managers. Second, the specific skills of an external manager have a positive effect on the product quality and business results. The econometric analysis conducted with panel data relating to 305 German wineries in the period 1996–1999 showed that private companies run by external managers produce better wines not only compared with private firms managed by the owner but also compared with cooperatives, foundations and state-owned companies.

## 4.2.2 Cooperatives

Wine cooperatives guarantee a lower degree of vertical integration compared with private companies since they directly control only the phase of vinification while the land remains the property of the member who cultivates it and then delivers the grapes to the winery. The member must obviously follow a series of indications and rules set by the cooperative, but the effort involved is a variable that cannot be completely controlled. As the members are remunerated on the basis of the quantities produced, the (individual) short-term advantages of opportunistic behavior aimed at maximizing quantities at the expense of quality tend to outweigh the (collective) long-term costs in terms of lower product quality and the reputation of the group (Pennerstorfer and Weiss, 2006, 2012; Fishman et al., 2008).[10] This is especially true when the number of members grows because it becomes more difficult to check the behavior of the various cooperative members and because the individual member views the damage he causes the group with his opportunistic behavior as insignificant. The problem of moral hazard is exacerbated when members are free to decide the quantities they deliver. The decentralization of decision making leads to excess production (Albæk and Schultz, 1998), negatively affecting quality.

Economic models generally consider affiliation to a cooperative as given and the loyalty to an association as high. In reality, members can differ in the quality of their land, their moral qualities, and their loyalty to the common cause. As cooperatives guarantee freedom of entry (the "open door" principle), the social base may, on average, be made up of "worse" individuals than those who produce their own wine or sell grapes to private wineries. In other words, other things being equal, it may well be that the most ambitious landowners or those who have the best land decide to produce on their own while the less resourceful or those who have no chance of excelling decide to join a cooperative. In contrast, Fulton (1999) underlined the importance of both the commitment of members and a cooperative's ideology for economic success where "cooperative ideology" means a preference for business types that can be controlled directly.

A series of empirical studies has led to rather consistent results indicating how cooperatives produce lower quality wines on average. Evidence of this is given in

Frick (2004) and Dilger (2004), who used German data, and Delmastro (2007), who used Piedmontese data on wine quality.[11] However, there exist illustrious examples of regions—such as the South Tyrol in Italy—that have, in fact, created prestigious cooperatives in terms of excellent quality while others—like Emilia Romagna and Trentino—have focused on quantity and ready-to-drink wines but guarantee excellent value for money. Both cases are success stories.

### 4.2.3   Bottlers

The lowest degree of vertical integration of production occurs with bottlers—private companies that do not produce their own wine but buy from other companies and then sell it under their own brand. In this case, the company relies entirely on suppliers both from the agronomic and the wine-making point of view. Although there is no empirical literature on the subject, these companies may be expected to have a minimum level of quality and to sell large quantities (usually) through mass market retailers. The very fact, however, that the wines produced by this type of company are not being reviewed in wine guides is indicative of their low quality.

### 4.3   Cooperatives in the Wine Sector: An Overview

In the previous section cooperatives were just fleetingly touched on. They were included in the classification of company types based on the degree of vertical integration that is then reflected in the quality of the product. This section will deal with them in a more in-depth, though not exhaustive, manner.

### 4.3.1   Beginning and Development of Cooperatives

Forms of spontaneous cooperation have always existed, but the origin of organized cooperation can be traced back to the creation of the first cooperative outlet in the English town of Rochdale in 1844 by twenty-eight textile workers and artisans. The aim was to open a business where even the poorest could buy basic necessities. It worked with modest contributions—about one pound each—from members. Unlike previous experiences that had failed immediately, this one was successful and expanded with the opening of a butchery and a mill. The greatest merit of this company, which has gone down in history with the name of the "Rochdale Society of Equitable Pioneers," is that it laid down the ideological foundations of the cooperative movement that is still today based on the principles established in the society's statute. Cooperatives later spread throughout Europe and then the world, moving from consumption to the production of goods and services, work, credit, building, and so on.

The growth of the cooperative movement in the wine sector was favored by the spread of phylloxera and powdery mildew, dramatically reducing the quantities produced toward the end of the nineteenth century with a consequent increase in average prices. When the remedies were finally found, production gradually returned to pre-crisis levels that, together with the widespread production of adulterated wines, led to a sharp contraction in prices. The profitability of wineries was put to the test even further by the increase in production costs to combat vine parasites and diseases (fertilizers, pesticides, and grafting of European grapevines onto US rootstocks). This was all taking place in a historical period that also saw the emergence of trade union struggles and rising wages. Simpson (2000) believes that these elements encouraged first informal cooperation and then the birth of cooperatives. In France it became easier to form cooperatives after the abolition in 1844 of compulsory governmental authorization for the establishment of associations of more than twenty people. To sum up, the need to reduce production costs by making collective purchases of raw materials and sharing technological knowledge to fight vine fungi and parasites between the end of the nineteenth century and the first half of the twentieth century played a fundamental role in the growth of cooperatives.

Cooperatives now hold significant, and at times dominant, market shares in various sectors of the economy, such as milk (100 percent in Malta, 99 percent in Sweden, 97 percent in Denmark and Finland, 90 percent in Uruguay, 84 percent in the Netherlands, 80 percent in Slovenia and Portugal), pork (100 percent in Malta, 90 percent in Denmark), beef (80 percent in Slovenia and Sweden), cotton (77 percent in Burkina Faso), fishing (90 percent in Malta), flowers (95 percent in the Netherlands), wood (73 percent in Canada, 70 percent in Slovenia, 38 percent in Finland), eggs (60 percent in Denmark), and fruit and vegetables (58 percent in the Netherlands) (Logue and Yates, 2005; Bogetoft, 2005; Pennerstorfer and Weiss, 2006). The list could go on, including credit unions, construction companies, and so forth. In the United States, thirty thousand cooperatives employ two million people while in Japan the turnover in the agricultural sector alone is around $90 billion dollars (ICA, n.d.b.). The important role played by cooperatives in the creation of income and employment was explicitly recognized by the International Labor Organization in Recommendation No. 127 of 1966, renewed in No. 193 of 2002, and later actively promoted within the framework of the World Employment Conference of 1976 (Logue and Yates, 2005). The actual number of cooperative members—not only in production but also in consumption, credit, and so on—is even higher (see table 4.3) for a global total of 718 million members in 2013,[12] while a less conservative estimate stretches this number to one billion individuals employing over 280 million people (ICA, n.d.b). Box 4.1 provides a brief description of the history of cooperatives in Italy.

Given their importance, it is surprising that the study of this type of company has practically disappeared from university courses and economics textbooks. Lynch, Urban, and Sommer (1989) examined the curricular programs of sixty-three American universities in 1977 and found that only twenty-four institutions, eighteen less

**Table 4.3**
Number of cooperative members in the world, 2013.

| Africa | | The Americas | | Asia | | Europe | |
|---|---|---|---|---|---|---|---|
| Kenya | 8,650,000 | USA | 225,901,137 | India | 93,755,144 | Italy | 12,555,533 |
| Nigeria | 5,490,825 | Canada | 18,620,000 | Japan | 77,034,387 | Great Britain | 10,019,000 |
| Rwanda | 1,624,032 | Brazil | 15,279,400 | Iran | 36,902,477 | France | 9,669,638 |
| Tanzania | 1,380,000 | Argentina | 4,894,400 | China | 31,000,000 | Poland | 8,100,000 |
| Zambia | 877,442 | Ecuador | 2,962,000 | Indonesia | 30,000,000 | Russia | 5,040,000 |
| Uganda | 509,000 | Mexico | 2,479,900 | Bangladesh | 30,000,000 | Finland | 4,211,781 |
| Mauritius | 150,000 | Colombia | 2,436,002 | Sri Lanka | 12,100,000 | Sweden | 3,961,475 |
| Lesotho | 90,000 | Costa Rica | 1,465,400 | Thailand | 10,552,839 | Germany | 3,327,727 |
| South Africa | 56,000 | Guatemala | 1,113,200 | Vietnam | 6,500,000 | Norway | 2,329,378 |
| Botswana | 51,400 | Uruguay | 909,598 | Malaysia | 5,819,170 | Denmark | 2,004,803 |
| Zimbabwe | 10,000 | Paraguay | 810,200 | Korea | 5,372,740 | Portugal | 2,000,006 |
| Guinea | 772 | Puerto Rico | 682,677 | Nepal | 3,206,100 | Belarus | 1,500,000 |
| | | Bolivia | 488,250 | Philippines | 2,000,000 | Turkey | 1,446,802 |
| | | Chile | 472,000 | Singapore | 1,400,000 | Cyprus | 1,193,982 |
| | | Honduras | 218,500 | Myanmar | 1,085,692 | Czech Republic | 722,205 |
| | | Panama | 216,000 | Pakistan | 921,999 | Spain | 589,848 |
| | | El Salvador | 136,700 | New Zealand | 600,000 | Ukraine | 482,700 |
| | | Peru | 133,560 | Kuwait | 444,753 | Austria | 477,000 |
| | | Dominican Republic | 118,900 | Israel | 153,000 | Bulgaria | 253,362 |

| Country | Value | Country | Value | Country | Value |
|---|---|---|---|---|---|
| Jamaica | 18,000 | Mongolia | 108,000 | Slovakia | 233,204 |
| Venezuela | 17 | Australia | 54,323 | Moldova | 200,200 |
| | | Tajikistan | 50,000 | Serbia | 135,367 |
| | | Kyrgyzstan | 20,000 | Hungary | 95,000 |
| | | Kazakhstan | 16,700 | Romania | 66,184 |
| | | | | Latvia | 50,000 |
| | | | | Holland | 50,000 |
| | | | | Georgia | 50,000 |
| | | | | Lithuania | 40,058 |
| | | | | Belgium | 40,000 |
| | | | | Croatia | 27,115 |
| | | | | Switzerland | 10,000 |
| | | | | Malta | 4,606 |
| | | | | Ireland | 2,500 |
| | | | | Armenia | 1,133 |
| **Total** | **18,889,471** | | **279,355,841** | | **70,890,607** |
| **Total combined** | | | | | **718,233,243** |

*Source:* ICA.

**Box 4.1**
Brief history of Italian cooperatives.

---

The first Italian cooperative was set up in Pinerolo in 1849 and pioneered the rapid development of the movement that took shape with the foundation of both the National Federation of Cooperatives (Federazione Nazionale delle Cooperative) in 1886—later to become the League of Cooperatives (Lega delle Cooperative) in 1993—and the Italian Cooperative Confederation (Confederazione Cooperativa Italiana), which was of Catholic origin, in 1919. The years of the fascist dictatorship, however, brusquely slowed down the development of the movement as it was seen as an obstacle to the totalitarian regime and considered a carrier of Bolshevik ideologies. Many cooperatives were forced to close and their offices destroyed while the League of Cooperatives and the Confederation of Italian Cooperatives were dissolved between 1925 and 1927 and replaced by the National Fascist Authority of Cooperation (Ente Nazionale Fascista della Cooperazione), founded in 1926.

The postwar years saw the revival of the movement; trade associations were reestablished in 1945 and the General Directorate of Cooperation at the Ministry of Labor was formed in 1946. The DLCPS No. 1577 (Basevi law) regulating cooperatives was passed in 1947 and the "social function of mutual aid cooperation without private speculation" was recognized in article 45 of the Constitution of the Italian Republic in 1948.[a] Today cooperatives are responsible for about 7 percent of gross domestic product (GDP), have twelve million members, provide over 1.1 million jobs, and hold leading positions in many sectors of the national economy (Fabbri, 2011).

*Note*: [a]The source of this excursus was a page on the Confcooperative website accessed in 2014, http://www.confcooperative.it/C9/La%20%20story%20cooperation/default .aspx (page no longer available).

---

than in 1960, had courses focusing on the study of cooperatives, and in any case, they were not compulsory. Hill (2000) checked twenty-five economics textbooks used in Canada and concluded that cooperatives are ignored or at the most just fleetingly mentioned,[13] while Kalmi (2007) analyzed the quality and quantity of space dedicated to the theme of cooperatives in economics textbooks used at the University of Helsinki from 1905 to 2005 and found that both had fallen drastically since World War II. The author explains this is due to the dominance of the neoclassical economic school of thought that includes only private company types in its models. Although this conclusion is plausible, the spread of McCarthyism and "Red Scare" around 1950 probably contributed to making the study of cooperatives unpopular in the United States and, in turn, in all other countries. Box 4.2 contains the definition, values, and principles of cooperatives as stated by the International Cooperative Alliance (ICA) while their strengths and weaknesses are discussed in the next paragraphs.

**Box 4.2**
The ICA definition, values, and principles of cooperatives.

---

### Definition

A cooperative is an autonomous association of persons united voluntarily to meet their common economic, social, and cultural needs and aspirations through a jointly owned and democratically-controlled enterprise.

### Values

Cooperatives are based on the values of self-help, self-responsibility, democracy, equality, equity, and solidarity. In the tradition of their founders, cooperative members believe in the ethical values of honesty, openness, social responsibility and caring for others.

### Principles

The cooperative principles are guidelines by which cooperatives put their values into practice.

1. Voluntary and Open Membership
Cooperatives are voluntary organizations, open to all persons able to use their services and willing to accept the responsibilities of membership, without gender, social, racial, political or religious discrimination.

2. Democratic Member Control
Cooperatives are democratic organizations controlled by their members, who actively participate in setting their policies and making decisions. Men and women serving as elected representatives are accountable to the membership. In primary cooperatives members have equal voting rights (one member, one vote) and cooperatives at other levels are also organized in a democratic manner.

3. Member Economic Participation
Members contribute equitably to, and democratically control, the capital of their cooperative. At least part of that capital is usually the common property of the cooperative. Members usually receive limited compensation, if any, on capital subscribed as a condition of membership. Members allocate surpluses for any or all of the following purposes: developing their cooperative, possibly by setting up reserves, part of which at least would be indivisible; benefiting members in proportion to their transactions with the cooperative; and supporting other activities approved by the membership.

4. Autonomy and Independence
Cooperatives are autonomous, self-help organizations controlled by their members. If they enter into agreements with other organizations, including governments, or raise capital from external sources, they do so on terms that ensure democratic control by their members and maintain their cooperative autonomy.

*(continued)*

**Box 4.2** (continued)

5. Education, Training, and Information
Cooperatives provide education and training for their members, elected representatives, managers, and employees so they can contribute effectively to the development of their cooperatives. They inform the general public—particularly young people and opinion leaders—about the nature and benefits of cooperation.

6. Cooperation among Cooperatives
Cooperatives serve their members most effectively and strengthen the cooperative movement by working together through local, national, regional and international structures.

7. Concern for Community
Cooperatives work for the sustainable development of their communities through policies approved by their members.

*Source*: ICA, n.d.a.

### 4.3.2   Advantages/Strengths

Cooperatives present a series of advantages and strengths that have made their development and diffusion in the world possible. The first is related to the increase in economies of scale for producers who would otherwise be too small to compete (Schroeder, 1992; Hansmann, 2012; Bijman et al., 2012). This is particularly important in countries which, for both historical and demographic reasons, have seen land ownership fragmented into a myriad of small plots. This is the case of Europe because the continent has a very long history and high population density. Table 4.4 shows the marked differences between wine-producing countries that have a population density in the Old World that are five, ten, or even one hundred times greater than in the New World, with the only exception of China (see box 4.3 for an explanation of why cooperatives are so important in Italy).

The second advantage of cooperatives is the opportunity for members to acquire an additional link in the value chain (Bijman et al., 2012)—that is, the transformation of grapes into wine and its marketing—materializing at the end of the process in the redistribution of profits to members in proportion to the quantities supplied.

This is particularly important in the wine sector, which is characterized by chronic excess supply, because the agronomic side is considered the weak link in the chain. In fact, grape producers have limited bargaining power due to the high perishability of the raw material, ferocious competition in the market, and domestic demand that has fallen in southern Europe since the 1970s. The cooperative, therefore, increases the bargaining power of the individual producer in line with the philosophy of "unity is strength," both when selling their products and buying raw materials, seeds, and

Table 4.4
Population density of main wine-producing countries.

| Country | Population growth rate (% year) 2005–12 | >Mid-year population estimate (thousands) 2012 | Area (km²) 2012 | Population density (inhabitants/km²) 2012 |
|---|---|---|---|---|
| Germany | −0.1 | 81,932 | 357,137 | 229 |
| Italy | 0.5 | 60,851 | 301,339 | 202 |
| China | 0.5 | 1,350,695 | 9,596,961 | 141 |
| France | 0.5 | 63,556 | 551,500 | 115 |
| Portugal | 0.0 | 10,542 | 92,212 | 114 |
| Austria | 0.4 | 8,466 | 83,871 | 101 |
| Spain | 0.9 | 46,163 | 505,992 | 91 |
| Greece | 0.2 | 11,290 | 131,957 | 86 |
| United States | 0.9 | 313,914 | 9,629,091 | 33 |
| Chile | 1.0 | 17,403 | 756,102 | 23 |
| Brazil | 0.8 | 193,947 | 8,514,877 | 23 |
| Uruguay | 0.3 | 3,381 | 176,215 | 19 |
| New Zealand | 1.0 | 4,433 | 270,467 | 16 |
| Argentina | 1.0 | 41,282 | 2,780,400 | 15 |
| Australia | 1.5 | 22,684 | 7,692,024 | 3 |
| Canada | 1.1 | 34,880 | 9,984,670 | 3 |

*Source*: Author's calculations using data from United Nations, Statistics Division (n.d.).

so forth (Schroeder, 1992; Bogetoft, 2005; Logue and Yates, 2005; Hansmann, 2012; Pascucci, Gardebroek, and Dries, 2012).

In a study on the US milk market, Cakir and Balagtas (2012) demonstrated that cooperatives use their market power to increase the sale price by up to 9 percent above the marginal cost, amounting to a transfer of income to their advantage of about $600 million a year. In a detailed historical excursus Simpson (2005) retraced the tumultuous events that upset the French wine market at the end of the nineteenth century. With the spread of phylloxera, grape production was drastically reduced after 1875, forcing wine producers to look for raw material abroad or to carry out real adulterations and frauds to the detriment of consumers. When phylloxera was defeated, domestic production returned to precrisis levels, but many manufacturers and traders continued to use foreign raw material and to "synthesize" artificial wines, thus reducing drastically the profitability of the sector. To solve the situation, grape growers in the regions of Midi, Bordeaux, and Champagne used their political influence to convince public authorities to intervene and fight fraud by establishing

**Box 4.3**
Why is the cooperative movement so important in Italy?

Cooperatives are very important in Europe because of its longer history and higher population density compared with the New World. This has led to fragmented land and small firms.

In Italy the situation is exacerbated by two further elements: 35 percent of the territory is covered by mountains and the 1950 reform intensified the dispersion of land ownership. Law No. 841/50, which contained rules for the expropriation, reclamation, transformation, and allocation of land to small farmers, tried to improve the dramatic economic conditions of agricultural workers immediately after World War II. Through the expropriation of large estates and the subsequent reallocation of the land in small units, it endeavored to create a less unequal distribution of wealth and income by giving laborers the basic instruments necessary to provide for themselves.

However, sometimes there is a trade-off between efficiency and fairness. The agrarian reform transformed large estates of thousands of hectares into a multitude of production units of such modest size (generally from two to twenty hectares) that they were often inefficient as a result of the reduced economies of scale. Joining forces in a cooperative was a way of circumventing the problem.

All this explains the leading role played by cooperatives in Europe and in Italy: as shown in table 4.5, eight of the top twenty-five Italian wineries (by turnover) were cooperatives in 2012 and 2013, and three among the top five were also cooperatives. The table also shows a strong propensity for these companies to internationalize, their share of foreign sales varying considerably but often exceeding 50 percent of total sales. In the wine sector in Italy the average size of plots of land owned by cooperative members is around a hectare, varying from a minimum of 0.1 to a maximum that generally does not exceed ten or twenty hectares. With the exception of the producers of the highest quality wines, the minimum size to cover operating costs is around twenty to thirty hectares, so the majority of these small owners would not be able to produce wine but only sell the raw material to other wine makers.

appellations and creating producer cooperatives to contrast the excessive power of producers and traders.

Both in advanced (Nilsson, 2001; Bijman et al., 2012) and developing (Becchetti, Castriota, and Conzo, 2015) countries, however, the most immediate and tangible advantage for members is the higher and more stable price paid by cooperatives than by private buyers for the raw material. The higher price is due to the reasons mentioned above—namely, economies of scale, bargaining power, and the appropriation of an additional link in the value chain. Further, if the cooperative establishes strict quality standards and is able to effectively monitor and motivate producers, the raw material may be of a higher quality than the members would have guaranteed to a private individual, with a positive reflection on the amount paid. Price stability is

**Table 4.5**
The top 25 wineries in Italy by turnover.

| Ranking | Firm name | Location | Turnover 2012 (millions of €) | Turnover 2013 (millions of €) | Foreign share (% of turnover 2013) | Type of firm |
|---|---|---|---|---|---|---|
| 1 | Cantine Riunite & Civ | Campegine (RE) | 512 | 534 | 62.5 | Cooperative |
| 2 | Caviro | Faenza (RA) | 284 | 327 | 27.2 | Cooperative |
| 3 | Campari (divisione vini) | Milan | 196 | 228 | n.d. | Family |
| 4 | Mezzacorona | Mezzocorona (TN) | 160 | 163 | 58.5 | Cooperative |
| 5 | Fratelli Martini Secondo Luigi | Cossano Belbo (CN) | 158 | 159 | 90.0 | Family |
| 6 | P. Antinori | Florence | 157 | 166 | 67.2 | Family |
| 7 | Cavit Cantina Viticultori | Ravina (TN) | 153 | 153 | 78.0 | Cooperative |
| 8 | Casa Vinicola Zonin | Gambellara (VI) | 140 | 154 | 76.2 | Family |
| 9 | Enoitalia | S. Martino Buon Albergo (VR) | 113 | 128 | 75.6 | Family |
| 10 | Giordano Vini | Diano d'Alba (CN) | 110 | 101 | 47.7 | Mixed |
| 11 | Cantina sociale cooperativa di Soave | Soave (VR) | 107 | 103 | 50.3 | Cooperative |
| 12 | Casa Vinicola Botter Carlo & Co. | Fossalta di Piave (VE) | 104 | 136 | 94.8 | Family |
| 13 | Gruppo Cevico | Lugo (RA) | 96 | 117 | 21.2 | Cooperative |
| 14 | Gruppo Santa Margherita | Fossalta di Portogruaro (VE) | 95 | 102 | 61.5 | Family |
| 15 | La Vis | Lavis (TN) | 92 | 85 | 76.2 | Cooperative |
| 16 | Compagnia De' Frescobaldi | Florence | 85 | 83 | 62.1 | Family |
| 17 | Schenk Italia | Ora (BZ) | 84 | 81 | 75.3 | Foreign |
| 18 | Collis Veneto Wine Group | Monteforte D'Alpone (VR) | 76 | 78 | 15.2 | Cooperative |
| 19 | Contri Spumanti | Cazzano di Tramigna (VR) | 71 | 93 | 39.3 | Family |
| 20 | Ruffino | Pontassieve (FI) | 71 | 72 | 93.2 | Foreign |
| 21 | Cantine Turrini Valdo & Figli | Riolo Terme (RA) | 67 | 85 | 81.6 | Family |
| 22 | Masi Agricola | S.Ambrogio di Valpolicella (VR) | 66 | 65 | 91.2 | Family |
| 23 | Fratelli Gancia & Co. | Canelli (AT) | 64 | 50 | 29.4 | Mixed |
| 24 | Casa Vinicola Caldirola | Missaglia (LE) | 64 | 59 | 28.3 | Family |
| 25 | MGM Mondo del Vino | Forlì | 64 | 66 | 66.6 | Mixed |

*Source:* Mediobanca (2014), Indagine sul Mercato Vinicolo, table 5.

the result of a cooperative's commitment not to follow fluctuations in the market, especially in times of crisis, favoring in this way its members' financial planning of revenue and expenses.[14]

Another significant benefit is the ancillary services provided to members, such as the supply of seeds and fertilizers, agronomic advice, and so on (Logue and Yates, 2005), which increase worker productivity and/or reduce costs, with obvious consequences for net income. Most of the time cooperative members underestimate or are completely unaware of the cost incurred to provide these services and the advantages they derive from them.

The separation of ownership from the effective control of the company is the source of what in economics is called the "principal-agent problem" (Rees, 1985; Milgrom and Roberts, 1992). In this respect cooperatives limit both the risk of moral hazard and the problem of the owner having to monitor the manager when he delegates power (Novkovic, 2008; Cook, 1994). In cooperatives, in fact, the manager is often a member (especially in smaller ones) while in private firms power is often delegated to an external manager who can pursue objectives that differ from those of the owner. The manager has a short-term perspective as he can be removed at any time and therefore often aims to maximize profits immediately (as year-end bonuses are linked to results), even if this behavior in the long run may jeopardize brand reputation or growth prospects.

In addition to the strictly economic advantages, cooperatives benefit from more favorable tax treatment than for-profit firms (Sexton, 1990; Tennbakk, 1995; Cook, 1995). They also have a psychological boost from the satisfaction of taking part, albeit indirectly, in the management of one's own business, which is based on the tenets of cooperatives—the ideals of justice, equity and reciprocity (Fehr et al., 2007; Pascucci et al., 2012).

From a more "macro" perspective, cooperatives have a greater survival rate, and their productivity is at least comparable with private companies (Logue and Yates, 2005; Defourney, Estrin, and Jones, 1985; Hall and Geyser, 2004; Simpson, 2005; see Valette, Amadieu, and Sentis, 2018 for a study on wine cooperatives). They also contribute to greater equity in distribution (Vanek, 1970) because members do not only act as a factor of labor but also have a share in the distribution of profits. Next, with a larger number of owners, cooperatives have a lower risk of frauds, and this is particularly important in the agri-food sector for their implications for human health. Last, but not least, cooperatives are very difficult to take over because the sale must be approved by an absolute majority of members. This ensures that the ownership of the company remains in the home country.

However, as Pascucci et al. (2012) pointed out, the advantages of cooperatives may be partially overestimated since many of them allow for less rigid relationships with their members and since many members are so only on paper.[15]

### 4.3.3   Disadvantages/Weaknesses

After reviewing the strengths of the cooperatives, we now move on to examine the weaknesses identified in the scientific literature. The first concerns an issue that has already been discussed—namely, the possible self-selection of members who could have poor land or less intrinsic motivation and initiative. In other words, it is argued that, all things being equal (for example, the number of hectares and geographic context), an owner of valuable land and/or someone who is strongly motivated is unlikely to join a cooperative and would probably prefer to create a company of his own.

Secondly, there are obvious problems of adverse selection and moral hazard. Adverse selection concerns the quality of the grapes delivered by members to cooperatives. As in the car market described by Akerlof (1970), the application of a uniform price to goods of differentiated, but not observable, quality can lead to the self-exclusion of the best suppliers or, hypothetically and theoretically speaking, even to the disappearance of the market. Besides, as some cooperatives do not insist on members delivering all their grapes and as there is a risk of a partial sale to third parties under the counter, the worst grapes could end up in the cooperative. Information asymmetries between members and the cooperative, therefore, lead to a moral hazard with respect to the quality of the grapes—this time as a consequence of the farmer's commitment rather than the quality of the land, even if having a share in company profits should mitigate the problem to some extent (Defourney, Estrin, and Jones, 1985; Jones and Pliskin, 1988).

Moreover, as members are remunerated on the basis of the quantities delivered, there is a strong incentive toward overproduction (Albæk and Schultz, 1998; Bogetoft, 2005), which has a negative impact on quality. As discussed above, the solution to the problems of adverse selection and moral hazard is quality control at the time of delivery, inspections, strict rules, and so on, but it is only partial. The scientific literature has shown how, on average, the quality of wine produced by cooperatives is lower than that of private companies.

While decision-making power is firmly concentrated in the hands of the owner or manager in private companies, it is exercised by a manager in cooperatives but should represent the will of all the members. This can cause difficulties both in organization (Defourney, Estrin, and Jones, 1985) and coordination, especially if the quality of the product is differentiated and members are heterogeneous.

It is often said (Furubotn and Pejovich, 1970; Porter and Scully, 1987) that cooperatives make fewer capital investments than traditional companies because of the lack of property rights that prevents them from selling shares or capital assets when members leave the cooperative. This would act as a deterrent to the use of equity and an incentive to rely on credit. In light of these considerations, it is reasonable to expect a higher level of indebtedness than in private companies (Soboh et al., 2009). Ferrier and Porter (1991) reached the conclusion that cooperatives are a suboptimal organization because of the limited time horizon of members; the nontransferability of shares; and control and management problems that lead to technical, allocation, and scale inefficiencies. The authors supported their statements by analyzing US data on the dairy sector in 1972. But, in fact, in the case of a member withdrawing from the cooperative, share capital is liquidated strictly at its nominal value, but it can also be reduced in proportion to the losses attributable to the capital and in any case on the basis of the criteria established in the articles of association. In line with this, Maietta and Sena (2008) used Italian data on a group of private companies and wine-producing cooperatives for the period 1996–2003 and did not find any under-capitalization in cooperatives in comparison with private companies.

Finally, cooperatives may be characterized by greater risk aversion and reluctance to adopt new technologies since the activity of the cooperative is the main source of livelihood for many members and often production cannot be diversified or unpro-ductive activities abandoned, as noted by Katz (1997) in a study conducted on 228 private companies and eighty-three cooperatives over the period 1988–1992. The author came to the conclusion that the company type plays a fundamental role in determining the corporate strategies of agricultural enterprises.

### 4.3.4   Conditions for the Success of Cooperatives

In order for a cooperative to be successful in terms of product quality and brand reputation as well as in economic terms, it needs to have a number of characteris-tics. The first is the homogeneity of members (Hanf and Schweickert, 2007a, 2007b; Capitello and Agnoli, 2009): in fact, if they are very different or pursue different objectives, this will lead to inefficiencies typical of when things are done randomly. To avoid the formation of heterogeneous groups of members, the objectives and operating rules of a cooperative must be established clearly from the beginning. Precise rules lead to the self-selection of members.

Second, the number of members should not be too small (to guarantee economies of scale and the visibility of the company brand) nor too large because the risk of opportunistic behavior will increase, as described above (Kollock, 1998; Rey and Tirole, 2007; Pennerstorfer and Weiss, 2007, 2013; Fishman et al., 2008; Castriota and Delmastro, 2015; Bonroy et al., 2018).[16] Although it may often be preferable to avoid the proliferation of members, the statute of many cooperatives forbids barri-ers to entry in accordance with the "open door" principle, even when there is a need to create "strategic groups" (Porter, 1980). However, if production or commercial capacity has reached saturation point, a temporary block on the entry of new mem-bers can be set up and possibly renewed.

The pursuit of shared statutory objectives depends on the commitment of members (Fulton, 1999) and is helped by the presence of large quantities of human capital and social capital. The first can be defined as an individual's set of skills, values, and state of health acquired during their life, which will affect future income (Becker, 1962). The second concerns the institutions, the rules, and norms that regulate the quality and quantity of social interactions (Putnam, 1993). Both vary from one region to another and influence the propensity to cooperate. In their study on a sample of Italian agricultural cooperatives, Pascucci et al. (2012) used binary variables to capture the less favorable attitude toward cooperation of people living in southern Italy, a circumstance that Menzani and Zamagni (2009) explained by the scarce presence of this type of organization in the south. The sustainability of the cooperative model depends on the existence of social preferences and cooperative rules compatible with low levels of opportunistic behavior and with stringent mutual monitoring (Ben-Ner and Ellman, 2013).

Strict rules, however, are useless if there is no effective system of controls and penalties (Castriota and Delmastro, 2015). The checks must be frequent enough to discourage opportunistic behavior by members—but not too frequent as to weigh down on the budget of the company—while sanctions in case of violations of the regulation must be severe. However, since it is the management of the company that is responsible for checks and penalties, and it may include members, there is a clear problem of conflict of interest.

## 4.4   Industrial Districts/Clusters

### 4.4.1   Definition of Cluster

An industrial district is an agglomeration of small- and medium-sized businesses operating in a circumscribed territory, specializing in one or more phases of a production process, and tightly integrated through a complex network of formal and informal economic and social relations. Alfred Marshall is considered the "father" of the theory of industrial districts as he had already defined the concept and its characteristics in the late nineteenth century. Marshall and Marshall (1879) stated that

> some of the advantages of division of labor can be obtained only in very large factories, but that many of them, more than at first sight appears, can be secured by small factories and workshops, provided there are a very great number of them in the same trade. The manufacture of a commodity often consists of several distinct stages, to each of which a separate room in the factory is devoted. But if the total amount of the commodity produced is very large, it may be profitable to devote separate small factories to each of these steps.[17]

Economies of scale, defined as the decrease in average production costs connected to the growth of the size of production, can be internal or external. The former refer to the size of the company and to the efficiency of its management while the latter are the result of the general development of the sector to which it belongs. In his studies with applications on the British economy of the late nineteenth century, Marshall (1890) showed how it is possible to pursue, with an interconnected network of small businesses, external economies of scale comparable with the internal ones present in large companies. In different contexts, therefore, the typical efficiency of large-scale internal production can be achieved by grouping a large number of small businesses in the same district and subdividing the production process into various phases, each of which can be performed in a small factory with maximum efficiency. Districts can have vertical forms (when businesses specialize in different stages of the production cycle) or horizontal forms (when they perform similar activities in the same production process), but most of the time there is a combination of the two.

Industrial districts are present in many countries of the world and have found ideal conditions for growth in Italy. The big industrial crisis of the 1970s forced large companies to relocate some stages of production and fostered the development of specialized small- and medium-sized enterprises in niche sectors, exploiting the craft traditions that had developed over the centuries. In Italy there are now more than one hundred industrial districts[18] employing approximately a quarter of all the workers in the country, with the greatest concentration in Lombardy and in the Marche (ISTAT, 2001). This has encouraged various scholars, first and foremost Giacomo Becattini, to investigate this situation, defined as "a socio-territorial entity characterized by the active presence of both a community of people and a population of firms in one naturally and historically bounded area. In the district unlike in other environments such as manufacturing towns, community and firms tend to merge. ... The fact that there is a dominant activity differentiates the district from a generic 'economic region'" (Becattini, 1990, p. 38).

Michael Porter has studied overseas business groups. He defined clusters as "geographical concentrations of interconnected companies, specialized suppliers, service providers, firms in related industries and associated institutions (e.g., universities, standards agencies, and trade associations) in a particular field that compete but also cooperate. ... Clusters are a striking feature of virtually every national, regional, state, and even metropolitan economy, especially in more economically advanced nations" (Porter, 2000, p. 15).[19] And again: "a cluster is a geographically concentrated group of interconnected businesses and associated institutions in a particular field, linked by commonalities and complementarities. ... Clusters also often extend downstream to channels or customers or laterally to manufacturers of complementary products or companies related by skills, technologies or common inputs" (Porter, 2000, p. 16).

### 4.4.2  Advantages of Clusters

In the same study Porter (2000) identified the competitive advantages of being located in an industrial cluster that may involve (1) productivity, (2) innovation, and (3) the creation of new businesses. When companies and production factors are concentrated in a delimited territory leading to agglomeration economies, competitive advantages tend to self-feed over time (Scott, 1988; Storper, 1989; Arthur, 1990; Krugman, 1991).

The location of a company near a cluster means skilled labor can be recruited and other production factors purchased at low prices while the proximity of companies of the same or different groups promotes institutional and personal relationships and, therefore, the exchange of information and ideas with spillover phenomena (Saxenian, 1994). Productivity is also boosted by the complementarities existing among companies that produce different goods or services—as in the case of the tourism sector involving many businesses that range from hotels to restaurants, from transport to airport services, and so forth—and reputation that spreads to all companies in an area when a number of operators achieve levels of excellence.

A series of services, such as specialized training programs or some types of infrastructure, is provided by public authorities only if a conspicuous number of companies justify the use of large public resources. Network economics has highlighted the role played by informal contacts in reducing transaction costs thanks to greater trust between parties (Mueller, Sumner, and Lapsley, 2006). Further, the agency problems that arise in vertically integrated companies can be avoided because clusters can guarantee greater efficiency and quality thanks to the competitive pressure of companies that have access to the same production factors and similar technologies and cost structures. Peer competition is heightened by the desire to excel in the local business community both in terms of economic results (extrinsic motivations) and prestige (intrinsic motivations of a selfish type). The presence of various companies that produce similar goods or services and have access to similar technologies and factors reduces monitoring costs since managers can easily compare the cost structure of their own company with competitors'.

The benefits of innovation are no less important than those of productivity. The network economy has always highlighted the importance of informal relationships in facilitating the dissemination of information (Hippel, 1994). Further, companies within a group are able to understand the new needs of customers and adopt process and product innovations more quickly than those operating in isolation. Watching the behavior of competitors enables them to continuously revise the benchmark they must aim at while taking part in research consortia and technology parks furthers interaction and information exchange with effects that filter down through the whole sector. The recruitment of specialized personnel from rival companies in the same district favors the transfer of technologies and knowledge, especially in areas where innovation is linked to learning-by-doing processes. As the presence of many companies with similar characteristics poses the risk of eroding profit margins, there is a powerful incentive for continuous innovation and product and/or cost differentiation.

Finally, clusters facilitate the creation of new businesses since entrepreneurs operating in the territories are more likely to hear about opportunities to do business in new products or services. The very existence of a cluster signals business opportunities and

profits, often attracting investments from neighboring areas. The entry of new companies is encouraged by low entry barriers: the presence of many companies with similar
production facilities, in fact, reflects economies of scale on the supply side, sustainable
capital requirements, and relatively easy access to distribution channels as well as the
absence of restrictive government policies. The risk of retaliation upon entry by companies already present in the market also decreases when the sector is fragmented. The
presence of a network of successful producers and suppliers reduces the perception of
potential new entrants as a threat, thus lowering barriers to entry even further.

### 4.4.3   Drawbacks of Clusters

The literature has, however, also identified some possible negative effects of clusters
on profitability and innovation. On the one hand, the concentration of many operators in a limited space can exacerbate competition (Mueller, Sumner, and Lapsley,
2006) and lead to the depletion of some factors of production (e.g., raw materials,
skilled labor, etc.). On the other hand, if companies have a common entrepreneurial
mind-set, the cluster can delay innovations, thereby perpetuating established habits
and approaches and rejecting new ones. A further disadvantage linked to industrial
districts is the lack of diversification in production. This exposes the area to the risk
of sudden impoverishment should the sector experience a phase of recession or even
begin to decline because of the emergence of radical process or product innovations
that benefit other districts.

Overall, in light of the studies and the arguments presented, the advantages seem
to far outweigh the disadvantages, so the state should support the development of
industrial districts.[20] Support can take place in four areas as follows (Porter, 2000,
p. 28, figure 3):

1. the context that influences company strategies and competition: the government
   should set up departments that can serve the cluster, work to attract foreign direct
   investments, favor exports, and eliminate barriers to local competition;
2. the conditions that determine demand: the state should establish a set of rules to
   reduce regulatory uncertainty as well as systems of classification and certification
   of quality and act as a sophisticated buyer of high-quality products;
3. the conditions that influence the supply of production factors: public authorities
   should set up study and training programs tailored to the needs of the cluster, finance
   research programs in local universities to promote technologies in the district, and
   provide the communication infrastructures and transport that companies need;
4. connected companies: the government should favor meetings (forums) that put all
   the companies in the cluster in contact with each other, work to attract suppliers
   of the surrounding areas, and establish industrial parks and free trade areas based
   on the characteristics of the district.

Clusters in the wine sector are geographic concentrations of interconnected companies belonging to sectors which serve each other, as well as the public institutions that provide public goods and services and trade associations. Figure 4.2, taken from Porter (2000), describes the structure and ramifications of the Californian wine cluster, but it applies to any region. On the agronomic side the production of wine involves companies supplying cuttings, chemicals, tools for the harvesting of grapes, and the irrigation of land while the enological side includes companies producing wine-making equipment, barrels, bottles, caps, labels, advertising, specialized magazines, tourism, and food. There are also the public bodies appointed to define and enforce the rules aimed at combating fraud and ensuring the quality and wholesomeness of the foods, and finally there are the public institutions that provide services like specific advanced training. The structure, therefore, is much

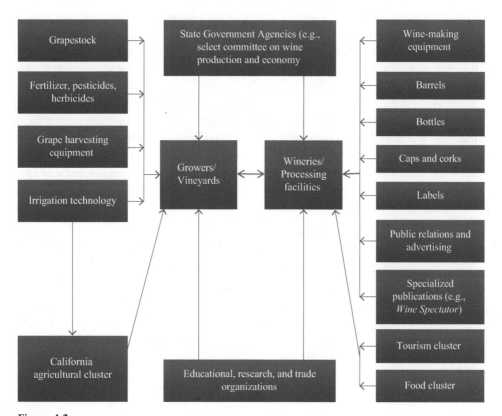

**Figure 4.2**
Structure of the wine cluster.
*Source*: Porter (2000), p. 17, figure 1. The figure describes the Californian wine cluster, but it fits any wine region in the world.

more complex than might be imagined and includes various professional sectors and figures.

The literature highlights the contribution made by wine clusters to local economies. Mueller, Sumner, and Lapsley (2006) focused on California and came to the conclusion that the geographic aggregation of companies producing grapes, wine, machinery, and services originates mainly from economies of scale in grape production and lower transport costs. Larreina (2007) analyzed the multiplicative effects of the development of the wine sector on the economy for the Spanish region of Rioja by constructing input-output tables that provide a complete picture of the flows of products and services in a given year. With this statistical tool, the direct, indirect, and induced contribution to GDP and employment of a certain economic sector—in this case the wine sector—can be reconstructed (Kurz, Dietzenbacher, and Lager, 1998).[21] The author, also in view of the indirect effect, came to the conclusion that a fifth of the Rioja economy is attributable to wine production, and public authorities should therefore support with appropriate policies. Larreina, Gómez-Bezares, and Aguado (2011) developed the study of the contribution of wine to the economy of this region with a series of different approaches. Other studies on wine clusters have been conducted by Doucet (2002) for Aquitaine in France, Porter and Sölvell (2003) in the Victoria region of Australia, and Williamson and Wood (2003) for Cape Town in South Africa, though the dispersion of companies in very extensive territories in the New World makes the measurement and classification of clusters more questionable. Finally, Francioni, Vissak, and Musso (2017) showed that in the wine sector network relationships help wine producers to expand internationally because they benefit from contacts with tourists, friends, relatives, and other partners.

# 5

# Wine and Finance

Whoever said that money can't buy happiness simply didn't know where to go shopping.
—A phrase attributed to Bo Derek

This chapter is divided into two sections. The first describes the opportunities to benefit, in terms of both expected return and portfolio diversification, from investments that include the purchase of bottles of wine and shares of wine companies listed on the stock exchange. Bottles of wine have a special nature because they have both investment and consumption value affecting the agent's utility. They are sold in the secondary market by auction firms like Christie's, Sotheby's, Zachy's, and Acker Merrall & Condit, and their sales amounted to $350 million in 2015. There are, in addition, sales over the internet (e.g., on eBay), which are difficult to quantify. (For a detailed description of how wine auctions work, see Cardebat, 2017, pp. 109–117.)

Looking at the sole expected return is not sufficient to compare assets because they carry different degrees of risk. The Sharpe ratio is given by the performance of an investment adjusted for its risk: $E(r)/\sigma$, where $E(r)$ is the expected return and $\sigma$ the standard deviation. When investing in a diversified portfolio of assets, the standard deviation is connected to two types of risk: market risk and firm-specific risk. The former cannot be eliminated because it depends on factors which affect the whole economy (gross domestic product growth rate, exchange rates, shocks to oil prices, etc.) while the latter can be reduced or even removed with an accurate diversification strategy. Therefore, one should also consider the correlation of the expected performance of an asset with that of the market. In fact, assets whose expected return is negatively correlated with that of the market are valuable because they are countercyclical. This insurance function comes at a cost. Assets with negative correlations with market returns should have lower expected returns. This model, known as the capital asset pricing model (CAPM), has been extended to include additional elements, as in Fama and French's three-factor model (1992, 1993, 1995). Appendix 5.1 provides a review of the theoretical concepts used throughout the section.

When comparing assets, we should take into account the peculiarities of markets that can generate additional sources of risk. The first one is the liquidity risk connected to the amount of time necessary for the sale and the uncertainty about the realized price; from this point of view, financial markets are more liquid than real estate and bottles of wine markets. The second one is the risk of counterfeiting, and in the case of bottles of wine it is always present but increases exponentially if performed by unknown private sellers.

The second section deals with the instruments to hedge risk, such as insurance and derivatives, to cope with catastrophic weather events and exchange rate fluctuations.

## 5.1   Investments

### 5.1.1   Investments in Bottles of Wine

There is a widespread belief that the purchase of collectibles, like works of art, stamps, and bottles of wine, are a valid alternative form of investment or can be complementary to traditional forms, like shares and debt securities. However, empirical studies have largely rejected these conjectures and demonstrated that collectible items have an unfavorable profile characterized by low yield and high risk (Burton and Jakobsen, 1999; Mandel, 2009). Baumol (1986) measured the real annual performance of paintings from 1652 to 1961. The data, collected by Reitlinger (1961), concerned auctions held predominantly in London and so until 1920 refer almost exclusively to sales at Christie's. Based on the author's calculations, the real yield rate was only 0.55 percent. Frey and Pommerehne (1989) extended Baumol's (1986) database to 1987 and included auctions in other countries. The authors came to similar conclusions, with an unattractive real rate yield of 1.4 percent between 1635 and 1949 and 1.6 percent between 1950 and 1987. Pesando (1993) analyzed the price of modern prints between 1977 and 1992 and found a real annual return of 1.51 percent, which is significantly lower than for shares and bonds in the same time span, while risk, defined as the standard deviation of portfolio returns, is the same or even higher.

More favorable results were obtained by Mei and Moses (2002) and Dimson and Spaenjers (2011). The first, with data relating to works of art sold in the period 1875–2000, found higher returns than for bonds, though lower than for shares, and a certain power of diversification. The second, instead, focused on stamps, which, like the works of art in Mei and Moses (2002), earn more than bonds but less than shares and can help to diversify risk, given the low correlation with equity returns.

Articles about the wine sector that are completely anecdotal, without any statistical basis, fantasizing about amazing returns and recommending purchases for investment purposes frequently appear. Hugh Johnson (1971),[1] for example, suggested buying bottles, letting them rest in the cellar for a few years, and then selling them because they will increase in value. An article written by Prial (1997) appeared in the

*New York Times* and reported on the boom in Bordeaux prices in 1996 (especially in cellars like Petrus and Chateau Ausone) and burgundy (Domaine de la Romanée-Conti). In the Italian newspaper *Corriere della Sera*, Ferraro (2014) compared the prices of the Ornellaia wine from 2007 to 2014 with those of some alternative investments (Standard & Poor's [S&P] 500, Financial Times Stock Exchange 100, gold and oil) and titled his article "Investing in Big Reds? Better than Gold and Oil." But clearly, no scientific conclusion can be drawn about the average profitability of investments in wine by taking into consideration only the bottle that gave the best results in the market in a limited period of time.

Indeed, academic studies (table 5.1) have provided conflicting results. A number of surveys have shown that wine is not an attractive investment because it does not offer higher returns and/or does not have a higher return-risk profile than other financial assets. Krasker (1979) used data from the Heublein auctions of Bordeaux red and Californian Cabernet Sauvignon wines between 1973 and 1977. To make the database homogenous, the author excluded the years prior to 1950 because wine stops maturing in a bottle after a certain number of years and, therefore, price changes can reflect, in part, the preferences toward antique goods that is a market in its own right. With a sample of only 137 observations, the expected return from stocking wine was no different to a risk-free asset. Wine storage costs, not detailed by any archive but econometrically estimated, amounted to $1.40 per year per bottle. Although a series of arguments were put forward to justify this figure, the author himself admitted that it seemed to be very high.

Weil (1993) adopted a different approach because it followed an investor's purchases from 1980 to 1992 rather than looking at auction prices. With an average yield of 9.5 percent and with 11 percent peaks for Bordeaux wines, the purchase of wine for investment purposes was less profitable than stocks in the New York Stock Exchange in the same period. Di Vittorio and Ginsburgh (1994), with a sample of about thirty thousand red wines from Bordeaux (vintages 1949–1989) sold at Christie's auctions, showed a price increase of 75 percent between 1981 and 1990 and a subsequent decrease of 15 percent in the following years. The average yield amounted to 4.2 percent, with strong variability between châteaux and a correlation with the weather conditions of the year.

Burton and Jakobsen (2001) studied auction prices of Bordeaux red wines in various houses (Christie's, Sotheby's, Davis & Company, etc.) between 1986 and 1996, narrowing the survey to vintages after 1960. Unlike the other investigations, this considered the costs of transaction, storage, and insurance and is therefore the most complete, but it did not take into consideration transport costs and taxes. The yield of wine proved to be lower than the Dow Jones and higher than treasury bills; however, it is very volatile, which makes it an even less attractive type of investment. Further, the most expensive wines have a below average expected return. Bentzen,

**Table 5.1**
Main results on investments in bottles of wine.

| Paper | Wines | Years | N | Methodology | Costs | | | | | Conclusions |
|---|---|---|---|---|---|---|---|---|---|---|
| | | | | | Transaction | Transport | Storage | Insurance | Tax | |
| Krasker (1979) | Bordeaux red wines and Californian Cabernet Sauvignon (years > 1950), Heublein auctions | 1973–1977 | 137 | Repeat sales | No | No | Estimated econometrically | No | No | Expected return from stocking wine was no different to a risk-free asset. |
| Jaeger (1981) | Bordeaux red wines and Californian Cabernet Sauvignon (years > 1950), Heublein auctions | 1969–1977 | 199 | Repeat sales | No | No | Used data from Freemark Abbey Winery | No | No | Positive return on wine compared with risk-free assets. Expected return and the risk for less expensive wines is greater than for more expensive wines. |
| Weil (1993) | French wines | 1980–1992 | 70 | Followed investor's purchases | n/a | n/a | n/a | n/a | n/a | Average return of 9.5%, with peaks of 11% for Bordeaux wines, much lower than the New York Stock Exchange in the same period. |

| Study | Data | Period | Sample | Method | | | | | | Comments |
|---|---|---|---|---|---|---|---|---|---|---|
| Di Vittorio and Ginsburgh (1994) | Bordeaux red wines (1949–1989), Christie's auctions | 1980–1992 | 29,901 | Hedonic regressions of price with dummy variable | No | No | No | No | No | The prices of wines analyzed grew 75% between 1981 and 1990 and then fell 15%. Average return of 4.2% with strong variation between châteaux and correlation with the yearly weather conditions. |
| Burton and Jacobsen (2001) | Bordeaux red wines (years > 1960), various auction houses (Christie's, Sotheby's, Davis & Company, etc.) | 1986–1996 | 10,558 | Repeat sales | Yes | No | Yes | Yes | No | The return on wine was lower than treasury bills. The return on wine is, however, very volatile, which makes this type of investment even less attractive. The more expensive wines had a lower than average expected return. |
| Bentzen, Leth-Sorensen, and Smith (2002) | Bordeaux red wines (years > 1950), Copenhagen Wine Auctions-Bruun Rasmussen | 1988–2002 | 48 auctions | Not very clear estimate method | No | No | No | No | No | Returns and volatility not reported. |

(continued)

**Table 5.1** (continued)

| Paper | Wines | Years | N | Methodology | Costs | | | | | Conclusions |
|---|---|---|---|---|---|---|---|---|---|---|
| | | | | | Transaction | Transport | Storage | Insurance | Tax | |
| Fogarty (2006) | Australian wines, Langton auctions (years > 1965) | 1989–2000 | 14,102 | Adjacent period hedonic regressions of price | No | No | No | No | No | The return on Australian wine is probably higher than on French wines while the yield-risk profile is comparable with Australian equities. |
| Fogarty (2007) | French wines (using data from Burton and Jakobsen, 2001) and Australian wines (using data from Fogarty, 2006) | 1986–1996 and 1989–2000 | 10,558 and 14,102 | Repeat sales plus adjacent period hedonic regressions of price | Yes | No | Yes | No | Yes | Studies on wine underestimate the real return on this form of investment because it is either exempt from tax or subject to limited taxation in many countries. If this factor is taken into account, the actual yield of wine is higher. Wine, therefore, offers interesting investment opportunities, also by virtue of the power to diversify a securities portfolio. |

| Study | Source | Period | N | Method | | | | | Findings |
|---|---|---|---|---|---|---|---|---|---|
| Sanning, Shaffer, and Sharratt (2008) | Bordeaux red wines (years 1893–1998), Chicago Wine Company auctions | 1996–2003 | 13,662 | Repeat sales | No | No | No | No | Average monthly return of 0.51%, which rose to 0.78% if just the best wines were included. The monthly return was 0.75% higher than predicted by estimating both the CAPM and the Fama-French three-factor model and was poorly correlated with risk factors. Wine, therefore, offers interesting opportunities for investment and diversification. |
| Fogarty (2010b) | Australian wines, Langton auctions (years > 1965) | 1990–2000 | 12,180 | Repeat sales | No | No | No | No | Return on wine is lower than on traditional shares. Nevertheless, wine does benefit slightly from diversification. |

(continued)

**Table 5.1** (continued)

| Paper | Wines | Years | N | Methodology | Costs | | | | | Conclusions |
|---|---|---|---|---|---|---|---|---|---|---|
| | | | | | Transaction | Transport | Storage | Insurance | Tax | |
| Masset and Henderson (2010) | Wines from various countries, Chicago Wine Company auctions (years 1981–2005) | 1996–2009 | More than 400,000 | Repeat sales | No | No | No | No | No | Fine wines had higher returns and lower volatility than equities, especially in times of crisis. Adding wine brought benefits in terms of diversification and average expected risk while the CAPM estimate showed a positive and significant alpha between 1996 and 2009 and a very low beta. |
| Cardebat and Figuet (2010) | French wines of Bordeaux; auctions in the United States, Great Britain, and France (131 years and 486 châteaux) | n.a. | 53,153 | Repeat sales | No | No | No | No | No | CAPM and Fama-French Three-Factor model estimates as in Sanning, Shaffer, and Sharratt (2008). Negative excess return (alpha null) but power of diversification as in Sanning, Shaffer, and Sharratt (beta null). |

| Study | Period | Sample size | Methodology | | | | | | Findings |
|---|---|---|---|---|---|---|---|---|---|
| Fogarty and Jones (2011) | 1988–2000 | 14,102 | Hedonic approach, repeat sales and hybrid model | No | No | No | No | No | Return on wine and its benefits in terms of portfolio diversification are influenced by the estimation methodology adopted (hedonic models, repeat sales, and hybrid models). |
| Devine and Lucey (2015) | 1996–2007 | 51,756 and 18,147 | Repeat sales | No | No | No | No | No | Wine offers higher returns than risk-free securities but with a more favorable return/risk ratio than shares. When the individual subregions are taken into consideration, the returns become more volatile so investment in wine should be limited to experts only. |
| Dimson, Rousseau, and Spaenjers (2015) | 1900–2012 | 9,492 | Arithmetic repeat-sales regression | No | No | Yes | Yes | No | The real financial return to wine investment is 4.1%, which exceeds government bonds, art, and investment-quality stamps. Wine appreciation is positively correlated with stock market returns. |

Leth-Sorensen, and Smith (2002) analyzed data on Bordeaux red wines of years after the 1950s (therefore excluding "antique goods"), which were sold in forty-eight Bruun Rasmussen auctions in Copenhagen between 1988 and 2002, but did not report on returns and volatility.

A series of other studies, however, came to opposite conclusions. Jaeger (1981) developed the Krasker study (1979) using the same database but extended the time horizon to the period 1969–1977 because the years 1973–1977 were particularly unfavorable and distorted the results. With an additional four years and sixty-two observations and using data from the Freemark Abbey Winery on the actual storage costs, the analysis showed a positive return on wine compared with risk-free securities. Fogarty (2006) collected data on Australian wines sold at Langton auctions between 1989 and 2000 and concluded that the yield on Australian wine is probably higher than on French wines while the yield-risk profile is comparable with Australian equities. Wine, therefore, appears to be a good form of investment. In the following year Fogarty (2007) broadened the survey by extending the database and considering the effects of taxation. The data on French wines were the same as those used by Burton and Jakobsen (2001) while the Australian data were the same as Fogarty (2006). The author emphasized how studies on wine underestimated the real performance of this form of investment since auction transactions are tax-exempt or subject to limited taxation in many countries. If this factor is taken into account, the actual yield of wine is higher. For this reason, wine offers interesting investment opportunities, also by virtue of its ability to diversify a securities portfolio.

Sanning, Shaffer, and Sharratt (2008) applied the CAPM and the Fama-French three-factor model to Bordeaux red wines of the years 1893–1998 sold at auction between 1996 and 2003 at the Chicago Wine Company and found an average monthly rate of return of 0.51 percent, which rose to 0.78 percent if just the best wines were included. The monthly return was 0.75 percent higher than predicted by estimating both the CAPM and the Fama-French three-factor model (alpha coefficient) and was poorly correlated with risk factors (beta coefficients). Fogarty (2010b), using data on Langton auctions during the years 1990–2000, found that the yield on wine is lower than on traditional shares. Nevertheless, wine provides a (modest) diversification benefit.

In Masset and Henderson's (2010) study, using data on wines from various countries sold at auction between 1996 and 2009 at the Chicago Wine Company (vintages 1981–2005), the best bottles (fine wines) had higher returns and lower volatility than equities, especially in times of crisis. Adding wine brought benefits in terms of diversification and average expected risk while the estimate of the CAPM showed a higher return than expected in the model (positive and significant alpha) and a poor correlation with market performance (very low beta). Fogarty and Jones (2011), using Australian Langton auction data from the period 1988–2000, demonstrated that the yield on wine and its benefits in terms of portfolio diversification depend on the estimation methodology adopted (hedonic models, repeat sales, and hybrid models were used in the study).

Cardebat and Figuet (2010) replicated the study by Sanning, Shaffer, and Sharratt (2008) with different data relating to the Bordeaux wines of 486 châteaux and found no excess return (alpha null) with both the CAPM and with the Fama-French three-factor model, but some power of diversification (beta null) as in Sanning, Shaffer, and Sharratt (2008). Devine and Lucey (2015), using data on the red wines of Bordeaux and Rhône sold at the auctions of the Chicago Wine Company between 1996 and 2007, concluded that wine offers higher returns than risk-free securities but with a more favorable return/risk ratio than shares. When the individual subregions are taken into consideration, the returns become more volatile, so only experts should invest in wine. Lastly, Dimson, Rousseau, and Spaenjers (2015) used data from Premier Cru Bordeaux over the period 1900–2012 and estimated a real financial return on wine investment of 4.1 percent, which exceeds government bonds, art, and investment-quality stamps.

From this quick review we can easily see how the scientific literature has not yet managed to reach some kind of consensus on the question of the opportunity and profitability of investments in wine. As in all empirical analyses, this may be due to the fact that using different databases and estimation methodologies can change the results significantly so that so-called "stylized facts" cannot be identified.[2] Moreover, as Fogarty and Jones (2011) pointed out, wine sales are not very frequent and require specific econometric methodologies that can produce variable results. An additional problem is that the estimated return of investment from wine bottles is influenced by the calculation method. Fogarty and Sadler (2016) applied six different methodologies to French data and showed that results change significantly. The comparison between financial assets and bottles of wine is made even more problematic by the following critical issues.

1. Wine and securities are not homogeneous in terms of costs, benefits, risk, and the degree of liquid assets.

   - *Costs*: Most studies on wine (see table 5.1) do not take into consideration transaction, transport, storage, and insurance costs (Fogarty, 2006). The estimated return on wine purchases is therefore overestimated compared with traditional financial assets which only have transaction costs. Further, unlike the transfer of securities, the cost of transporting wine depends on the final destination while storage costs are influenced by climate so the same purchase by investors in different countries may present different net returns. The same applies to the direct participation in the auction that involves direct (monetary) and indirect (opportunity) costs.

- *Benefits*: Calculating the performance of an asset must take into consideration all costs and benefits produced, not just those of an economic nature. Unlike normal financial assets, collectibles—such as, for example, works of art, stamps, and also wine—generate utility for those who own them. Art works and stamps can be enjoyed by the owner and their guests, and possession alone can be a source of pride (Burton and Jakobsen, 2001; Mandel, 2009). If these psychological benefits are added to the expected return and risk in the investor utility function, then it is easier to justify the low returns on investment in art found in the literature (for a review, see Mandel, 2009). Wine, however, is different from other collectibles since to be enjoyed it has to be drunk and, therefore, destroyed. The only flow of benefits before its destruction is the possible gratification of possessing a cellar full of prestigious labels, but this would hardly seem likely to have a significant impact on the investor's utility function and make them ready to sacrifice part of the expected return from an alternative investment.
- *Risk*: The literature has not studied how the type of wine storage, whether at home or in specialized companies, can affect the resale price at auction. It may well be that wine held in a personal cellar provides fewer guarantees for the conservation of the product under optimal conditions, and therefore, it may be more difficult to resell.
- *Liquidity*: A comparison of the returns corrected for risk takes into account only the volatility of the asset price of financial resources but not the greater liquidity of securities. These can be sold in real time, whereas the liquidation of a winery usually takes four to five months (Burton and Jakobsen, 2001).

2. *Taxes*: A comparison between the return on traditional financial assets and bottles of wine is complicated by the different tax treatment of the two investment forms. In many countries, in fact, the sale of bottles is tax free (Burton and Jakobsen, 2001; Fogarty, 2007). This tends to underestimate the performance of wine.

3. The purchase of bottles of wine for investment purposes involves valuable products that are sold mainly in auctions in London, New York, Chicago, and a few other cities. The number of lots and participants is very limited. Bidders can also take part in auctions by telephone—a very common occurrence in the wine sector. With few buyers, a physical presence at the auction can make the difference as the number of participants can be seen, which is an indicator of the interest of buyers in the good sold. Therefore, anyone who participates in the auction by telephone loses out on precious information and often ends up paying a higher price (Ginsburgh, 1998). Financial markets, by contrast, have millions of buyers participating in the online sales and purchases.

Basically, comparing the return and risk profiles of traditional financial assets and bottles of wine is like comparing apples and pears: they are both fruits, both give juice, but they are not the same thing. The comparison is further complicated by the different nature of wine which is a consumer good while stocks and bonds are investment assets. Consumer goods such as oil, copper, and meat are purchased primarily for consumption while investment assets such as gold and silver are purchased mainly for investment purposes. These two metals have multiple industrial and commercial uses and, therefore, a double value, but they are universally regarded as investment goods because there are a substantial number of people holding them for this second purpose. A high number of buyers for investment purposes ensures that there are no long-lasting opportunities for arbitrage. For consumer goods, however, this is not the case. The return on the physical possession of an asset (convenience yield) can keep a productive process active or exploit temporary local shortages of the goods (see Hull, 2009, chapter 5).

### 5.1.2   Investments in Winery Securities

While there is an abundance of studies on the opportunity to include bottles of wine in security portfolios, the scientific literature has completely ignored investments in shares of publicly listed wineries. The only exception is Baldi et al. (2010), who used a nonlinear cointegration model to study the long-term relationship between the price indices of winery shares (Global Wine Industry Share Price Indexes) and equity indices in five countries: France, United States, Chile, China, and Australia.[3] The presence of cointegration between different stock market indices indicates that, apart from short-term deviations from a common equilibrium, they move in the same direction in the long term (Masih and Masih, 1997; Patra and Poshakwale, 2008). In the presence of cointegration, the two indices move together, which reduces the power of diversification, while the absence of cointegration reduces risk by leaving the expected long-term return unchanged (Berument, Akdi, and Atakan 2005; Ratner, 1996). The presence of cointegration also implies a causal effect in at least one direction which allows the use of the short-term returns of an index to predict the movements of the other.

The results of the study by Baldi, Vandone, and Peri (2010) showed that, in the more mature markets—that is, France and the United States—there is Granger causality from the share index to the wine sector index and that the adjustment speed of the latter to the long-term equilibrium is lower than the equity index. This implies that deviations from the equilibrium of the wineries index last longer than those of the composite index. Investors, therefore, can make profits by anticipating short-term changes in the index of wine prices based on the movements of the composite index. In less mature markets like China and Chile, the two indices always present a nonlinear cointegration but this time with the same speed of adjustment to the long-term equilibrium.

## 5.2   Risk-Hedging Instruments

Wineries are exposed to a number of risks that influence the investor's utility and the expected return. The two most important factors are the variability of weather conditions and that of exchange rates due to the increasing share of exports in total production. These two risks can be managed ex ante or ex post. Reserves may be accumulated to tackle difficult times before damage actually occurs, thus reducing exposure to risk factors as far as possible (e.g., making irrigation systems or selecting clones of plants that are more resistant to pests) through insurance contracts and derivative instruments.[4] Insurance and derivatives can both be used as hedging instruments against risk, but they are different from a regulatory, accounting, fiscal, and legal point of view (World Bank, 2011, p.18). An insurance contract, moreover, guarantees an indemnity that is proportional to the damages actually suffered, conditional on their verification and postponed in time. A derivative, instead, may depend on the value of some underlying condition (the weather, the price of goods, or currencies) regardless of the actual damage suffered or the possession of the good. The ex post solutions consist in the request for support from the state when none of these measures have been taken.

### 5.2.1   The Benefits of Covering Risks

This section deals with the topic of risk coverage through the use of financial instruments. Covering against risk—as, for example, through the purchase of an insurance policy—means the inclusion of a new security in the portfolio at the same time as lowering both risk and the expected return. Risk is reduced because the yield on the policy is negatively correlated with at least one portfolio asset that is insured. At the same time, the expected return decreases as a result of the premium payable to the insurer in exchange for data collection and processing, administrative costs, the opportunity costs of holding reserves, and any compensation for damage. Taking out insurance therefore entails a trade-off between reducing risk on the one hand and maintaining the expected return on the other. The willingness to pay a premium to the insurer to reduce risk is a direct function of the investor's risk aversion.

The presence of an insurance and derivatives market at affordable prices can stimulate economic growth as individuals with greater risk aversion will invest in activities characterized by greater uncertainty (and therefore a higher yield) only if they can transfer the uncertainty onto third parties (Skees, Barnett, and Collier, 2008).[5] This point has often been used to advocate public intervention in favor of the insurance market, especially against natural disasters and catastrophic events. This, however, would only affect sectors subsidized by the state, draining resources from others and creating a phenomenon of displacement. Further, if support for the insurance market is not well organized, it can also generate strong inefficiencies favoring very risky investments—as, for example, when building is encouraged in developed countries in plains prone to the risk of flooding (Mileti, 1999) or when cultivation of crops that need a lot of water are encouraged in areas subject to drought (Skees, 2001).

### 5.2.2   Insurance Against Damage

Is it always possible to cover a risk by transferring it to a third party? The answer, unfortunately, is no. The scientific literature suggests five criteria for determining the insurability of a risk:[6] (1) the distribution of the event is known, or it is possible to calculate expected future losses; (2) the damages are observable; (3) the risk of adverse selection is limited; (4) the risk of moral hazard is limited; and (5) the extent of losses is not excessive.

*Known distribution.* The first is an actuarial condition. To predict expected damage, a circumstance must belong to a sufficiently homogeneous category of events that are subject to the same sources of risk and their distribution must be reliably estimated. According to the law of large numbers (Bernoulli, 1713), this requires an appropriate sample size whose quality, however, is not always the same. If losses are independent and identically distributed (IID), then the prediction is much simpler and more accurate and only a small sample is needed. The more the distribution moves away from the IID, the more observations are required because tail events become difficult to forecast. If the distribution approaches IID, the risk of catastrophic events with a large number of subjects affected by significant damage or even devastation decreases. A loss is considered catastrophic when it is unexpected and extraordinarily large compared with the amount of assets in the insured portfolio. Catastrophic losses have two properties: they are difficult to predict, and they are concentrated geographically. Hurricanes are less catastrophic than earthquakes because they occur more regularly and frequently and are easier to predict. Apart from the unpredictability of the type of event, the possibility of modeling the risk can also depend on the lack of databases with long and reliable time series.[7] The factors that determine an insurance premium are the average expected loss, the operating costs associated with the creation and management of the portfolio (insurance pool), reserves for unexpected losses, and the profit of the insurer. The first three are costs, and the fourth is a gain. If the distribution of events is known, then both the operating costs and the reserves necessary to deal with unexpected events decrease. The insurance of events that are difficult to predict may involve such high premiums as to discourage the applicant from covering against the risk.

*Observable damage.* The second insurance criterion is quite an obvious condition: if losses are not verifiable, the insured can declare greater damage than was actually suffered, which can prevent the market from even being set up. Further, the cost of measuring the damage must be reasonably limited; otherwise, the insurance premium rises.

*Adverse selection.* In the insurance sector, adverse selection occurs when the insurer cannot assess the degree of risk for the various subjects requiring coverage and is forced to apply a single premium to all. This can push the best customers out of the market, increasing the average risk of a portfolio and pushing the insurer to increase further the premiums in line with Akerlof's (1970) vicious circle, which has already been described. Insurers make enormous efforts to assess the risk associated with each customer so as to differentiate and personalize the premium, but sometimes this is difficult or even impossible.

*Moral hazard.* Moral hazard occurs when the insured intentionally causes a loss to obtain compensation, claims damages greater than those actually suffered, or reduces efforts to avoid ill-fated events (Rothschild and Stiglitz, 1976; Gollier, 2005). The absence of moral hazard implies that losses are defined, measurable, and accidental and are uninfluenced by the behavior of the insured. On the contrary, the presence of opportunistic behavior leads to the start of vicious circles.

*No excessive losses.* The last condition is related to the size of potential losses, which (if excessive) can make insurance by a single agent impractical. A terrorist attack with bacteriological weapons or a large-scale earthquake would cause losses in terms of lives and physical capital that no insurance company could compensate. Risk sharing can alleviate the problem, but the size of potential losses could make the accumulation of such big reserves necessary as to make the insurance premium unacceptable. Pollner (2001) pointed out how the natural catastrophes of the early 1990s (e.g., earthquakes and hurricanes) caused a sharp contraction in the insurance market, leading to increases in premiums of between 200 percent and 300 percent in the Caribbean.

An almost perfect example of insurable risk is the automotive market, characterized by the presence of a huge number of circulating vehicles (millions in each country) and a high number of accidents but relatively randomly distributed and limited damage (at most a few million euros or dollars for one accident). Losses are verifiable, although sometimes it can be difficult to distinguish previous damage from that of the accident. The problems of adverse selection and moral hazard are present but are mitigated by regular official vehicle inspections and the introduction of incentive mechanisms, such as points on a driver's license, the "bonus-malus" system, and black boxes in cars.

### 5.2.3  Weather Risk

Of the two main risks affecting viticulture, weather is undoubtedly more difficult to insure against. Skees, Barnett, and Collier (2008) underlined how agricultural insurance is one of the most difficult to develop since damage is not independent (geographic correlation) and there are marked information asymmetries between the insured and the insurer. There are two types of agricultural insurance:

1. *Multiple Peril Crop Insurance* (MPCI) is related to crop yield and was created and subsidized by the US government (Barnett, 2000; O'Boyle, 2018). If productivity (e.g., quintals or tonnes per hectare) is lower than the historical average of the insured land by a certain percentage (e.g., 50 percent or 70 percent), the farmer obtains compensation proportional to the missing gain (Skees and Barnett, 2004). Since the yield can depend on a number of variables, which range from climatic conditions to management, it is impossible to identify which factor(s) is responsible and to insure the harvest against one or more specific elements. For this reason, the insurance compensates the owner of the land if the yield falls below a preestablished threshold, regardless of the cause. This type of insurance is difficult to develop because of the obvious information asymmetry problems, the amount and accuracy of the necessary data, the premium that grows together with the number of risks covered, and the correlation between the damage of farms located in the regions affected by the same calamity. The empirical evidence presented by Hazell (1992) was based on seven countries[8] in the 1970s and 1980s and showed that the total compensation paid and administrative costs of MPCI far exceeded the value of the premiums collected (in all countries, the outflows were more than double the revenue). The author concluded that this model of risk coverage, even taking into account the social benefits, is economically unsustainable. In fact, it was maintained only through public subsidies, and over the years has been withdrawn or considerably reduced in many countries (Skees, Barnett, and Collier, 2008);

2. *Index insurance.*[9] In this case compensation is a function of an index that constitutes a proxy for the damage suffered by the individual producer and is provided by an independent third party (Barnett and Mahul, 2007). The two most important types of indexes are the average productivity of a certain area, which requires a great deal of data, and the climatic conditions (e.g., temperature, rain, humidity, wind) recorded at a particular weather station. This type of index insurance is easier to provide because long and reliable time series of meteorological data are also available for the least developed countries. The administrative costs are, in addition, very low because data does not have to be collected ex ante or damage measured in each individual farm ex post, thus eliminating the problems of adverse selection and moral hazard and making compensation faster. For this tool to be effective, the index and individual damage must, of course, be strongly correlated. The difference between the loss suffered and compensation constitutes the basis risk, which represents the greatest obstacle to the development of index insurance. Further, there is not enough data for an evaluation of the actuarial performance of these instruments at the moment, and the consequences of climate change raise questions about the possibility of developing and managing them (Hellmuth et al., 2009).

In any case, derivatives on weather conditions exist and are strongly supported by both the World Bank and the International Fund for Agricultural Development (IFAD) as instruments of risk coverage, especially for farmers in developing countries.[10] With particular reference to the wine sector, three different studies have developed price models for weather derivatives and have concluded that these tools can offer valid coverage against weather risks (see Turvey, Weersink, and Chiang [2006] and Cyr and Kusy [2007] for an application to the Ontario ice-wine; and Zara [2010] for a simulation on the Controlled Designation of Origin [DOC] Oltrepò Pavese Bonarda).

### 5.2.4  Currency Risk

Exporting and importing companies are also exposed to currency risk. Let us imagine a European winery that signs a contract (at time $t = 0$) with a US importer for the delivery of one thousand bottles of wine within six months (at time $t = T$) at the agreed price of €5 for each unit. At time $t = 0$, the exchange rate of dollars to euros ($/€) is 1.30. In the absence of currency fluctuations, the importer will have to pay the sum of $1,000 \times 5 \times 1.30 = \$6,500$ at time $t = T$. If, however, the exchange rate increases to 1.40, the company will have to pay $1,000 \times 5 \times 1.40 = \$7,000$. To cover against risk, the importer can adopt various strategies (see Björk, 1998, pp. 1–3), such as the following two examples.

1. Buy €5,000 today for the price of $6,500 and keep it in a checking account for six months. This eliminates the currency risk completely. The problem is that this blocks a substantial amount of money for six months or that the company may not even have this sum available.

2. A second and more sophisticated solution consists in buying a European call option for €5,000 with a strike price of $K$ $/€ at time $t = 0$ to be exercised eventually at time $t = T$. The currency option gives the purchaser the right, but not the obligation, to buy €5,000 at the predetermined exchange rate $K$ in six months' time, and this will only happen if the dollar has devalued in the meanwhile. For example, the option confers the right to buy €5,000 at the $/€ rate of 1.30. If during the six months the rate increases (e.g., 1.40), then at time $t = T$ the option is exercised and the company pays $5,000 \times 1.30 = \$6,500$ to buy €5,000, which at the current rate is worth $5,000 \times 1.40 = \$7,000$. If, on the contrary, the rate remains unchanged or decreases, the option is not exercised. From a conceptual point of view, currency options do not differ much from index insurance, in which the premium is paid only if an indicator exceeds or falls below a certain threshold.

The options therefore protect companies from price decreases or increases above a preestablished threshold and may have any type of goods or securities traded on the market—from currencies to cereals and so on—as underlying securities. European options differ from American options because the right to buy or sell at the preestablished price can only be exercised in $t = T$ and not at any time as in the US options.

Derivatives in general can be used to pursue two opposite objectives—namely, speculation and risk hedging. They can be traded on the stock market, in which case they have standardized characteristics, or informally and unregulated (over the counter, or OTC), adapting the instrument to the needs of the individual customer. According to data released by the Bank for International Settlements, there were $96.5 trillion of derivatives traded on the regulated market and $640 trillion on the OTC market still outstanding in the first half of 2019.[11] Most OTC securities consist of derivatives on interest rates (interest rate contracts, 81.8 percent), but those on exchange rates also play a significant role (foreign exchange contracts, 15.4 percent).[12]

## Appendix 5.1: Notions of Investment Theory

When faced with investment choices,[13] a person's temperament can take on infinite nuances ranging between the two extremes of a risk-averse individual and a risk lover, with the figure of a risk-neutral individual somewhere in the middle. The first case constitutes normality:[14] one of the basic principles of finance is that investors facing a higher level of risk demand greater returns. The difference between the performance of two investments, one with and one without uncertainty, is called "risk premium." The second case is a gambler who, on the contrary, enjoys uncertainty. The third profile is a person who is indifferent to risk and considers only the expected return in investment choices. A gamble, however, must not be confused with speculation: in a gamble the person derives pleasure from the random situation, even if this does not involve any increase in the expected return, whereas in speculation the investor is willing to accept a higher level of risk only in exchange for an adequate expected return. Aversion to risk and speculation are, therefore, not incompatible.

A utility function which is widely used in literature and by the Association of Investment Management and Research formalizes and summarizes these different profiles with the following utility function:

$$(5.1) \qquad\qquad U = E(r) - 0.005A\sigma^2,$$

where $U$ represents the utility of the individual, $E(r)$ and $\sigma^2$ the expected yield and the variance of the investment respectively, and $A$ the index of subjective aversion to risk. The number 0.005 is a convention to scale the standard deviation $\sigma$ and to express it as a percentage (e.g., 20 percent) instead of in decimal numbers (0.20), as with the expected return. Equation 5.1 represents the three types of investors. Assigning a positive (or negative) value to $A$ describes the behavior of the subject who is averse to risk (or a risk lover) while utility coincides with the simple expected return when $A$ is equal to zero (indifferent). Graphically the indifference curves for the risk-averse subject are positively inclined and have a rising slope (figure 5.A.1). In this case $p$ refers to a single security, but the same concept applies to a portfolio.

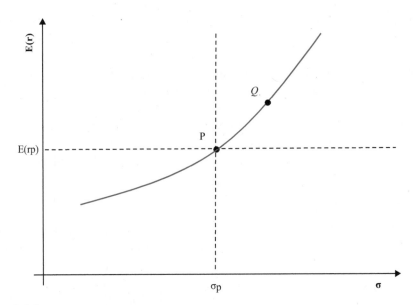

**Figure 5.A.1**
The indifference curve.

If we consider investment $p$, which has an expected return $E(r_p)$ and a standard deviation $\sigma_p$, the expected utility is $U(p)$. All the investments placed top left of point P (quadrant II) have higher expected returns and/or lower risk and are therefore preferred in accordance with the mean-variance criterion. The opposite is true of investments placed in the lower right corner (quadrant IV) while in the other two areas utility can be greater than, equal to, or less than point P. From this simple explanation, therefore, we can see the importance of not limiting the comparison of two alternative investments simply to the expected return and, as a consequence, the need to consider the degree of risk.

When we move from considering a single security to analyzing the risk of an entire portfolio, we have to take into consideration the interactions that exist between the different investments. A security that has returns negatively correlated with other portfolio assets can be used as a risk-hedging instrument in the same way as an insurance policy. In finance there are two sources of risk: (1) market risk, which is attributable to factors such as the performance of the economy and the interest and exchange rates, affects the returns of all companies and, as a consequence, is not diversifiable; and (2) firm-specific or idiosyncratic risk, which instead concerns the management of a single company. If the returns on the securities do not have a perfect positive correlation ($\rho < 1$), diversification by adding securities to the portfolio reduces the firm-specific risk, but market risk cannot be eliminated. Figure 5.A.2

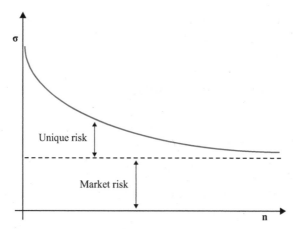

**Figure 5.A.2**
Portfolio risk and number of assets.

shows the benefits of diversification obtained by adding identical shares (e.g., units) of different securities. This example proposes a very simple technique for risk diversification that assigns the same weight to each security in the portfolio.

Let us now analyze the case in which the weight varies, and consider only two securities for the sake of simplicity. The return on the portfolio is given by the average of the returns of the individual investments $r(s)$ weighted for its respective share $w(s)$ with $w_1 + w_2 = 1$:

$$(5.2) \qquad\qquad r_p = \sum_s \mathrm{w}(s)\, r(s).$$

The variance is given by the sum of the variances $\sigma^2$ of the two weighted securities for the squares of the respective weights $w$ plus a component that includes the covariance between the two returns:

$$(5.3) \qquad\qquad \sigma_p^2 = w_1^2 \sigma_1^2 + w_2^2 \sigma_2^2 + 2 w_1 w_2 Cov(r_1, r_2),$$

where the latter is given by

$$(5.4) \qquad\qquad Cov(r_1, r_2) = \sum_s \Pr(s)[r_1(s) - E(r_1)]\,[r_2(s) - E(r_2)],$$

where $s$ represents a possible scenario. A positive covariance implies that the yields of the two securities move on average in the same direction (positive correlation) and vice versa (negative correlation). From this simple formula we can immediately understand the role played by covariance in determining portfolio risk: a positive covariance between the two securities increases portfolio risk while a negative covariance reduces it. In the latter case an instrument is considered a hedge, following the old proverb "do not put all your eggs in the same basket," and therefore its function is similar to that of an insurance.

Efficient diversification involves trying to minimize risk for any given expected level of return by buying shares that are not necessarily the same as the securities characterized by negative correlations. Now, to what extent can the portfolio risk decrease? For analytical simplicity, only two securities will be considered (1 and 2),[15] but the results also hold for portfolios with a large number of securities. Since

(5.5)                              $Cov(r_1, r_2) = \rho_{12}\sigma_1\sigma_2,$

if the two securities have a perfect negative correlation ($\rho = -1$), equation 5.3 becomes:

(5.6)                              $\sigma_p^2 = (w_1\sigma_1 - w_2\sigma_2)^2.$

At this point it is sufficient to choose portfolio shares $w_1$ and $w_2$ so that the portfolio standard deviation is zero (no risk):

(5.7)                              $w_1\sigma_1 - w_2\sigma_2 = 0.$

The portfolio securities that solve this equation by eliminating risk are:

(5.8)                              $w_1 = \dfrac{\sigma_2}{\sigma_1 + \sigma_2}$, and

(5.9)                              $w_2 = 1 - w_1 = \dfrac{\sigma_1}{\sigma_1 + \sigma_2}.$

If short sales are allowed, the same results can be obtained in the presence of a perfectly positive correlation ($\rho = 1$). Perfectly positive and negative correlations are purely theoretical; in real life they are much lower. In any case, the purchase of securities with $-1 < \rho < 1$ constitutes a valid alternative to those without risk since they allow a reduction in portfolio variance at the same time as achieving higher returns, thus moving the efficient frontier to the upper left (figure 5.A.3).

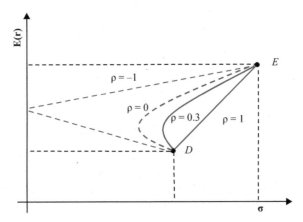

**Figure 5.A.3**
Portfolio expected return and standard deviation.

From an operational point of view, the investor has to take three separate steps to choose the best portfolio (figure 5.A.4). The first, as formalized by Markowitz (1952), consists in the identification of the efficient frontier, given by the set of portfolios that minimizes the variance for each given expected level of return through diversification, and the selection of securities characterized by negative covariances. The second involves the definition of the highest capital allocation line (CAL). This line, which represents combinations of risk and return on portfolios, including one part of risk-free securities and one of risky securities, has the maximum slope when it is tangent to the efficient frontier. This point is called the "optimal portfolio" since it maximizes the expected return per unit of risk. All points that are placed on the CAL maximize the yield expected per unit of risk. The third step consists in deciding at which point the investor wants to position themselves and therefore the risk profile. Once the CAL has been selected, individuals will try to maximize their utility, which depends on the expected performance and portfolio risk, in relation to their risk aversion. People with a higher degree of aversion will choose a less aggressive portfolio, positioning themselves on the left side of the line and vice versa.

If we now consider individual securities, what kind of performance should we expect? The CAPM proposed by Sharpe (1964) and developed by Lintner (1965) and Mossin (1966) is an equilibrium model of the financial market that establishes a relationship between the expected return of a security and its riskiness. The hypotheses underlying the model are as follows:

1. "atomistic" investors: each investor holds too few securities to influence the market, so they are price-takers;
2. investment plans concern only a single period, even if this short-sightedness is a suboptimal strategy;
3. investments only concern securities traded in the market (e.g., shares, bonds, and debt securities);

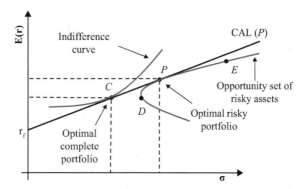

**Figure 5.A.4**
Identification of the optimal portfolio.

4. there are no taxes or transaction costs;

5. investors aim to maximize the relationship of expected return on risk (mean-variance optimizers) and therefore use Markowitz's (1952) portfolio selection model;

6. expectations are homogeneous. Since all investors use Markowitz's (1952) model with exactly the same predictions about the expected return on securities, all obtain the same efficient frontier and the same optimal portfolio.

If all these conditions are satisfied, it follows that (a) all investors hold a portfolio of risky securities that replicates the market portfolio, (b) the market portfolio is on the efficient frontier at a tangent to the CAL, which is therefore maximized, and (c) the risk premium of each security is a function of the risk premium of the market portfolio and of the beta coefficient, measuring to what extent the security and the market move in the same direction:

$$(5.10) \qquad \beta_i = \frac{Cov(r_i, r_M)}{\sigma_M^2},$$

while the risk premium of the single security that can be expected is

$$(5.11) \qquad E(r_i) - r_f = \frac{Cov(r_i, r_M)}{\sigma_M^2}[E(r_M) - r_f] = \beta_i[E(r_M) - r_f].$$

This expression is often represented only in terms of expected performance:

$$(5.12) \qquad E(r_i) = r_f + \beta_i[E(r_M) - r_f].$$

The expected return on an asset is the rate on risk-free securities plus the risk premium of the market portfolio multiplied by beta, which measures the riskiness of the security. With $\beta = 1$ we have an expected return that is perfectly correlated with the market portfolio, and thus there should be the same risk premium as for a market portfolio. A security that has $\beta > 1$ is considered aggressive since its performance amplifies fluctuations in the market. A security that has $\beta < 1$ is considered defensive as it may be used as a hedging instrument. Since there is no free lunch, this insurance function involves a cost to the investor which leads to a lower expected return.

The CAPM is a theoretical model based on very restrictive and unrealistic assumptions. In practice, there are big investors (for example, investment funds like Black Rock) capable of influencing markets, and there are long-term strategies, taxes, transaction costs, uneven expectations, and so on. Moreover, the expected returns are not observable, and we have to know the composition of the real portfolio market and not that of its approximation as in the S&P500 Index to be able to verify the model correctly (Roll, 1977). It is, nevertheless, often used to obtain rough indications about whether it is desirable to buy securities. The security market line (SML) (figure 5.A.5) relates the expected and beta yield. When the beta yield is equal to

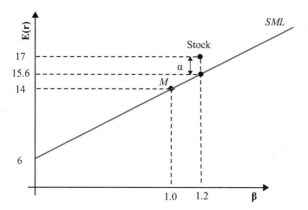

**Figure 5.A.5**
SML and an asset with positive $\alpha$.

zero, the CAPM does not provide for any risk premium; whereas when it is equal to one it provides for the risk premium of the market portfolio. When the expected return (positive alpha) on securities that are above the SML is too high, these securities will have too low a price. Instead, the securities that are below (negative alpha) will have too low an expected return, and therefore these securities are too expensive. The estimation of the alpha coefficient can, therefore, evaluate the presence of abnormal returns while the beta coefficient is the risk profile of securities (aggressive, neutral, or defensive).

Since expected returns are not observable, the following single index model expressed in terms of excess return on the rate of risk-free securities ($R_i = r_i - r_f$ and $R_M = R_M - R_f$) can be estimated:

$$(5.13) \qquad R_i = \alpha_i + \beta_i R_M + \varepsilon_i,$$

where $\varepsilon_i$ is the specific risk linked to the single security. $Cov(R_i, \varepsilon_i)$ is zero while the covariance between $R_i$ and $R_M$ is

$$(5.14) \quad Cov(R_i, R_M) = Cov(\beta_i R_M + \varepsilon_i, R_M) = \beta_i Cov(R_M, R_M) + Cov(\varepsilon_i, R_M) = \beta_i \sigma_M^2,$$

in which alpha disappears because it is constant. It follows that the coefficient beta is equal to

$$(5.15) \qquad \beta_i = \frac{Cov(R_i, R_M)}{\sigma_M^2}.$$

The beta coefficient obtained by estimating this single index model is identical to that of the CAPM with expected returns, with the exception of the market portfolio that is replaced by an observable market index. More recently Fama and French (1992,

1993, 1995) have extended the CAPM by introducing two additional risk factors based on the widespread belief that securities of small companies with a strong book-to-market ratio present higher returns than expected with the Sharpe (1964), Lintner (1965), and Mossin (1966) model. The two risk factors added to the model are small minus big (SMB) and high minus low (HML).[16] The first is given by the difference between the expected returns of portfolios of small capitalization stocks and large capitalization stocks while the second is given by the difference between the expected returns of portfolios of securities with a strong book-to-market ratio and portfolios of securities with a low book-to-market ratio:

$$(5.16) \qquad E(r_i) - r_f = a_i + b_i \left[ E(r_m) - r_f \right] + s_i E(SMB) + h_i E(HML).$$

If the price of the securities is correctly established, the coefficient $a$ should be zero. As for the CAPM, an econometric estimate can once again be made with the following model:

$$(5.17) \qquad r_i - r_f = a_i + b_i(r_M - r_f) + s_i E(SMB) + h_i E(HML) + \varepsilon_i.$$

The authors verified the model using US stock data and found that the inclusion of two additional risk factors increases the explained variance (R2) from 70 percent to over 90 percent. Fama and French's model, as indeed the CAPM, can assess whether wine investments guarantee too high or too low returns compared with the model (and therefore the price is too low or too high, respectively) as well as their correlation with the specific risk factors.

# 6

# Asymmetric Information

To know the vintage and quality of a wine one need not drink the whole cask.
—A phrase attributed to Oscar Wilde (1854–1900)

This chapter is divided into three sections. The first discusses the concept of asymmetric information, which is relevant in many fields besides economics; it analyses the causes that give rise to the two different situations of adverse selection and moral hazard and discusses possible solutions for these market failures. The second describes the advantages in terms of higher sales and price to be gained from the three sources of reputation—namely, individual, collective, and institutional. Finally, the main differences between the Old World and the New World are discussed.

## 6.1 Information Asymmetry: Problems and Possible Solutions

"Asymmetry" of information is when traders do not all have the same (complete) information in a transaction and can give rise to two different situations: adverse selection and moral hazard. Adverse selection is when one of the two parties, the "principal," cannot know of one or more exogenous characteristics of the "agent," the object of the transaction or the situations in which they may find themselves. It is important to underline that these characteristics preexist the decision to carry out the transaction and, therefore, are called exogenous. In contrast, moral hazard occurs after the decision to carry out the transaction when the delegating party cannot see the actions performed by the agent or the characteristics of a good they have supplied. In this case the actions and characteristics are subsequent to the decision to carry out the transaction.

The literature on information asymmetry has received great impetus from the pioneering work of Akerlof (1970) on adverse selection. In this article the author presented a theoretical model applied to the used car market in which the owner-seller knows exactly the characteristics and the degree of wear of the vehicle being sold while the buyer has trouble in establishing the actual quality of the vehicle because of their limited knowledge of mechanics or in detecting defects that cannot be easily seen. Assuming that the price quotes on the market refer to cars with average characteristics and qualities in terms of mileage and damage suffered, it follows that only owners with cars of equal or below average quality will gain by proceeding with the sale. The market responds to a decrease in the average quality of the cars with a parallel decrease in the average price offered, leading to a further lowering of quality down to only lemons. Theoretically, this downward spiral can continue until the complete disappearance of the market.

A situation of this kind is inefficient from a Pareto point of view since the will of the two parties, the principal and the agent, to conclude a mutually beneficial transaction is hampered by asymmetric information. In line with these intuitions, Shapiro (1983) showed that in the presence of a consumer's incomplete information, companies produce goods and services of lower quality because of the incentive to make short-term gains. All of this, of course, can continue until customers understand the real quality of the product that will then hamper purchases and in this way distort the market.

In the wine sector, information asymmetries are very strong.[1] Wine is, in fact, a classic example of an "experience good" (Cardebat, 2017, p. 32; Thornton, 2013, p. 38).[2] With the exception of repeat purchases or those that take place after tastings, bottles are generally bought sealed and when the wine has not yet been tasted. The principal discovers the quality only after purchase at the time of consumption. As the quality and the characteristics of the product are preexisting, this is a typical adverse selection situation. Moral hazard, instead, does not exist since, once the transaction has been completed, the producer can no longer influence in any way the quality of the drink because it has already been produced and bottled. Adverse selection in the wine sector does not lead to the disappearance of the market as a whole (people will not stop drinking wine because of uncertainty about the quality of the product on the shelf), but it can reduce the buyer's willingness to pay. It may also weaken the correlation between actual quality and the price paid to the detriment of those who find themselves in a negative spiral and to the advantage of others who may invest in effective advertising campaigns and marketing.

There are four possible solutions to the problem of adverse selection, two of which are private and two, public. Their purpose is to signal quality to consumers, thereby reducing the information asymmetry (Cardebat, 2017, pp. 32–35). Private solutions emerge spontaneously on the initiative of companies operating in the wine sector and consist in building a solid reputation for individual companies (corporate reputation) or business consortia (collective reputation). Public solutions are provided by national or supranational public authorities and take the form of wine classification systems (leading to "institutional" reputation) and quality control. In the next section the three forms of reputation will be discussed (table 6.1) while the role of controls will be briefly mentioned in the context of collective reputation.

**Table 6.1**
Individual, collective, and "institutional" reputation.

| | |
|---|---|
| Individual reputation | Refers to a single company and is built up by the firm by investing in quality, advertising campaigns, etc. |
| Collective reputation | Refers to a group of companies that have joined together in a consortium, creating a collective brand (appellation—e.g., Chianti Classico, Barolo, etc.), and that are committed to following strict rules on standards and production procedures of the product specifications. |
| "Institutional" reputation | Refers to the classification of wine established by public authorities (VdT, IGT, DOC, and DOCG in Italy). The state sends a signal to the consumer ordering the wines according to their quality and production standards. |

*Notes*: VdT = Vino da Tavola (table wine); IGT = Indicazione Geografica Tipica (wine typical of a region); DOC = Denominazione di Origine Controllata (Controlled Designation of Origin); and DOCG= Denominazione di Origine Controllata e Garantita (Controlled and Guaranteed Designation of Origin).

## 6.2 Reputation

Reputation is the expectation about the quality of an asset or the abilities or behavior of one or more agents (Bar-Isaac and Tadelis, 2008)[3] and depends on the quality and behavior observed in the past.[4] Quality and reputation are two concepts that are connected but that do not necessarily coincide because reputation obviously depends on quality but is also influenced by other factors such as advertising campaigns, word of mouth,[5] and so on. It follows that reputation can be better or worse than the actual quality.

In the economic field, the presence of information asymmetries makes reputation a valuable tool to increase sales both in value and in volume. Building a reputation requires significant short-term investments to obtain long-term returns (Wilson, 1985). Not only do certain costs in the short term correspond to uncertain and delayed benefits over time, but reputation can suddenly be damaged by intentional actions or accidental events that risk undermining a multiyear commitment (Fombrun and Shanley, 1990). However, an increase in sales and/or the price premium resulting from reputation can compensate for the efforts made by the company (Barney, 1991).

In the last thirty years there has been a proliferation of literature on the causes and consequences of reputation. Among the theoretical articles that suggest a positive impact of firm reputation on selling price are Klein and Leffer (1981), Shapiro (1983), Rogerson (1983), Allen (1984), and Houser and Wooders (2006) while empirical studies focusing on e-commerce include Melnik and Alm (2002), Keser (2003), Resnick et al. (2006), and Cabral and Hortaçsu (2010). Businesses that enjoy a better reputation are able to achieve above average profits in the long run, as demonstrated by Roberts and Dowling (2002).[6]

In the wine sector, reputation plays a very important role in consumer choices, especially for high-end products (Heijbroek, 2003). As already mentioned in chapter 2, the positive relationship between the three forms of reputation and the price of bottles sold has been widely demonstrated using data for the French Bordeaux region (Combris, Lecocq, and Visser, 1997; Landon and Smith, 1997, 1998; Cardebat and Figuet, 2004; Ali and Nauges, 2007), Australia (Oczkowski, 1994, 2001, 2018), the United States (Costanigro, McCluskey, and Mittelhammer, 2007; Costanigro, McCluskey, and Goemans, 2010; San Martín, Brümmer, and Troncoso, 2008; Cross, Plantinga, and Stavins, 2011), the Italian region of Piedmont (Benfratello, Piacenza, and Sacchetto, 2009; Corsi and Strøm, 2013), Germany (Frick and Simmons, 2013), and four countries of the New World (Schamel, 2000).[7]

Given that reputation helps to increase both the volume of sales and the average price that a consumer is willing to pay, we now move on to an analysis of its determinants based on the studies of Castriota and Delmastro (2012, 2015) using Italian data. The two studies use ratings from wine guides that collect and publish information by assigning votes to companies or appellations. The function of wine guides is to reduce information asymmetries in a complex market where consumers have too much information and limited skills (Marks, 2015, p. 120) and can therefore be compared in all respects with that of rating agencies (Hay, 2010). Nowadays wine guides, journalists, gurus, and bloggers are able to influence market prices with their own assessments as demonstrated by Ali, Lecocq, and Visser (2008) in the study of the American critic Robert Parker and French en primeur wines.[8]

One may rightly ask whether the number of stars assigned by wine guides to companies and appellations is a distorted proxy of reputation. In principle, there is nothing to guarantee that the opinion of experts reflects that of consumers who are on average less experienced—not to mention the possible risk that judges are paid under the table to give flattering evaluations. However, as emphasized by Costanigro, McCluskey, and Goemans (2010),[9] a number of studies have found a positive correlation between expert ratings and the price of wine regardless of the country, guide, or expert. Since price is nothing but a consumer's willingness to pay, this proves the correspondence between the opinions of critics and buyers.

### 6.2.1   Individual Reputation

The wine sector is a perfect candidate for an analysis of the determinants of a company's reputation because many producers are moved by strong intrinsic motivations (Scott Morton and Podolny, 2002) to pursue qualitative excellence regardless of the possible return on investments.

This section refers to Castriota and Delmastro (2012), who used data from a sample of 581 companies located in northwest Italy (figure 6.1). All 581 wineries have won a national reputation (appearing in the *Espresso* guide), but only sixty-seven have achieved notoriety at an international level (present in Hugh Johnson's guide).

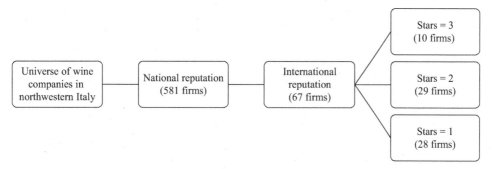

**Figure 6.1**
Structure of the sample of firms used in the study of firm reputation.
*Source*: This figure is from Castriota and Delmastro (2012), p. 59, figure 2.

International prestige is therefore a privilege reserved for few companies. Corporate reputation is built up chronologically from (1) no reputation, followed by (2) notoriety at a national level, and finally (3) at an international level.[10]

The theoretical literature[11] has identified a series of variables among the determinants of corporate reputation that can be largely attributed to the two pillars of the information economy: the innate or acquired characteristics of a company—such as ability and skills—and its actions—such as commitment, seriousness, and honesty. However, over the years, the literature has added other factors not related to these two macro-categories. For a review, see appendix 6.1.

The analysis carried out by Castriota and Delmastro (2012) shows that in general, determinants that influence national reputation will also influence international reputation. As reported in the theoretical literature, age plays a significant role; this determinant reflects the importance of the learning process both for the consumer and the entrepreneur. The involvement of the owner as a wine maker influences reputation positively through greater commitment and pursuit of excellence or through specific skills acquired over the decades. Company size is positively correlated with reputation since it ensures greater visibility and the adoption of large-scale technologies.[12] Company form does not seem to be relevant nor indeed does the recruitment of famous external wine makers or horizontal differentiation.[13] Finally, the collective reputation of the most prestigious appellations can positively influence the reputation of an individual company.

After having considered the similarities, let's now move on to the differences. The first concerns yields per hectare that have a negative effect on national reputation but zero effect on international reputation. This can be explained by the greater amount of knowledge and information a national observer can find compared to one in a distant country. Geographical proximity and contacts can also allow small niche businesses that pursue excellence but with very low yields to emerge and be known nationally. The second difference concerns the role of the institutional brand Controlled and Guaranteed Designation of Origin (DOCG) that is recognized by the state and appears to be relevant only at an international level whereas the prestige of the most important appellations matters most (collective reputation) at a national level. Here once again, less detailed knowledge about context and producers forces the international observer to rely more on the institutional signals provided by public authorities (recognition of DOCG). The appellation system and the classification system, therefore, can both be useful for building a reputation, even if they act through different channels: the first being national and the second, international.

### 6.2.2 Collective Reputation

The analysis of the determinants of collective reputation draws on Castriota and Delmastro (2015). As reported in the previous section, the presence of strong information asymmetries between producer and consumer, the nature of "experience goods," the dispersal of land ownership, and the need for combating fraud have all encouraged the creation of producer consortia and collective brands. When there is a very large number of products, consumers often buy goods of the more prestigious collective brands to save time (Andersson, 2002). Buyers must decide the type of information and the level of detail to collect (Costanigro, McCluskey, and Goemans, 2010), and so they generally start with geographic names (e.g., Italian or French wines are considered good), then move on to collective brands (e.g., Champagne and Barolo) and finally with the individual brands, with the best vintages generally reserved for just a few experts (Fleckinger, 2007). Last but perhaps not least is the role that some consumers attribute to regional traditions, for which they are willing to pay a price premium (Vogel, 1995; Grebitus, Lusk, and Nayga, 2013; Balogh et al., 2016).

A good collective reputation benefits sales volumes and prices, and this is particularly useful where companies are small (as in Italy) and it becomes impossible to build a reputation at an individual level (as, for example, happens in the New World). Indications of origin are so important that they were given protection by the World Trade Organization in the Marrakesh Agreement of 1994 (see box 6.1). This is the reason why economists started to study how collective reputation is formed and what factors contribute to it. Collective reputation is defined as the aggregation of the individual reputations of all the associated companies (Tirole, 1996; Landon and Smith, 1998) or its most famous members (Gergaud and Livat, 2004), In Tirole's model (1996) the new members of a group "inherit" the reputation of the older ones, thereby benefiting from it or paying the price for it, even long after the senior members have left. Collective reputation is thus history dependent and creates stereotypes. As for the determinants, many of the variables that influence a company's prestige do so in exactly the same way as for collective brands. See appendix 6.2 for a review of the theoretical determinants of collective reputation.

**Box 6.1**
Annex 1C: Trade-Related Aspects of Intellectual Property Rights, Marrakesh Agreement establishing the World Trade Organization, signed in Marrakesh, Morocco on 15 April 1994.

---

**Section 3: Geographical Indications**

**Article 22**

Protection of Geographical Indications
1. Geographical indications are, for the purposes of this Agreement, indications which identify a good as originating in the territory of a Member, or a region or locality in that territory, where a given quality, reputation or other characteristic of the good is essentially attributable to its geographical origin.
2. In respect of geographical indications, Members shall provide the legal means for interested parties to prevent:
    (a) the use of any means in the designation or presentation of a good that indicates or suggests that the good in question originates in a geographical area other than the true place of origin in a manner which misleads the public as to the geographical origin of the good;
    (b) any use which constitutes an act of unfair competition within the meaning of Article 10*bis* of the Paris Convention (1967).
3. A Member shall, *ex officio* if its legislation so permits or at the request of an interested party, refuse or invalidate the registration of a trademark which contains or consists of a geographical indication with respect to goods not originating in the territory indicated, if use of the indication in the trademark for such goods in that Member is of such a nature as to mislead the public as to the true place of origin.
4. The protection under paragraphs 1, 2 and 3 shall be applicable against a geographical indication which, although literally true as to the territory, region or locality in which the goods originate, falsely represents to the public that the goods originate in another territory.

**Article 23**

Additional Protection for Geographical Indications for Wines and Spirits
1. Each Member shall provide the legal means for interested parties to prevent use of a geographical indication identifying wines for wines not originating in the place indicated by the geographical indication in question or identifying spirits for spirits not originating in the place indicated by the geographical indication in question, even where the true origin of the goods is indicated or the geographical indication is used in translation or accompanied by expressions such as "kind," "type," "style," "imitation," or the like.
2. The registration of a trademark for wines which contains or consists of a geographical indication identifying wines or for spirits which contains or consists of a geographical indication identifying spirits shall be refused or invalidated, *ex officio* if a

*(continued)*

**Box 6.1** (continued)

> Member's legislation so permits or at the request of an interested party, with respect to such wines or spirits not having this origin.
> 3. In the case of homonymous geographical indications for wines, protection shall be accorded to each indication, subject to the provisions of paragraph 4 of Article 22. Each Member shall determine the practical conditions under which the homonymous indications in question will be differentiated from each other, taking into account the need to ensure equitable treatment of the producers concerned and that consumers are not misled.
> 4. In order to facilitate the protection of geographical indications for wines, negotiations shall be undertaken in the Council for TRIPS concerning the establishment of a multilateral system of notification and registration of geographical indications for wines eligible for protection in those Members participating in the system.
>
> *Source:* WTO (1994).

The empirical analysis of Castriota and Delmastro (2015) begins with the static and then continues with the dynamic. As in the case of individual reputation, age has a positive and significant coefficient. Both compulsory and optional quality standards are strongly significant, showing that the sacrifices made by members are then repaid in the form of group reputation. The frequency of controls and the size of sanctions are important to ensure the rules are observed and to build the prestige of the appellation. In line with Fishman et al.'s (2008) intuitions, when the size of the group (the number of producers) increases, the reputation first grows as a result of greater visibility and then, having reached a peak, decreases because of the incentive for opportunistic behavior and difficulty in controlling members' work (see figure 6.2). Free entry of new members, therefore, is not optimal.

Without considering the minimum quality standards, the DOCG is associated with a strongly and significantly better reputation. The inclusion of new regressors progressively weakens the explicative power of the DOCG variable, which is no longer significant. This means two things. First, once all the regressors are included, collective reputation does not depend on mere formal recognition but on real intrinsic qualities. Second, the wine classification system is still important as it is correlated with the reputation of collective brands, quality standards, and controls and can act as an (imperfect) substitute for information that is more detailed but difficult to collect.

The dynamic analysis confirms the persistence hypothesized in Tirole's theoretical model (1996). There is, however, a certain variability in time shown by the fact that on average, 21 percent of the appellations increase or decrease the number of stars received over a period of five years, a percentage that rises considerably when time is extended to thirty years. It is interesting to observe how greater persistence is found

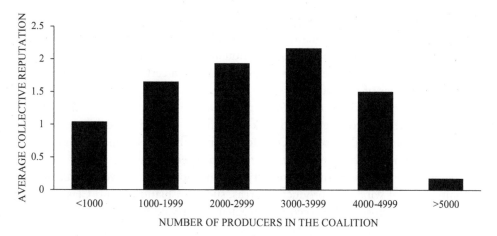

**Figure 6.2**
Average collective reputation by number of producers.
*Note:* Collective reputation is measured with the number of stars assigned by the Hugh Johnson's wine guide.
*Source:* Corresponds to figure 1 in Castriota and Delmastro (2015).

at the minimum reputation level. Therefore, there is a sort of "bad reputation trap" from which, once in, it is difficult but not impossible to get out of. For this reason, it is important for a consortium to fix high minimum quality standards when it is first established to avoid actions that could seriously damage the brand over the years.

To summarize, building a collective brand needs time, high-quality standards that are both compulsory and optional, strict compliance with the rules obtained with frequent checks and high penalties, a number of members that is neither too small nor too large, and finally, if possible, a favorable socioeconomic context. In light of these considerations it is not surprising therefore that, of the 1,424 subappellations existing in Italy in 2008, none had four stars and only six had 3.5 stars. While individual reputation is difficult to build, collective reputation is even more so because it is based on the autonomous choices of numerous operators who aim for the maximization of their individual well-being and certainly not of the collective well-being. The creation of prestigious collective brands over the centuries as a private response to market failures has, therefore, something miraculous about it, and they must be protected in an intelligent and careful way.

The main risks in recent times have been as follows.

1. *An excessive number of appellations (Colman, 2008, pp. 60–62)*: As shown by Delmond and McCluskey (2018), as the number of geographic indications present in the agricultural markets expands, the returns to each region's collective reputation increase to a peak and then start decreasing. All over the world the number of appellations continues to increase rapidly, confusing consumers and thus becoming ineffective. The number of subappellations rose from 686 in 1978

to 1,424 in 2008. Many subappellations (about 16 percent in 2008), however, are not even produced while many others are so small as to be unknown to almost all consumers. A possible solution is rationalization, leading to a reduction in the number of those subappellations that do not reach a certain level of prestige or a certain number of producers.

2. *An excessive number of members in the groups*: Some appellations have such a large number of producers as to make it difficult, if not impossible, to focus on quality (Colman, 2008, p. 64). Therefore, a limit should be imposed on the entry of new members when an optimal number has been reached.

3. *Overlapping names*: In time some very similar appellations have been created by name, type of wine, and production areas, such as Chianti/Chianti Classico (both DOCG) and Prosecco/Prosecco di Conegliano and Valdobbiadene (the first being Controlled Designation of Origin [DOC] and the second, DOCG) for historical reasons. Chianti is produced throughout nearly all of northern Tuscany and has many producers and average quality standards while Chianti Classico is an older appellation and has more limited borders (six municipalities between Siena and Florence), fewer producers, and much more stringent quality standards. The problem is that only a small minority of experts knows these differences and recognizes the quality of the Chianti Classico. The overwhelming majority of people do not know that they are two almost identical but actually very different wines, to the detriment of Gallo Nero producers,[14] who have difficulty in transmitting an image of excellence in the middle of this sea of wine where nobody fully understands provenance and quality. A very similar situation applies to Prosecco, a wine that was only produced in the areas around Conegliano and Valdobbiadene until 2009. In that year the designation DOC Prosecco was extended to eastern Veneto and Friuli-Venezia Giulia. Two DOCGs, Conegliano Valdobbiadene-Prosecco and Colli Asolani-Prosecco, were created to compensate the old producers for their loss of exclusivity. Once again, the existence of almost the same—but actually different—appellations is known only to a few enthusiasts, with the risk that the average price will decrease due to the sudden excess supply and a progressive and permanent damage to reputation if the new producers (followers) should try to draw a short-term advantage from the cumulative investments of historic companies (leaders).

### 6.2.3  Institutional Reputation

The third form of reputation originates from recognition by national or supranational authorities. The classification system of wines was started in France in 1855 by order of Napoleon III, who wanted the vineyards of the Bordeaux region to be classified in order of quality for the *Exposition Universelle de Paris*. In the same year the recognition of the cru classé was attributed to sixty wines (from the *Premiers Crus* to the *Cinquièmes Crus*).

The need to create a classification system of wines that clearly and simply identified the best products became even more pressing in the first two decades of the twentieth century when buyers were confused by frauds, phylloxera, and Algerian wine that was passed off as French. In 1935 the French government created the *Appellation d'Origine Contrôllée* (Controlled Designation of Origin, or AOC), and it established the *Institut National des Appellations d'Origine* (National Institute of Origin and Quality, or INAO) with the task of regulating the AOC. The system of appellations adopted the rules established by Baron Pierre Le Roy for the production of his wines. The Baron had, in fact, marked the boundaries within which his wines could be produced and controlled the permitted vines, the rules on pruning and vinification, and minimum alcohol content.

In Italy the need to regulate the production of wines, especially quality wines, dates back to the post–World War I period. In 1921, the Honorable Arturo Marescalchi, founder of the current Assoenologi (Association of Oenologists), presented the first proposal in parliament for the production of "typical wines" that was approved by royal decree in 1924 and converted into law in 1926. It was subsequently amended and modified in 1930 and 1937 but never came into force because no implementing decree was passed. A long period of silence followed, and Europe took up the initiative. France, which was highly influential in the Common Organization of Agricultural Markets, gradually molded the wine sector more and more in its image and likeness in time.

"Wines with Designation of Origin" (modeled on the French Appellation d'Origine) were discussed for the first time at the 1957 Treaty of Rome conference, thereby laying the foundations for a common classification system of products that took shape in a 1962 Community Law. Italy quickly adopted this approach with the DPR 930/63 to be followed with the recognition of the first DOC for Vernaccia di San Gimignano in 1966 while the approval of the first three DOCG—Barolo, Brunello di Montalcino, and Vino Nobile di Montepulciano—dates back to 1980. All the countries in the European Community, including those who joined later (e.g., Spain and Portugal) or even more recently (e.g., Eastern Europe), have created similar classification systems following the 1935 French system (Meloni and Swinnen, 2013). The classification system that was in force before the Council Regulation (EC) No. 479/2008 reform is shown in table 6.2 and has been modified into the four slightly different categories: Denominazione di Origine Protetta (Protected Designation of Origin, or DOP), Indicazione Geografica Protetta (Indication of Geographic Protection, or IGP), varietal wine, and generic wine, which member states can waiver.

The hierarchical structure of wines was intended to resemble a pyramid (see figure 6.3) and has been adopted in all EU countries, though with different names (see table 6.2). A few prestigious products sit at the top and are made to very strict standards. As we move down toward the base of the pyramid, the quantities of products gradually increase, but they have a poorer (or zero) reputation as a result of far less

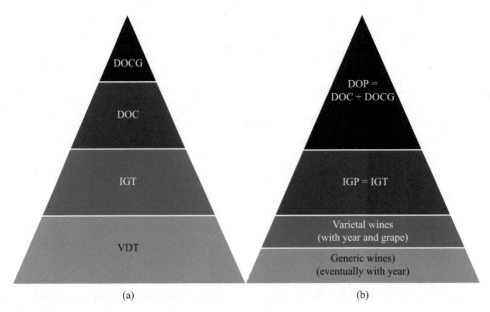

**Figure 6.3**
The pyramid of institutional reputation in the old (a) and new (b) wine classification systems.

stringent rules. Up until 2008, Italian law provided for four levels in order of increasing quality (see figure 6.3[a]):

- Table wines: can be produced in any region, with any vine and any vintage. The geographic area, the vines used, and the vintage cannot be shown on the label. Quality standards are aimed at ensuring the mere wholesomeness of the drink.
- IGT: the grapes must come from at least 85 percent of land that falls within delineated boundaries of the production area, but it generally covers quite a large surface (sometimes whole regions like the Tuscan and Sicilian IGTs). Minimum quality standards exist but are very bland (e.g., very high yields per hectare).
- DOC: the grapes must come entirely from land that falls within delineated boundaries of the production area and are generally quite circumscribed, and the minimum quality standards are quite strict. Analyses are performed by the competent authorities during production.
- DOCG: the grapes must come entirely from land that falls within delineated boundaries of the production area which is very restricted, and the minimum quality standards are even more stringent. By law the DOCG can only be assigned to wines that have had DOC recognition for at least five years and have reached a high level of prestige. Analyses are performed by the competent authorities both during production and bottling.

Table 6.2

Old wine classification systems in Europe in descending order of quality.

| France | Italy | Spain | Portugal | Germany | Austria |
|---|---|---|---|---|---|
| AOC: Appelation d'Origine Contrôlée | DOCG: Denominazione di Origine Controllata e Garantita | Denominación de Origen Calificada | DOC: Denominacao de Origem Controlada | Qualitätwein mit Praedikat or Kabinett | Qualitätwein mit Praedikat or Kabinett |
| VDQS: Vins Delimité de Qualité Superieure | DOC: Denominazione di Origine Controllata | Denominación de Origen | IPR: Indicacao de Proveniencia Regulamentada | Qualitätwein Bestimmter Anbaugebiete | Qualitätwein |
| Vin de Pays | IGT: Indicazione Geografica Tipica | DenominaciónEspecifica | Vinho regional | Landwein | Landwein |
| Vins de Table | VdT: Vino da Tavola | Vino de la Tierra | Vinho de mesa | Deutscher Tafelwein | Tafelwein |
| Vin de consummation courante | | Vino de mesa | | EWG | |

In Castriota and Delmastro's (2012, 2015) studies the authors show how—despite the controversy that in Italy often accompanies the system of appellations which are considered inefficient—the international reputation of Italian wineries and appellations appears to be significantly influenced and correlated with the production of DOCG wines. Notwithstanding the numerous imperfections, the system of appellations has been tested and the econometric results demonstrate how it brings indisputable benefits to consumers and companies, especially for new and/or small companies and/or those that specialize in the medium- and high-quality segments of the market. In recent times, however, the national system of classification of wines has, in some cases quite rightly, been criticized because of the sudden increase in the number of appellations, the increase in their share of the total national production and the imperfect correlation between type of designation (table wines, IGT, DOC, DOCG) and the actual quality of products. Therefore, a reform has been called for, but national authorities have taken no action so far.

Instead, the European Community has taken up the initiative once again and issued Council Regulation (EC) No. 479/2008 that doubles the number of acronyms. In fact, companies can freely decide whether to use the old acronyms DOC and DOCG or the acronym DOP (which incorporates both) and the old IGT or IGP. Table wine is now simply called wine and can optionally report the year of harvest[15] and, for varietal wines, also the name of the vine[16] on the label (see figure 6.3b), which was previously prohibited for this category of wine and was allowed only for better quality wines. Recognition of Geographical Indications (IGT and IGP) and of the Appellation-Designations of Origin (DOC, DOCG, and DOP) are no longer given by national but rather by European authorities while the minimum time necessary to progress from DOC to DOCG has been extended to ten years.

There are at least three problems with this current classification system.

1. *Two parallel systems.* In spite of the provisions of the Community Regulation, member states can continue to use their traditional terms (IGT, DOC, and DOCG in Italy). The first problem consists, therefore, in the potential confusion resulting from two parallel quality reporting systems and in the flattening of the pyramidal structure that incorporates DOC and DOCG within the DOP. A buyer faced with a DOCG wine—whose vineyard now has decided to put the DOP mark on the label—could mistakenly think that it is a DOC (only the last letter in the initials is different from DOC) or, believing it is something completely different, not understand whether the acronym "DOP" is better or worse than the other two. Likewise, a varietal wine, which can carry the vintage and vine on the label (e.g., Chardonnay 2014), can be mistaken for a higher quality wine (e.g., DOC Alto Adige Chardonnay 2014), whereas this could not happen before the EU reform of 2008 as the generic wines were classified as table wine and therefore could not state the vine and vintage. Besides being potentially misleading, the new pyramid

structure is also unbalanced in favor of quality wines (see figure 6.3[b]). The top, where "DOP" absorbs both DOC and DOCG, has become larger and dilutes two qualitatively differentiated levels into a single brand. If a part of the base, the varietal wines, was wrongly associated with IGP, then the base of the pyramid would be limited to only generic wines, with an intermediate level given by varietal and IGP wines and a higher level given by DOC and DOCG together. Thus, we risk improving the image of a part of the less prized wines (generic and varietal wines) and decreasing the reputation of those at the highest level (DOCG), with a net flattening of the signals transmitted to consumers. Instead of going toward a more refined naming system, following the model of some historical French appellations that identify hierarchically and qualitatively categories of land, Europe is moving in the opposite direction to a less selective system, leaving more and more room for marketing investments of large international groups and less and less to collective and institutional reputation (Castriota and Delmastro, 2009).

2. The appellation mechanism is now managed by the European Commission after being checked for conformity by national authorities and on the proposals of producer associations. This bureaucratic and cumbersome mechanism is likely to slow down the procedure of assigning new appellations, giving the final word to supranational institutions that do not know the real situation either of the areas or of the sector. Before handing over to European Community bodies, however, there has been a real "gold rush" to get higher level appellations from the EU Ministry of Agriculture and Forestry because it would have become more complicated later.

3. For the pyramid modeling system to be effective, the most prestigious awards have to be granted to a small part of the production. If, by definition, the share of quality wines (DOC) was 0 percent in 1963 (the year the DOCs were created), then it had risen to 41.5 percent in 2019 (DOC and DOCG).[17] Similar paths have been observed in other EU countries. If this trend were to continue, most wine would be classified as quality (*todos caballeros*, "everyone's a winner") in a few years, and this recognition would become useless. A possible solution, though probably impractical from a political point of view, could be to grant the more prestigious recognition to a fixed share of national products (e.g., DOCG to 5 percent or 10 percent of the wine) after which one or more appellations would have to be relegated to the lower level (e.g., to DOC), similar to soccer (football) championships where every year the worst teams of a tournament are downgraded and replaced by the best teams of the lower category. This solution would involve the problem of who should decide in an objective manner and without conflicts of interest which appellations should be downgraded, but it is also true that some DOCGs are much less prestigious than some DOCs and have been recognized in a far from meritocratic way.[18]

In general, both before and after the reform of the Council Regulation (EC) No. 479/2008, the qualitatively superior categories have always had more stringent production rules. This generally leads to an increase in quality and reputation, but it also imposes constraints that limit producers' ability to adapt to technological innovations and consumers' changing tastes (Shepherd, 2006). For this reason, some wine makers, especially when they reach a high individual reputation, decide to produce a significant share of wine that is bottled as IGT or VdT. They are free to move and experiment in a way that the rigid disciplinary rules of appellations/designations of origin do not allow.

## 6.3   Differences Between the Old and New World

Reputation is a very important variable for producers in all countries, but there are fundamental differences in the importance of the three forms between the Old and the New World. As mentioned above, the sector in Europe and especially in Italy is composed of small companies, and this has made it difficult for most companies to build a solid reputation on an individual basis and has encouraged the creation of collective brands and a hierarchical classification system. In contrast, the New World focuses on the recognizability of company brands.

In recent times, however, even non-European countries are becoming aware of the need to protect producers of especially prestigious areas that have been able to build a collective reputation in time.[19] For example, the American Viticultural Areas (AVAs) have been established in the United States and are halfway between the IGT and generic local wines. The AVAs, in fact, simply represent a geographic boundary within which wine can be produced,[20] but unlike the European appellations, they do not impose any minimum quality standards. The vines, the agronomic and wine-making techniques, or the minimum quality standards to reduce information asymmetries between producer and consumer are not specified. Only the distinctive characteristics of the area (soil, climate, etc.) compared with the surrounding areas are described.[21] The sole requirement is that at least 85 percent of grapes should be produced within the AVA. It is, therefore, a matter of collective reputation, even if the very fact of producing wine within an AVA is a recognition, which, in theory, can confer institutional reputation to some extent. However, this aspect has not yet been studied in the literature. Currently there are 242 AVAs,[22] with some applications awaiting approval. Other New World countries have also set up geographical indications to protect their most famous areas (table 6.3). The *Indicación Geográfica* (geographical indications, or IG) and the *Denominación de Origen Controlada* (Denomination of Origin, or DOC) have been established in Argentina and Uruguay; the *Denominación de Origen* (Denomination of Origin, or DO) has been established in Chile; the geographical indications, in Australia; and the Wines of Origin (WO), in South Africa.

**Table 6.3**
Appellations in the Old and New World, 2013.

| Country | IG | IGP | DOP |
|---|---|---|---|
| Old World | | | |
| Italy | – | 129 | 476 |
| France | – | 75 | 376 |
| Spain | – | 44 | 97 |
| Portugal | – | 10 | 46 |
| Germany | – | 26 | 13 |
| New World | | | |
| USA | 227 | – | – |
| South Africa | 153 | – | – |
| Australia | 78 | – | – |
| Chile | 61 | – | – |
| Argentina | n.a. | – | – |
| China | n.a. | – | – |
| New Zealand | n.a. | – | – |
| Russian Federation | n.a. | – | – |

*Source*: Data for all countries excepting the United States were downloaded from the European Commission's website on September 18, 2013. Data for the United States were sourced from the Government Printing Office, accessed on November 21, 2014.

## Appendix 6.1: Theoretical Determinants of Firm Reputation

A first group of variables includes the characteristics of the entrepreneur and his company. The age of a company should have a positive effect on reputation (Thornton, 2013, p. 177) as the entrepreneur and the wine maker learn from experience (learning by doing) while consumers learn about the company in time through repeated purchases. Intrinsic motivation is very important for the quality of wine produced and consequently for the reputation of the company. It is influenced in part by the work of the wine maker that is done personally by the owner. Indeed, the family business structure can influence the quality of products since an external manager may be driven by objectives that differ from those of the owner and aim at short-term profitability (Cadbury, 2000). In any case, after having spent their whole life in the family business, an internal manager may have accumulated specific knowledge of the company that is invaluable (Donnelley, 1964). However, family disagreements can lead to serious management problems in a company with an internal manager (Christiansen, 1953; Levinson, 1971; Barnes and Hershon, 1976; Lansberg, 1983), and even more importantly, the selection of managers will be made from a small group of individuals and not based on meritocratic criteria (Burkart, Panunzi, and Shleifer, 2003; Pérez-González, 2006).[23]

Company size, measured by the number of bottles produced, can generate positive effects because a greater number of regular customers, combined with the phenomenon of word of mouth, makes large companies disproportionately[24] more visible in the eyes of the market (Rob and Fishman, 2005). Further, greater resources mean new technologies can be adopted and massive advertising and promotional campaigns made. Belonging to an industrial group can also bring advantages in terms of visibility while the cooperative form, as illustrated in the chapters 2 and 4, may be associated with a poorer quality and reputation because the incentive to opportunistic behavior can increase if there is a very large number of members (Winfree and McCluskey, 2005; Fleckinger, 2007; McQuade, Salant, and Winfree, 2008; Fishman et al., 2008).

The maximum yields per hectare are fixed for superior wines (IGT, DOC, and DOCG) but not for VdTs. Therefore, for this last type of wine, the wine grower is absolutely free to choose. The decision to join a consortium to produce quality wines is also free. As for the grapes purchased externally, quality is generally associated with control of the entire production chain. Ultimately, every manufacturer is faced with a trade-off between quantity and quality and must decide how much to sacrifice of the first for the second or vice versa.

Other potentially relevant variables include horizontal/vertical differentiation and the stretching of reputation. The production of many types of wine can help to satisfy the tastes of a diversified clientele with reputation being transferred from products of a higher level to those of a lower level (Wernerfelt, 1988). Hiring famous oenologists[25] as external consultants can help to increase the reputation of a business both directly by providing useful knowledge to improve the quality of products and indirectly by "transferring" part of the oenologist's reputation to the business they are working for (Kreps, 1990; Bar-Isaac and Tadelis, 2008, section 6).

Given the uncertainty surrounding the purchase of wine, businesses and public authorities have established collective brands (appellations) and classification systems (in Italy: VdT, IGT, DOC, and DOCG) to reduce information asymmetries between firms and consumers. In line with Tirole's (1996) model, in which individual and collective reputation influence each other, the production of wines belonging to prestigious consortia can help to increase the reputation of a company; conversely, Winfree and McCluskey (2005) showed that when a collective brand is shared but there is no traceability, companies have an incentive to choose a suboptimal quality level for the group, which makes the adoption of minimum quality standards desirable (see also Winfree, McIntosh, and Nadreau, 2018; Delmond, McCluskey, and Winfree, 2018).

Although closely linked, the two concepts of collective and institutional reputation must be kept distinct. Each collective brand (e.g., Toscana, Torgiano, or Greco di Tufo) is, in fact, associated with a level of quality recognized by the state (in the three cases mentioned: IGT, DOC, and DOCG, respectively). DOCG is, all things being equal, more prestigious than DOC. Within the same segment, however, there are groups of companies that have managed, in time, to create more famous appellations for a number of reasons (climate, soil, native vines, minimum quality standards, virtuous behavior of members, etc.) and others that have failed in their intention. Therefore, DOCs may include a lot of collective reputations that differ greatly from each other (some with zero and others with three stars) or even DOCs with a better reputation than some DOCG.

## Appendix 6.2: Theoretical Determinants of Collective Reputation

From a theoretical point of view, the determinants of the reputation of groups of companies can be summarized in four categories: (1) the general characteristics of the group, (2) quality standards and horizontal differentiation, (3) the control system, and (4) the geographic context.

A collective brand can be recognized by its status, as DOC or DOCG in Italy. DOCG has higher quality standards, and therefore a higher quality can be expected. Apart from this, the very fact of having obtained recognition from public authorities could positively influence the reputation of the group. The age of companies is an important variable since consumers and producers learn from experience accumulated over time. The size of the group can also play a crucial role because larger groups have more resources to allocate to advertising campaigns and have a larger customer base, which, combined with the phenomenon of word of mouth, makes them more visible in the eyes of the market (Rob and Fishman, 2005).

On the other hand, when groups get too big, the incentives for opportunistic behavior grow and social norms become ineffective (Kandori, 1992; Saak, 2012). Moreover, as in cooperatives, every company pursues the maximization of individual profit in the absence of centralized planning. The sum of these behaviors leads to excess production (Albæk and Schultz, 1998) with potential damage to the reputation of the whole group. Fishman et al. (2008) reconciled these two opposing views with a theoretical model in which members have both an incentive to invest in the reputation of a group and behave in an opportunistic way. When the group is small, the incentive for virtuous behavior prevails while the opposite happens when the group becomes excessively big. In this case the cost of investment, which falls entirely on the single enterprise, has a minimal impact on overall prestige and generates an expected return that must be shared with all the other partners.[26]

The minimum quality standards (Winfree and McCluskey, 2005) are the rules and requirements set by the group or by authorities to ensure a minimum quality level to protect the consumer and to promote the formation of prestigious brands. Many businesses and professions are subject to the issuing of authorizations or licenses precisely for this reason, though the system is often accused of actually wanting to regulate the market and especially to prevent the entry of competitors and to artificially keep prices high. However, establishing very strict rules is completely pointless if they are not observed.

We come, thus, to the third group of variables in which the traceability of a producer makes a system of frequent checks and fines sufficiently burdensome to discourage possible opportunistic behavior (Allingham and Sandmo, 1972).[27] However, it must not become oppressive; otherwise it generates distrust and resentment, thus undermining intrinsic motivation and the commitment of the subject being checked (Frey, 1993; Bénabou and Tirole, 2003). In wine making, controls can take place either in the vineyard (e.g., vineyard surface area, number and type of variety, company documents, etc.) or in the winery (e.g., state appellations with the relative approval by the certifying body, cellar handling, chemical analysis on wine samples, etc.).

Finally, the geographic context plays a fundamental role in the development of the economy and the creation of prestigious collective brands through public policies, the construction of infrastructures, the crime rate, and so forth (for an example, see Abrams and Lewis, 1995). This aspect is particularly important in Italy in light of the huge differences between the north and south in the indicators of economic and social progress. Mentality, which influences the preparation for the setting up and for the management of a company, as well as compliance with rules is the result of thousands of years of history and varies enormously from one region to another. The literature on the role of social capital in economic development began with Banfield (1958) and continued with Putnam (1993) and Guiso, Sapienza, and Zingales (2004) (for further information, see appendix 6.3). In all these studies the geographic area of reference is Italy, which is particularly suitable for the purpose for the reason given above. Finally, some climatic and soil characteristics in agriculture (the so-called primitives) influence production both quantitatively and qualitatively (Cross et al., 2013).

**Appendix 6.3: Social Capital**

Although the literature on wine economics has not yet explored its role and benefits, social capital constitutes a fundamental variable for the creation of a prestigious appellation or a solid production cooperative. A brief review of the main contributions of scholars in the various disciplines is therefore appropriate.

In *The Moral Basis of a Backward Society*, Banfield (1958) first argued that the underdevelopment of a country (in his study it was a small community in southern Italy) may, in part, be due to a lack of trust that individuals have toward people outside their family nucleus. This theory, which was very innovative at the time, did not fit well into the models of economic development proposed by researchers of that period, and therefore it was not given the attention it deserved. The only exception was Arrow (1972), who wrote: "It can be plausibly argued that much of the economic backwardness in the world can be explained by the lack of mutual confidence. See Banfield's remarkable study of a small community in southern Italy." But over the last decades Banfield's ideas have been picked up again and investigated in greater depth.

The term "social capital" was coined by sociologists who referred to the advantages and disadvantages of belonging to a certain community (Bordieu, 1985). Coleman (1990) defined social capital as a resource for individuals that result from social relations. The source of this capital lies in the people to whom a person relates. Sociologists identify two reasons why individuals should want to make their resources available for other people without getting anything in return. The first is mentality: people pay their debts, comply with the law (e.g., obey traffic rules), and perform acts of charity because they feel a moral duty to behave correctly and civilly. The second can be traced back to less noble, instrumental reasons: the costs of transactions can be reduced by saving on legal and insurance costs if they are made with people who they trust. In this case people do not behave correctly for ethical reasons but rather because a good reputation can produce economic returns. For this reason, Coleman (1990) defined social capital as the extension of horizontal relationships among members of a community, and its role is to increase society's resources and allocate them more efficiently.

In recent years, however, the concept of social capital has been taken up and adapted by political science scholars such as Putnam (1993) and Fukuyama (1995). In their studies social capital becomes the property of extended communities, even nations, rather than small groups. Putnam (1993) defined social capital as a characteristic of social life (habits, norms, trust) that allows individuals to act together more effectively to achieve common goals. The difference between the sociologists' and political scientists' notions of social capital is in the size of the group of reference: sociologists focus on small groups while scholars of political sciences are interested in larger groups. For this reason, they adopt the average turnout at the polls or participation in volunteer associations as indexes to measure social capital. Guiso, Sapienza, and Zingales (2004) adopted the definition of social capital formulated by scholars of political science and examined the relationship that these social indicators have with one of the elements that affect economic growth: namely, financial development. In fact, social capital increases the level of trust of individuals who complete a transaction. In communities where social capital is high, people have a greater degree of trust for two reasons. The first is that people believe more readily in the promises of others, and this is a result of a moral attitude formed by education and experience. The second is that the social system guarantees a more effective punishment for those who do not fulfill their obligations. Given that financial contracts require a high level of trust, social capital should have considerable effects on the development of financial markets.

Generally, there are substantial differences in social capital between one country and another but only moderate differences from one region to another in the same state. An important exception among industrialized countries is Italy. Despite the fact that unification was completed about a century and a half ago and that the same administrative, judiciary, legal, and tax systems apply throughout the country, there are huge differences between the social capital of the northern and southern regions. Guiso, Sapienza, and Zingales (2004) measured the social capital of Italian provinces with two indicators: electoral participation and the average number of blood donations per inhabitant. Once these types of variables were obtained, the authors studied the effect of social capital on the allocation of a household portfolio, the use of checks, the possibility of obtaining finance, and the use of informal loans (by subjects other than financial intermediaries—mainly family or friends). The analysis of the data showed that in areas with high social capital, households keep a larger part of their resources in the form of shares rather than money, make a greater use of checks, and access finance more easily when they request it. Further, the importance of social capital is greater in the regions where the judicial system is less efficient and people are less educated as they have to rely on trust because of their limited understanding of bargaining mechanisms. These conclusions can easily be extended to most of the emerging and developing countries that have similar characteristics. It is important to note that the level of social capital depends on both the area in which people live and the area in which people were born.

This theory, however, deserves to be developed with further studies of both a theoretical and empirical nature as many illustrious economists like Solow (1995) have shown a certain skepticism about the link that exists, from a conceptual point of view, between social capital and economic development. Putnam (1993) also acknowledged that the mechanisms through which social capital contribute to economic development need to be studied in more depth.

# 7

---

# Economic and Social Externalities

I tried to drown my problems in alcohol, but I realized that they float.
—Anonymous

This chapter deals with the economic and social externalities arising from the production and consumption of wine. The first section provides a theoretical review of the concept of externality, its consequences, and possible solutions. It then describes the positive and negative externalities of the production and consumption of alcoholic beverages in general and wine in particular. Although most people consume a moderate amount of alcohol and benefit from it, alcohol abuse has severe consequences on drinkers and society as a whole. For this reason, the review puts particular emphasis on this latter aspect and draws heavily on medical literature. The last section discusses the policies adopted across countries and over time in order to tackle alcohol abuse and its negative consequences.

## 7.1 Externalities: Definition, Consequences, and Possible Solutions

### 7.1.1 Definition of Externalities
Externalities are advantages or disadvantages for either producers or consumers that are created by the activity of an operator who does not receive or pay a price for them. Externalities have the following features.

- They may result from production (for example, pollution generated by the chimneys of industrial companies or soil pollution caused by pesticide treatments) or consumption (for example, use of cars).
- They can be positive (for example, research and development, the planting of an orchard near a beekeeping business, protection of the landscape) or negative (for example, pollution or loud music).
- They are reciprocal: when the right of one party to produce or consume infringes the rights of others, should we stop them? Doing so, however, will harm the

producer/consumer. Consider nightclubs that inevitably end up generating noise—who should exercise their rights: those who enjoy themselves (the right to have fun) or those who impose silence (the right to have quiet)? If a restriction is imposed on nightlife, the first party is damaged to protect the second and vice versa. The choice of which party to protect is subjective and depends largely on sociodemographic factors, such as age and gender, as well as the mentality and customs of a country and the historical period. As an example, in 1905 the governor of Pennsylvania vetoed a law prohibiting public spitting to protect the public from spreading contagions, but this practice was considered an inalienable right of any gentleman.

• Normally a zero level of pollution is not desirable: there is a need for the right balance between benefits and social costs, which requires positive quantities of goods produced and, consequently, of pollution.

### 7.1.2   Causes and Consequences of Externalities

The onset of externalities is not linked so much to the imperfection of markets as to two other factors. The first is the absence of exchange markets and property rights. When assets belong to the community, operators are encouraged to overexploit resources (e.g., hunting, fishing, using water), causing difficulties and costs to others. The second is the emergence of a good (e.g., noise and air pollution or the dissemination of information technology through worker training) as a result of the production and/or consumption of another good. In the case of pollution, the damage falls on both people (damage to health) and companies (e.g., pollution from sulphur dioxide during the production of iron and steel damages fishing enterprises through acid rains).

The problem of externalities arises because a producer or consumer does not consider the costs or benefits that their choices will have for other individuals. In the case of negative externalities, such as pollution, the balance between supply and demand will occur at point E, where demand and supply are equal (figure 7.1). However, this kind of solution is inefficient from a social point of view. If social costs are taken into due consideration, the supply curve will shift upward and there will be a new equilibrium at point E' with a higher price and a lower quantity. In the presence of positive externalities, we have the opposite case, with a suboptimal production/consumption level and a high price. In this case, public subsidies are necessary to restore a socially efficient equilibrium. If externalities exist, markets are no longer able to guarantee an efficient allocation of resources.

### 7.1.3   Solutions to Externalities

There are both private and public solutions to solve the problem (for a more detailed analysis, see Katz et al., 2011, chapter 18). The private solutions include the following.

1. *Codes of behavior.* Rules are one way to persuade individuals to take account of the externalities they produce (e.g., do not throw paper on the floor, cough into

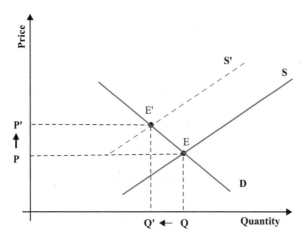

**Figure 7.1**
Supply curve with negative externalities.

the inside of your elbow, etc.). They correspond to the precept "don't do unto others what you don't want others to do to you," which less elegantly can be translated as "before doing something, consider its external costs and benefits." An extreme example (more theoretical than real) is when companies decide spontaneously not to pollute for reasons of business ethics.

2. *Mergers.* In the case of pollution, if the two parties involved (e.g., a steel mill and a fishing company) merge, the new single enterprise will internalize the negative externality by maximizing the new profit function. If the steel mill buys the fishing company, it will be willing to reduce pollution so as not to damage the fishing activity too much (and vice versa). Another example similar to mergers is joint ventures and consortia in the field of research and development, which aim to avoid underinvestment caused by the presence of positive externalities (e.g., the Sematech organization of microchip companies).

3. *Bargaining between parties.* If the cause of the discussed inefficiencies is the failure to assign property rights, then the simplest solution is to complete the assignment. This creates a market for the good in question. Property rights actually mean the right to use an asset, especially regarding natural resources. Coase (1960) elaborated a theoretical model,[1] which shows that, if the costs of negotiation and transaction are null, bargaining between economic agents leads to solutions that are efficient from a social point of view, even in the presence of externalities and regardless of who initially owns the property rights. If, in fact, the right is assigned to the polluter, they can reduce the quantities produced if compensated by the injured party, thus restoring a Pareto-efficient equilibrium. If, on the contrary, the right is assigned to the damaged party, they can claim compensation

from the polluter and reach the same result. The *allocation* of resources remains the same; only the *distribution* changes: in the first case the polluter is favored over the damaged party. The assignment of property rights is a political and not an economic choice. The Coase (1960) theorem, however, was criticized because of its too restrictive assumptions. Bargaining can, in fact, fail if: (a) negotiations are burdensome—this happens whenever active participation involves costs (documentation, meetings, travelling, legal fees, etc.) or the parties involved are too numerous, which leads to opportunistic behavior; (b) if there are problems in identifying the cause of or responsibility for damages (one example is the difficulty in identifying the pesticide responsible for the death of bees; in the case of air pollution, there are millions of companies that contribute to this problem); and (c) if information asymmetries exist, making it impossible to know how willing the counterparty is to pay. This can lead to the request for an excessive price, thus causing negotiations to fail.

Public solutions, which mainly concern diseconomies related to pollution, include the items below.

1. *Taxation (subsidy)* on negative (positive) externalities proposed by Pigou (1920). The difference between private and social marginal cost can be removed by introducing a positive tax proportional to quantities produced. Efficiency is restored by adding a tax equal to the social marginal cost to the private marginal cost. The opposite happens if the company creates an external economy and benefits from a subsidy.

2. *Incentives to eliminate external economies*. Instead of imposing a variable tax, the socially optimal quantity can be produced by granting a fixed subsidy for the loss of production. The value of the subsidy in this case is equal to the value of the diseconomy from an optimal social point of view. In this way the curve of private marginal costs is shifted parallelly upward, restoring efficiency.

3. *Negotiable rights to create diseconomies*. These have been introduced to solve the problems of environmental pollution. In this case the "optimal" level of diseconomy has to be established for each company. The right to pollute up to a specified limit is then assigned (e.g., through an auction), and the companies can either use the rights or adopt new technologies, reduce emissions, and sell the rights. This type of solution was adopted in the Kyoto Protocol for the reduction of greenhouse gases. It is important to note that, unlike taxes, the effect seems more certain, given that the level of pollution is prefixed while it is only hypothesized with taxes.

4. *Regulation*. It consists of introducing specific laws that impose limits, the adoption of technologies, and so on (e.g., an obligation to install filters and/or purifiers; an obligation to clear the snow from the sidewalk in front of one's own home; etc.). This solution automatically eliminates external diseconomies, but the effective compliance with the laws has to be checked and the offenders punished.

As we will see later, there is a much wider range of options available for the consumption of alcoholic beverages that are implemented all over the world with different levels of coercion and effectiveness. In some cases, the negative effects of alcohol abuse fall on consumers themselves and not on third parties, making it very different from a situation where a third party is damaged.

## 7.2   Externality of Production in the Wine Sector

Wine making in the main wine-producing countries has a turnover of billions of euros, without considering the allied activities linked to the production of rooted cuttings, agricultural machinery, fertilizers, pesticides, industrial machinery, bottling materials, and so forth. As already discussed in chapter 4, various industrial districts have grown in connection with wine production over time. In addition to their multiplicative effects on the regional economies, however, there also exist positive externalities of production, such as (1) the protection of the countryside, (2) the protection of the region, and (3) wine tourism.

The land cultivated with vineyards generates positive production externalities, first of all benefiting everybody who can enjoy the view of the magical landscape. Second, agriculture generally plays a crucial role in safeguarding the countryside. In many countries agricultural land has been "cannibalized" by real estate speculators to make way for houses, industrial warehouses, and large-scale solar parks. Indeed, in countries like Italy these solar parks have benefited from generous incentives granted over a long period for the production of "green" energy and have appeared in many regions. The roots of vines hold the soil in place when it rains, reducing the risk of landslides. Hydrogeological instability and the frequent emergencies that have hit many countries are the result of nefarious policies that have favored the real estate sector to the detriment of the countryside and stability of the land. Last, wine-growing activity generates positive effects on the hospitality industry and trade when consortia successfully promote a wine road.

A series of studies have analyzed the characteristics and demonstrated the positive impact of wine tourism on the economy in countries like Australia (Charters and Ali-Knight, 2002), South Africa (Bruwer, 2003), Spain (López and Martín, 2006), Greece (Karafolas, 2007), Italy (Asero and Patti, 2009; Nunes and Loureiro, 2012; Francioni, Vissak, and Musso, 2017), Chile (Hojman and Hunter-Jones, 2010), and Washington State (Storchmann, 2010).[2] Wine tourism has enormous potential that requires the concerted efforts of many people and institutions to express itself at its best.

These three positive externalities are closely connected to the cultural content of wine. Article 1 of the 1970 UNESCO Convention on the Means of Prohibiting and Preventing the Illicit Import, Export and Transfer of Ownership of Cultural Property defines "cultural property" as "property which, on religious or secular grounds, is specifically designated by each state as being of importance for archaeology, pre-history, history, literature, art or science" (UNESCO 1970). For cultural goods, "the creation of economic value is not their only *raison d'être*" because through them "human beings express their identity and work out ways of living together" (Throsby, 2001, p. 134). Even though wine does not explicitly appear in the long list provided in the same article, UNESCO has named the Champagne and Prosecco hills and wineries as World Heritage Sites (Cardebat, 2017, pp. 38–39).

Wine has been an important element of conviviality and well-being since ancient times. Many wine areas have excelled in achieving high quality, expressing the local identity, and preserving the landscape. The wine "properties, traditions, varieties, and indeed its effects, are a part of European culture and identity as it is to walk our streets or fish our rivers. ... And it is our duty to conserve such a cultural asset" (Hugh Johnson, 2009, p. 9). A useful approach could be to apply the concept of cultural good to wine by asking a simple question: is wine closer to a painting or to a screwdriver? For most wine consumers, the answer is clear: it is close to a painting. Further, while bottles of wine are not nonrival and some consumers are easily excluded, the vineyard landscape is more similar to a public good that everybody can freely enjoy without preventing others from doing so.

Some cultural economists believe that cultural goods carry not only an economic value but also a cultural one that should considered separately (Throsby, 2001). A cultural good—especially if it is public like the landscape—justifies governmental support based on market failure (Towse, 2010, pp. 51, 171–174); left to the market-place, people underestimate the positive externalities, and production turns out to be lower than socially optimal (Marks, 2015, pp. 178, 183, 194). Further, for other cultural goods like theaters and classical music, the Baumol and Bowen's (1966) "cost disease" argument supports the need for public intervention in favor of the arts sector.[3] Finally, public subsidies can have a positive influence on the quality of cultural goods and, in turn, on demand, as shown with data on Australian, Austrian, and French theaters by, respectively, Throsby (1990), Krebs and Pommerehne (1995), and Urrutiaguer (2002).

The policy measures government can adopt range from lowering taxes on certain items to subsidizing or directly owning organizations that promote culture (Towse, 2010, pp. 32–34).

### 7.3  Consumption Externality in the Alcoholic Beverages Sector

The consumption of alcohol produces a series of positive or negative effects depending on the quantities taken on a daily basis. For "moderate consumption," US authorities mean a glass a day for women and two for men (US Department of Agriculture and US Department of Health and Human Services, 2010, p. 16).[4]

### 7.3.1   Benefits from Moderate Consumption

Pearl's (1926) seminal population study showed the lower mortality of moderate drinkers. The author cautiously avoided attributing benefits to lighter drinking, but he concluded that such drinking was probably not harmful and suggested that the higher mortality among abstainers in some cases might be due to "constitutional" weakness, which pushed people to avoid alcohol. Since then, the medical literature on the effects of alcohol has expanded and has consistently shown the positive effects of light drinking. The main benefits of moderate consumption are as follows.

- *A reduction in the risk of cardiovascular disease.* Over a hundred scientific studies have shown the existence of an inverse relationship between a moderate consumption of alcohol and the onset of cardiovascular disease (Harvard School of Public Health, n.d.). These surveys use extensive databases, with up to hundreds of thousands of individuals, for periods of time that can exceed fifteen years and record very significant decreases in the main cardiovascular diseases (in some cases up to 80 percent) compared with teetotalers (Stampfer et al., 1988; Klatsky, Armstrong, and Friedman, 1990; Thun et al., 1997; Camargo et al., 1997a, 1997b; Renaud et al., 1999, Mukamal et al., 2003). Moderate consumption of alcohol increases high-density lipoprotein levels, also known as HDL or "good cholesterol," which is said to have protective effects on heart disease. Leger, Cochrane, and Moore (1979) analyzed deaths due to heart disease in eighteen developed countries and found a strong negative relationship with per capita consumption of alcohol and especially wine. Klatsky et al. (2003) also found a J-shaped effect between alcohol consumption—and in particular wine—and mortality. Chiva-Blanch et al. (2013) confirmed the beneficial effects of a moderate daily consumption of alcohol which, however, are even more evident in drinks rich in polyphenols like wine and beer. The effect is heightened with red wine. These studies, therefore, seem to give credence to the so-called "French paradox" that was proposed by the French scientist Serge Renaud in 1991 on the TV show *60 Minutes*, in which he attributed the low incidence of cardiovascular disease among French people (approximately just a third compared with Americans, despite having a similar diet in terms of saturated fat) to the daily consumption of two glasses of red wine (Colman, 2008, p. 83). This led to a rapid increase in the sales of this drink in the United States. Other studies, however, attribute the beneficial effects to alcohol in itself and do not find significant differences among beer-, wine-, or spirit-drinking countries (Rimm et al., 1996; Mukamal et al., 2003).
- *Reduction in the incidence of tumors.* The incidence of kidney tumors (Rashidkhani et al., 2005; Greving et al., 2007) and lymphatic tumors (Morton et al., 2005) is lower among moderate drinkers than chronic and occasional drinkers or teetotalers and is not significantly different for tumors of the colon (Shrubsole

et al., 2007), ovaries (Rota et al., 2012) and other female sexual organs (Hjartåker, Meo, and Weiderpass, 2010).

- *Other beneficial health effects.* Moderate consumption of alcohol leads to a reduction in the probability of catching a cold (Cohen et al., 1993), gallstones (Grodstein et al., 1994; Leitzmann et al., 1999), and type 2 diabetes (Conigrave et al., 2001; Koppes et al., 2005; Djousse et al., 2007). In general, while excessive consumption leads to a reduction in life expectancy due to its harmful effects on health, the potential for domestic and road accidents, and so on, moderate consumers of alcohol, perhaps surprisingly, have a life expectancy that is on average higher than teetotalers. As demonstrated by Doll et al. (1994) using a sample of over twelve thousand male British doctors interviewed in 1978 and studied for the following thirteen years up to 1991, there was an inverse U-shaped effect between alcohol consumption and life expectancy, even though there is a risk of endogeneity due to omitted variables, such as the previous state of health. Teetotalers could, in fact, be such because they do not like the taste of alcohol or because of their precarious health conditions attributable to other diseases, in which case it is not surprising if teetotalers have a shorter life expectancy. Holahan et al. (2010) reviewed empirical studies on this topic and the methodological problems that can skew results. The authors studied the relationship between moderate consumption of alcohol and total mortality (all-cause mortality) net of a series of other sociodemographic variables, previous illnesses, state of health, and so forth, with a database of 1,824 individuals observed for over twenty years: the same positive and significant effect of moderate consumption was still present, though reduced, even after the inclusion of possible confounders in the statistical analysis. Fueller (2011) had similar results.
- *Improvement of sex life.* Moderate consumption of alcohol causes an increase in libido in both men and women (Harvey and Beckman, 1986; Beckman and Ackerman, 1995) and a reduction in anxiety as well as a decrease in erectile dysfunction problems compared with both teetotalers and hardened drinkers (Chew et al., 2009).
- *Effects on mental efficiency.* Moderate consumption of alcohol increases cognitive performance (Galanis et al., 2000; Rodgers et al., 2005) and reduces the risk of developing degenerative brain diseases, such as Alzheimer's disease and other forms of dementia (Cupples, 2000; Sabia et al., 2018), although it does not seem to affect the incidence of Parkinson's disease (Checkoway et al., 2002).
- *Effects on psychological well-being*: Baum-Baicker (1985) reviewed the benefits of alcohol consumption on the human psyche, which include an anti-stress function, increased happiness, euphoria, conviviality, and emotional expressiveness (alcohol as "a social lubricant"). The nonlinear effect of alcohol consumption on happiness has been demonstrated empirically with Russian data by Massin and Kopp (2010) and Krekhovets and Leonova (2013). Monahan and Lannutti (2000) showed that

the consumption of alcohol makes women with less self-esteem more uninhibited and less anxious on a first date with a man, whereas nervous tension and depression decrease with moderate consumption. Pernanen (1991) and Hall (1996) demonstrated its relaxing effect while Skogen et al. (2009) found a U-effect between alcohol consumption on the one hand and anxiety and depression on the other. These results were taken up and developed in an extensive study by Peele and Brodsky (2000), confirming the psychological benefits and emphasizing how scientific literature tends to overemphasize damage.[5] When we move to heavier drinking, the situation changes. Yörük and Yörük (2012) applied a discontinuity methodology to the consumption of alcohol in twenty-one-year-olds and showed that when the sample subjects reached the legal age for buying alcoholic beverages, their consumption of alcohol increased 1.5 times, but it did not result in greater psychological well-being. The very fact, however, that the majority of citizens in democratic regimes vote freely in favor of selling alcoholic beverages, despite all potential damages that will be described below, may indicate (1) that people underestimate its social costs or (2) that this study underestimated the psychological benefits.

### 7.3.2   Damage from Abuse

Moderate alcohol consumption is fine, but when it becomes excessive, everything goes wrong. The harmful consequences of alcohol come not only from too much intake but also as a result of the mode of consumption (moderate daily consumption or binge drinking; Rehm et al., 2003, 2004). Alcohol abuse is responsible for 4.5 percent of illnesses and accidents and causes about 2.5 million deaths every year—about 4 percent of the world total. This is higher than diseases such as HIV/AIDS and tuberculosis, making it one of the first causes of death, especially for young people and men (WHO, 2011).

To these we can also add other types of more or less serious harm caused by three mechanisms: (1) long-term toxic effects on internal organs and tissues, (2) short-term intoxication, and (3) dependence (Rehm et al., 2003). The effects are even more harmful when alcohol is produced either at home—illegally—or, in any case, outside government controls, which according to some estimates account for almost 30 percent of the total (WHO, 2011, p. xi). In general, negative consequences can affect the consumer (self-regarding), third parties (other-regarding), both, and society as a whole. Among the negative effects for the drinker are the following items.

- *Physical health*: Decades of medical studies have shown a very strong positive correlation between alcohol abuse and the occurrence of diseases in the cardiocirculatory and gastrointestinal apparatus (e.g., cirrhosis and pancreatitis), diabetes, and tumors of various organs (larynx, esophagus, oral cavity, liver, colon, breast, etc.; see Baan et al., 2007). Since alcohol is a substance that has many calories and interferes with metabolic functions by urging individuals to eat more, a diet that includes a high consumption of this type of beverage may lead to a greater risk of being overweight or even obese. De Castro (2000) also showed that the amount of food consumed increases with the number of diners and the presence of alcoholic beverages on the table, and it is proven that alcohol reduces the capacity of the organism to burn fat (Leibowitz, 2007; Stewart et al., 2006). French et al. (2010) found, however, rather limited effects of alcohol consumption on Americans' weight. The possible causes are a parallel reduction of food consumption to compensate for the increase in calories or increased sports activity by drinkers of alcohol (French and Zavala, 2007).

- *Mental health and work*: Excessive consumption of alcohol is associated with neuropsychiatric disorders, such as epilepsy (Samokhvalov et al., 2010) and dementia (Sabia et al., 2018), as well as poorer performances at school (Carrell, Hoekstra, and West, 2011; Balsa, Giuliano, and French, 2011) and work with repercussions on productivity, wages, and the probability of becoming and remaining unemployed. The belief that alcohol consumption and/or abuse decreases worker productivity is so ingrained that there are regulations in all countries that prohibit its consumption in the workplace, especially in those jobs where the consequences could be catastrophic (e.g., police officers, soldiers, airline pilots, etc.). On this point, however, the scientific literature has not provided clear evidence against alcohol. In a pioneering work, Fisher (1926) concluded that the daily consumption of three glasses of beer led to a 10 percent reduction in productivity while Pidd et al. (2006) found a strong positive relationship between alcohol consumption and absenteeism in the workplace. But a series of other studies produced contrasting results with negative, positive, parabolic, or null effects (see Berger and Leigh, 1988; Manning et al., 1991; French and Zarkin, 1995; Heien, 1996; Mullahy and Sindelar, 1996; Hamilton and Hamilton, 1997). Renna (2008) showed that negative effects disappear when switching from a one-stage to a two-stage ordinary least squares estimate, in which the total wage decreases because of the reduction in hours of work performed while hourly pay is the same. As pointed out by Kenkel and Ribar (1994), it is difficult to identify the relationship between alcohol consumption and productivity because of the possible presence of (1) errors in self-declarations about consumption, (2) reverse causality between consumption and income, and (3) omitted variables that influence simultaneously consumption and salary. As demonstrated by Dave and Saffer (2008), for example, risk tolerance significantly influences alcohol consumption. The preferences of individuals, however, can determine many other choices about behavior, making it difficult to isolate the effect of the specific variable on the physical and mental health of citizens.
- *Suicides*: The tendency to commit suicide can be influenced by the consumption of alcohol (Cook and Moore, 2000). Blood samples extracted from cadavers of

suicides often have high levels of alcohol (Hayward, Zubrick, and Silburn, 1992) while Skog and Elekes (1993) demonstrated the correlation between alcohol consumption and suicides using Hungarian data. In this case, however, the problem of the reverse causality is very important: while the consumption of alcohol can induce depression, it is also true that depressed people seek consolation or stupor in the abuse of this substance. From an empirical point of view this problem was overcome with a natural experiment carried out by Wasserman, Varnik, and Eklund (1994). The authors, in fact, studied the relationship between these two variables in the Soviet Union during the perestroika period (1985–1990), when the authorities imposed much more restrictive policies on alcohol consumption than in previous years. The study showed that a drastic reduction in suicides and violent deaths was associated with a decrease in alcohol consumption, with –68 percent and –85 percent respectively compared with 1984 levels.

- *Accidents*: Over the decades, ample evidence has been produced about the fact that the loss of clarity of mind and coordination in movements can favor accidents at home and at work as well as road accidents involving cars, motorcycles, and so on (Wechsler et al., 1969; Cook and Moore, 2000; Skog, 2001; Borges, Cherpitelb, and Mittlemanc, 2004; Quinlan et al., 2005; WHO, 2011; Rickard, Costanigro, and Garg, 2013).

The main damage caused to other people can be found below.

- *Damage to a fetus*: Fetal alcohol syndrome is the most serious of the fetal pathologies caused by the consumption of alcohol in pregnancy and is the result of an intake of eighty grams of alcohol a day. The negative consequences on the fetus can be physical, with facial dysmorphologies and growth and/or psychological and neurological deficits, with a series of disorders and deficiencies. With lower intakes— for example, ten grams per day—the risk of serious damage is reduced but remains; in this case it is called fetal alcohol effects (Waterson and Murray-Lyons, 1990).
- *Episodes of violence and crime*: Alcohol abuse causes loss of control over actions and undermines reasoning ability. It becomes more difficult to appreciate the consequences of one's own actions when highly intoxicated: parents can react violently to their children's provocations or whims, men can insist or even force women to have sexual intercourse or turn a heated discussion into a physical fight, fans at the stadium can look for a fight to avenge the defeat of their own team, the victims of robberies can stupidly rebel in front of a pointed gun, and so forth (Cook and Moore, 2000). A state of intoxication reduces the ability of individuals to negotiate peaceful solutions to disputes that arise inside or outside the family unit as well as exacerbating conflicts about financial or marital difficulties. Some people become particularly violent when they drink (Fagan, 1990). In a study using US data from 1979 to 1988, Cook and Moore (1993) demonstrated

the positive relationship between per capita alcohol consumption and the incidence of rape, armed robbery, and theft. While there is a correlation with murders, it is weak. French and MacLean (2006) came to similar conclusions after having appropriately checked possible endogeneity. Abbey (2002) reviewed studies on sexual assaults in US colleges that considered both the attacker's and the victim's state of intoxication and highlighted the methodological limitations of the literature. A positive relationship, sometimes minimal or moderate, between alcohol abuse on the one hand and the frequency and intensity of domestic violence on the other has been found in several studies (e.g., Leonard and Quigley, 1999; Brecklin, 2002; Testa, Quigley, and Leonard, 2003; Foran and O'Leary, 2008). This is, however, rather controversial because of the problem of omitted variables. While people who abuse alcohol are known to have greater marital conflict that often results in domestic violence, what remains to be demonstrated is whether it is a direct causal relationship and not a mere correlation caused by other variables. Zubretsky and Digirolamo (1996), for example, claimed that alcohol simply acts as an "excuse" for physical attacks and that they do not cease with a reduction in consumption. A treatment sample (those who frequently and badly beat up their partners) is not the same as a control sample because they often live in a context of social degradation and have themselves been victims of violence.

Some of the consequences of alcohol abuse can fall both on those who drink and on completely unrelated third parties. The following fall into this category.

- *Unwanted pregnancies and sexually transmitted diseases*: A state of intoxication means the risk of unwanted pregnancy or the transmission of infectious diseases connected to unprotected sex is often underestimated. Apart from the moral aspects, the problem of children born outside marriage to very young parents is linked to the affective and socioeconomic conditions that they will grow up in. A single mother may have to interrupt or postpone her studies or give up important job opportunities. Sexually transmitted diseases (e.g., HIV/AIDS, gonorrhea, syphilis, chlamydia, etc.) have dramatic consequences on people's health and cause significant costs for the national health system. The incidence of these diseases is highest among young people, homosexuals, drug addicts, and some ethnic groups (e.g., African Americans in the United States). An extensive literature has illustrated the correlation between alcohol consumption and unprotected sex (Donovan and McEwan, 1995) without, however, showing the direction of causation for the usual problem of omitted variables. The most recent studies have tried to isolate the direction of causation by taking into account the heterogeneity of people. In one study on data for US adolescents, Markowitz, Kaestner, and Grossman (2005) found alcohol consumption had no effect on the probability of having sexual relations but identified a negative impact on the probability of taking precautions.

- *Road accidents*: The literature on the effects of alcohol consumption on road accidents has unequivocally shown its ill-fated effect on the ability to control a vehicle (Levitt and Porter, 2001; Baughman et al., 2001; Carpenter and Dobkin, 2011). Levitt and Porter (2001) estimated that the probability of drivers with traces of alcohol in their blood causing fatal accidents is seven times higher than for those who are sober, with this risk rising to thirteen times higher for those who are drunk from a legal point of view. The risk of causing harm to third parties grows with the size of the vehicle involved: with a motorcycle it falls almost exclusively on the driver, but with sports utility vehicles and trucks it falls on third parties (Gayer, 2004).

### 7.3.3  Costs of Abuse for Society

This review clearly shows the econometric problems involved in measuring the benefits and costs of alcohol consumption that often risk skewing results, though the WHO (2011) seems to ignore them in its report. It is reasonable to think—and it is widely agreed—that moderate consumption of alcohol is acceptable (if not desirable) while abuse, as always, must be condemned for the negative consequences it brings on individuals and other people.[6] The costs for society are, in fact, enormous. Figure 7.2 reports the WHO world estimates for the percentage share of disability-adjusted life years (DALYs) caused by alcohol according to the type of harm. This index puts the years of life lost and life lived with disabilities due to illnesses and accidents caused by alcohol together in a single indicator, thus measuring the total cost in terms of years of full health lost (Murray and Lopez, 1996; Murray et al., 2002). Clearly, alcohol abuse is responsible for a variable but significant percentage of a series of diseases, disorders, and actions.

Figure 7.3 shows the percentage of deaths attributable to alcohol abuse around the world in 2016. This share is minimal in Muslim countries where consumption is actually close to zero for cultural and religious reasons and highest in the former Soviet Union. Table 7.1 gives the World Health Organization's (WHO's) estimate of the years of life corrected for disability per one hundred thousand inhabitants for major harm to themselves. Alcohol abuse takes on the characteristics of a real social scourge in the Russian Federation, followed by South Africa, Brazil, and China.

Economic quantification or "data monetization" of the damages is not without methodological problems. Rehm et al. (2009) reviewed 247 summaries and forty-seven complete texts of articles and reports with considerable methodological differences that often make it impossible to compare results. The most relevant divergences concern the choice of the discount rate used to calculate the present value of future costs attributable to premature death (usually 6 percent, but it can vary between 3 percent and 10 percent), the inclusion or otherwise of the benefits from moderate consumption, and the types of costs considered. These can, in fact, include the costs

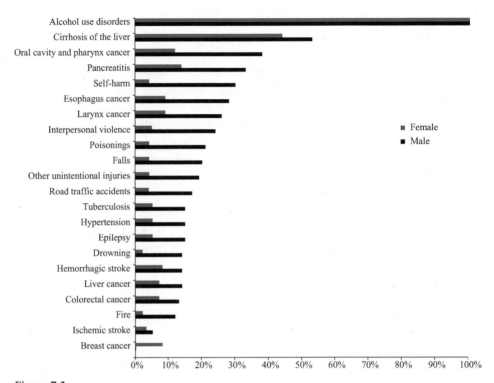

**Figure 7.2**
Percent share of disability-adjusted life years attributable to alcohol in the world, 2012.
*Source*: Author's calculations using data from WHO.

borne by the national health and judicial systems, administrative costs, social services, lack of income, decreased productivity, and psychological pain arising from the death of a spouse or their disability. In the same study the authors quantified total costs net of the benefits of moderate consumption in four high-income countries (France, the United States, Scotland, and Canada) and two medium-income countries (Korea and Thailand), which turned out to be between 1.3 percent and 3.3 percent of GDP per capita per year.

## 7.4   Policies to Combat Alcohol Abuse

The negative effects of alcohol abuse can be countered by a range of policies that vary according to country and to the historical period considered (see table 7.2). The results achieved depend not only on the efforts made but also on the combination of instruments adopted. The WHO (2011) underlined how governments pay little attention to the issues of public health and security policies despite the fact that millions of people die each year from causes related to alcohol abuse.[7]

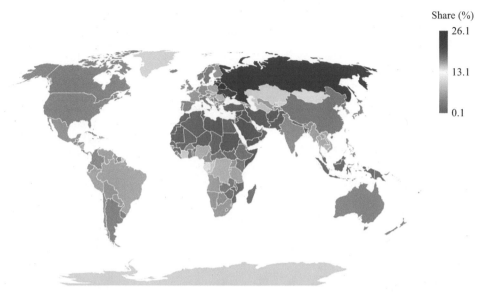

**Figure 7.3**
Percentage of deaths attributable to alcohol abuse around the world, 2016.
*Source*: Author's calculations using data from WHO.

Policies can be aimed at either reducing alcohol consumption or mitigating the harmful consequences of abuse. The first, in ascending order of restriction, are (1) prevention campaigns, (2) public and private transport policies, (3) taxes, (4) limits on the availability of alcohol, and (5) prohibition while the second are mainly aimed at discouraging drunk driving and providing cars with adequate active and passive safety features.

### 7.4.1   Prevention Campaigns

The first tools to discourage abuse are prevention campaigns put in place by governments ("social marketing"), restrictions on alcohol advertising, and medical advice. Campaigns aimed at increasing awareness of the harm caused by alcohol depend entirely on the sensitivity, commitment, and organizational ability of public authorities (see WHO, 2011, p. 52) while restrictions on advertising can reduce or prevent the overall media bombardment or limit it in some contexts (e.g., during time slots when there is a predominantly young audience). Suggestions from a doctor are a very precise tool that only targets people at risk without burdening the entire population as in the case of taxes. This measure can also be less expensive compared to other policies, such as shakedowns and the arrest of violators, that are aimed at discouraging drunk driving. The effectiveness of a doctor's advice in reducing abuse has been demonstrated in a series of studies using econometric analyses that take into account

**Table 7.1**
Harm from alcohol abuse (years of life corrected for disability every 100,000 inhabitants), 2004.

| Country | Intoxication | Breast cancer | Colorectal cancer | Liver cancer | Oral and pharyngeal cancer | Esophagus cancer | Cerebral vascular disease | Diabetes | Ischemia |
|---|---|---|---|---|---|---|---|---|---|
| Argentina | 15 | 316 | 142 | 36 | 29 | 45 | 527 | 467 | 666 |
| Australia | 80 | 337 | 160 | 28 | 28 | 35 | 228 | 201 | 492 |
| Austria | 10 | 259 | 142 | 48 | 53 | 24 | 242 | 261 | 520 |
| Belgium | 37 | 388 | 153 | 27 | 56 | 38 | 350 | 182 | 596 |
| Brazil | 6 | 336 | 103 | 26 | 66 | 50 | 836 | 509 | 951 |
| Chile | 11 | 166 | 76 | 38 | 11 | 30 | 396 | 397 | 431 |
| China | 89 | 138 | 81 | 270 | 34 | 184 | 1,072 | 232 | 416 |
| Denmark | 86 | 393 | 209 | 27 | 43 | 49 | 358 | 226 | 551 |
| Finland | 216 | 245 | 101 | 34 | 18 | 20 | 394 | 190 | 730 |
| France | 23 | 346 | 165 | 65 | 85 | 55 | 242 | 184 | 279 |
| Germany | 21 | 311 | 161 | 33 | 48 | 37 | 289 | 213 | 593 |
| Greece | 89 | 272 | 99 | 57 | 20 | 11 | 522 | 256 | 727 |
| Iceland | 55 | 295 | 128 | 5 | 22 | 24 | 239 | 96 | 576 |
| Ireland | 10 | 357 | 173 | 27 | 26 | 58 | 280 | 146 | 701 |
| Italy | 10 | 277 | 139 | 67 | 33 | 17 | 287 | 276 | 426 |
| Luxembourg | 85 | 294 | 135 | 37 | 66 | 35 | 376 | 162 | 513 |
| Holland | 22 | 383 | 175 | 20 | 30 | 55 | 291 | 195 | 417 |
| Norway | 197 | 301 | 183 | 12 | 19 | 24 | 283 | 175 | 482 |
| New Zealand | 33 | 387 | 197 | 26 | 26 | 27 | 305 | 209 | 634 |
| Portugal | 19 | 271 | 182 | 38 | 63 | 39 | 681 | 324 | 507 |
| Russian Federation | 891 | 312 | 179 | 44 | 60 | 35 | 1,776 | 204 | 3,051 |
| Spain | 38 | 237 | 164 | 11 | 50 | 33 | 276 | 243 | 401 |
| South Africa | 30 | 285 | 92 | 48 | 74 | 168 | 1,284 | 839 | 990 |
| Sweden | 56 | 230 | 131 | 38 | 18 | 19 | 281 | 204 | 543 |
| Switzerland | 7 | 291 | 121 | 5 | 35 | 28 | 184 | 181 | 368 |
| UK | 47 | 351 | 152 | 67 | 28 | 65 | 348 | 168 | 674 |
| USA | 160 | 329 | 144 | 17 | 23 | 33 | 327 | 374 | 715 |

*Source*: Author's calculations using data from WHO.

| Cirrhosis | Alcohol abuse disorders | Suffocation | Falls | Fires | Self-inflicted wounds | Other unintentional wounds | Low birth rates and pre-term births | Violence | Road accidents | Total |
|---|---|---|---|---|---|---|---|---|---|---|
| 309 | 550 | 47 | 82 | 32 | 217 | 504 | 284 | 365 | 328 | 4,961 |
| 220 | 571 | 24 | 175 | 18 | 248 | 253 | 76 | 59 | 265 | 3,498 |
| 328 | 426 | 18 | 151 | 9 | 298 | 198 | 166 | 29 | 292 | 3,474 |
| 243 | 265 | 14 | 148 | 19 | 372 | 180 | 62 | 56 | 387 | 3,573 |
| 69 | 627 | 111 | 161 | 24 | 129 | 439 | 376 | 1,256 | 656 | 6,731 |
| 136 | 543 | 73 | 104 | 32 | 248 | 311 | 90 | 322 | 379 | 3,794 |
| 85 | 495 | 281 | 314 | 20 | 295 | 426 | 411 | 74 | 602 | 5,519 |
| 400 | 514 | 22 | 134 | 17 | 237 | 221 | 105 | 34 | 254 | 3,880 |
| 161 | 687 | 44 | 266 | 29 | 450 | 385 | 67 | 72 | 275 | 4,384 |
| 96 | 520 | 30 | 160 | 18 | 332 | 264 | 52 | 32 | 323 | 3,271 |
| 222 | 519 | 11 | 103 | 11 | 220 | 134 | 106 | 27 | 229 | 3,288 |
| 277 | 365 | 48 | 136 | 10 | 55 | 159 | 110 | 21 | 447 | 3,681 |
| 303 | 291 | 17 | 134 | 21 | 273 | 196 | 53 | 32 | 230 | 2,990 |
| 44 | 470 | 31 | 120 | 16 | 307 | 134 | 93 | 30 | 226 | 3,249 |
| 94 | 80 | 19 | 130 | 8 | 113 | 157 | 74 | 27 | 345 | 2,579 |
| 104 | 485 | 18 | 183 | 14 | 267 | 258 | 116 | 46 | 384 | 3,578 |
| 33 | 499 | 15 | 90 | 6 | 180 | 109 | 67 | 37 | 163 | 2,787 |
| 72 | 969 | 40 | 157 | 25 | 256 | 261 | 28 | 31 | 227 | 3,742 |
| 579 | 284 | 43 | 170 | 17 | 309 | 268 | 140 | 56 | 368 | 4,078 |
| 67 | 413 | 37 | 154 | 16 | 155 | 217 | 88 | 45 | 487 | 3,803 |
| 132 | 1,277 | 251 | 473 | 212 | 789 | 1,699 | 149 | 845 | 933 | 13,312 |
| 530 | 106 | 28 | 109 | 10 | 137 | 173 | 66 | 41 | 314 | 2,967 |
| 114 | 287 | 77 | 104 | 248 | 359 | 609 | 468 | 2,031 | 1,138 | 9,245 |
| 91 | 766 | 23 | 107 | 16 | 255 | 146 | 53 | 42 | 206 | 3,225 |
| 171 | 399 | 15 | 139 | 9 | 281 | 173 | 77 | 36 | 205 | 2,725 |
| 185 | 663 | 9 | 97 | 14 | 169 | 131 | 150 | 61 | 203 | 3,582 |
| 70 | 600 | 32 | 114 | 28 | 242 | 202 | 144 | 221 | 449 | 4,224 |

**Table 7.2**
Main policies to reduce alcohol abuse in the world, 2012.

| Country | Restricted advertising on TV | Restricted sponsorship on TV | Minimum age | Blood alcohol content (%) | Random breath testing | Educational guides | Restricted sales (days) |
|---|---|---|---|---|---|---|---|
| Argentina | partial | no | 18 | 5 | yes | no | no |
| Australia | prohibition | no | 18 | 5 | yes | yes | no |
| Austria | no | partial | 16 | 5 | yes | yes | no |
| Belgium | no | no | 0 | 5 | yes | yes | no |
| Brazil | partial | . | 18 | 2 | yes | no | yes |
| Chile | no | no | 18 | 5 | no | no | . |
| China | partial | no | 0 | 2 | yes | yes | no |
| Denmark | partial | prohibition | 16 | 5 | . | yes | no |
| Finland | partial | partial | 18 | 5 | yes | yes | yes |
| France | prohibition | prohibition | 16 | 5 | yes | no | yes |
| Germany | partial | prohibition | 16 | 5 | no | yes | no |
| Greece | . | . | . | 5 | yes | . | . |
| Iceland | prohibition | prohibition | 20 | 5 | yes | no | yes |
| Ireland | partial | partial | 18 | 8 | yes | no | yes |
| Italy | partial | voluntary | 0 | 5 | yes | yes | no |
| Luxembourg | . | . | . | . | yes | . | . |
| Holland | voluntary | no | 16 | 5 | yes | yes | yes |
| New Zealand | no | no | 18 | 8 | yes | yes | yes |
| Norway | prohibition | no | 18 | 2 | yes | yes | yes |
| Portugal | partial | partial | 16 | 5 | yes | yes | no |
| Russian Federation | partial | partial | 18 | 3 | yes | no | yes |
| South Africa | no | no | 18 | 5 | yes | yes | no |
| Spain | partial | no | 16 | 5 | yes | yes | no |
| Sweden | prohibition | prohibition | 20 | 2 | yes | no | yes |
| Switzerland | prohibition | prohibition | 16 | 5 | yes | yes | no |
| UK | partial | prohibition | 18 | 8 | yes | no | no |
| USA | voluntary | no | 21 | 8 | yes | . | no |

*Source*: Author's elaborations using data from WHO.

| Restricted sales (hours) | Restricted sales (shops) | Restricted sponsorship of sport | Restricted sales promotion | Mono-poly | License | Excise tax (hl) | VAT % | Restricted consumption in public | Warnings on labels |
|---|---|---|---|---|---|---|---|---|---|
| yes | . | voluntary | voluntary | . | . | . | . | prohibition | Yes |
| no | no | no | partial | no | yes | . | 10.0 | partial | no |
| no | no | no | no | no | no | 0.0 | 20.0 | no | no |
| no | no | no | no | no | yes | 47.1 | 21.0 | voluntary | Yes |
| no | yes | no | no | no | yes | . | 25.0 | partial | Yes |
| . | . | no | no | no | . | . | 19.0 | no | no |
| no | no | no | no | no | Yes | . | 17.0 | no | no |
| no | no | no | no | no | yes | 82.3 | 25.0 | no | no |
| yes | yes | partial | prohibition | yes | yes | 257.0 | 22.0 | partial | no |
| yes | yes | prohibition | no | no | yes | 3.5 | 19.6 | no | yes |
| no | no | no | no | no | no | 0.0 | 19.0 | no | no |
| . | . | . | . | . | no | 0.0 | . | . | . |
| yes | yes | partial | no | yes | yes | . | 24.5 | voluntary | no |
| yes | yes | voluntary | no | no | yes | 328.1 | 21.0 | voluntary | no |
| no | no | partial | partial | no | yes | 0.0 | 20.0 | voluntary | no |
| . | . | . | . | . | . | 0.0 | . | . | . |
| yes | yes | voluntary | voluntary | no | yes | 68.5 | 19.0 | no | no |
| no | yes | voluntary | no | no | yes | . | 12.5 | voluntary | no |
| yes | yes | prohibition | prohibition | yes | yes | 517.7 | 25.0 | partial | yes |
| no | yes | partial | no | no | yes | 0.0 | 21.0 | no | yes |
| yes | yes | partial | partial | no | yes | . | 18.0 | prohibition | no |
| yes | yes | no | no | no | yes | . | 14.0 | no | no |
| no | yes | no | no | no | no | 0.0 | 16.0 | partial | no |
| yes | yes | prohibition | prohibition | yes | yes | 222.0 | 25.0 | voluntary | no |
| no | no | partial | partial | no | no | 0.0 | 7.6 | partial | yes |
| yes | . | voluntary | partial | no | yes | 265.0 | 17.5 | no | no |
| yes | . | . | . | no | no | . | 8.0 | . | yes |

the endogeneity linked to the higher levels of consumption by people receiving treatment or, to avoid this problem, with randomization assigned to samples of individuals at risk (see Fleming et al., 1997; Kenkel and Terza, 2011). The literature about restrictions on advertising has produced discordant results with a series of studies that did not find significant effects on the demand and consumption of alcoholic beverages (Smart, 1988; Fisher, 1993; Calfee and Schergata, 1994; Bang, 1998; Nelson, 1999, 2010) and others, on the contrary, that found negative effects (Saffer, 1991; Tremblay and Okuyama, 2001). As emphasized in this last study, however, the fact that advertising does not influence the demand for alcohol does not mean that it does not have any impact on consumption. Even if advertising does not encourage people to increase consumption, the removal of restrictions can generate price competition that leads them to consume greater quantities but with an unchanged demand curve.

### 7.4.2 Public and Private Transport Policies

Taxes acting on the price of other goods or services needed to consume alcohol are excise taxes on fuel, the cost of public transport, and taxi fares. Chi et al. (2011) demonstrated using Mississippi data that when gasoline prices rise, accidents due to drunkenness decrease. These measures can, to some extent, affect the number of outings with friends and therefore the consumption of alcoholic beverages (income effect), but above all they influence the choice of the mode of transport (substitution effect) by encouraging people to use the cheapest means available, whether public or private. The income effect measures the impact of increased purchasing power on consumption while the substitution effect shows the impact of changing relative income and prices on consumption. However, there is wide consensus that creating an extensive and efficient network of buses and subways and subsidizing or minimizing the cost of taxis can help reduce drunk driving. The decision to drive under the effect of alcohol has two implications: it is dangerous to drink excessively and criminal to drive while intoxicated (Jackson and Owens, 2011). "A person commits an offense if the expected utility to him exceeds the utility he could get by using his time and other resources at other activities" (Becker, 1968). In this comparison of expected utilities, scholars have focused mainly on policies that increase the costs of illegal behavior (e.g., taxes on alcoholic beverages, laws that make it difficult to buy them, and penalties for those who are positive on the breathalyzer). As pointed out by Jackson and Owens (2011), however, there is no empirical literature demonstrating the positive impact of public transport development on road deaths caused by drunk driving for two reasons. First, these policies are accused of combating road accidents but not alcohol abuse, which would still not be discouraged. Second, the infrastructure is "given"; it changes very little over time and therefore from an econometric point of view generates identification problems. The authors tried to fill this gap by using data on the Washington, DC, city metro network, whose opening hours at weekends was progressively extended between 1999 and 2003 from midnight to three in the morning. The results demonstrated that offering a safer alternative reduces the percentage of people driving while intoxicated, with the effects tending to vanish as the distance increases from areas served by a metro station.[8] The effect of ridesharing on public health is debated. A number of studies have found that ridesharing services, such as Uber and Lyft, reduce intoxicated driving and fatal accidents (Uber and MADD, 2015; Greenwood and Wattal, 2016; Dills and Mulholland, 2017), even though Brazil and Kirk (2016), using US data from 2005 to 2014, find no association between the timing of the deployment of Uber in US metropolitan counties and the number of subsequent traffic fatalities. The effect of ridesharing on road accidents, therefore, depends on the study analyzed. Furthermore, Burgdorf, Lennon, and Teltser (2019) find that the spread of UberX across the United States has increased alcohol consumption and abuse, as well as employment and total earnings at drinking establishments. The results of these studies suggest that the economic and social effects of ridesharing are complex and deserve further investigation.

### 7.4.3 Taxes

Taxes generate an income effect that influences the total consumption of alcohol and a substitution effect that affects the choices related to the types of drinks bought (e.g., wine, beer, and spirits). In their extensive review of seventy-two studies, Elder et al. (2010) found a systematic negative relation between tax or price of alcohol and indexes of excessive drinking or alcohol-related health and criminal outcomes. Durrance et al. (2011) showed that taxes on alcohol are negatively correlated with the murders of women while Cook and Durrance (2013) demonstrated in a larger survey with US data that the sudden and substantial increase in federal excise taxes on all alcoholic beverages approved by the Bush administration in 1991 led to a reduction in total alcohol consumption and, as a consequence, a reduction of fatal road accidents and nine types of offenses. In 2004 Finland reduced alcohol duties by one third on average; this was followed by a 10 percent increase in alcohol consumption and a 46 percent rise in liver disease deaths (Mäkelä and Osterberg, 2009).[9]

Taxes on alcoholic beverages consist of excise taxes and value added tax (VAT). From a formal point of view excise taxes are "indirect" as they fall on the producer who then passes them on to the buyer while VAT is "direct" because it weighs directly on consumption. But, from a practical point of view, excise duties are applied to quantities whereas VAT as a percentage of the price.[10] Two similar products (e.g., two bottles of wine) of equal alcohol content but different prices will, therefore, have the same excise duties but different VAT.

The policy of taxing alcoholic beverages implies two important choices. The first concerns the distribution of the burden between taxes on quantities (excise taxes) and taxes on value (VAT). Based on the assumption that the best products are more expensive, economic theory (see Barzel, 1976) argued that quantity taxes encourage an increase in the quality of products because their relative weight decreases for higher range goods. Ad valorem taxes, on the contrary, do not change the relative prices between products of different quality levels. Ljunge (2011) confirmed these hypotheses in an empirical study of US data showing that the market share of quality wines is a direct function of excise tax on alcohol while ad valorem taxes are irrelevant. In light of these results, if the objective of economic policy is to raise the quality of products to increase competitiveness on international markets, then the shift of the tax burden from VAT to excise duty can be one feasible strategy.

The second choice concerns the possible differentiation of the level of taxation to hit more heavily those drinks that are associated with the greater number of accidents and offenses or are produced in greater quantities abroad. Rickard, Costanigro, and Garg (2013) found, for example, that with the same amount of alcohol consumed, a greater share of wine as part of the total amount is associated with a reduction in road deaths while the opposite happens with beer and spirits. This is perhaps due to the different profile of wine consumers compared with those who drink beer and spirits. A series of studies has shown that wine drinkers have a higher socioeconomic level and adopt healthier lifestyles (Baltieri et al., 2009),[11] whereas beer and spirits are the typical drinks of the working class, students (Siegel et al., 2011), and people who try to dull their senses quickly or cheaply. In the United States wine is almost considered a status symbol and is consumed in important places or on special occasions (dinners with friends and relatives, restaurants, wine bars, etc.). If the consumption of alcohol is to be reduced, it seems appropriate to start by hitting the products consumed by the categories most at risk and avoid going against the drinks bought by the more educated and sophisticated part of society that usually considers moderation as its behavior model.

In addition, Saffer (1989) found using US data that the greatest reduction in alcohol consumption occurs with an increase in taxes on spirits, followed by taxes on beer, and finally on wine. The author concluded that one tax on all alcoholic beverages is inefficient whereas higher taxes should be levied on beer and spirits and lower taxes should be levied on wine to reduce the total consumption of alcohol. In Europe, however, the application of a differentiated taxation aimed at favoring the consumption of wine to the detriment of beer is not allowed by EU laws.

Table 7.3 shows excise duties and VAT in Europe: there are very marked differences in the absolute tax burden, with much higher levels in the countries of northern Europe (Cook and Moore, 2000). The Sixth Council Directive 77/388/EEC of 17 May 1977 on the harmonization of the laws relating to turnover taxes (sixth VAT directive) forced member states to impose a single rate of at least 15 percent on goods and services (in Italy it stands at 22 percent). Reduced rates (in Italy, of 10 percent and 4 percent) may be applied to some necessary goods or goods that are of interest to disadvantaged groups, including drinks but not alcoholic ones. It is therefore not

**Table 7.3**
Excise tax (€/hectoliter) and VAT (%) on alcoholic beverages in Europe, 2013.

| Country | Beer Excise tax (degrees Plato) | Beer Excise tax (alcohol) | Beer VAT | Wine Excise tax (alcohol) | Wine VAT | Other alcoholic beverages Excise tax (alcohol) | Other alcoholic beverages VAT |
|---|---|---|---|---|---|---|---|
| Belgium | 1.84 | | 21 | 56.97 | 21 | 56.97 | 21 |
| Bulgaria | 0.76 | | 20 | | 20 | | 20 |
| Czech Republic | 1.24 | | 21 | | 21 | | 21 |
| Denmark | | 7.51 | 25 | 147–197 | 25 | 147.68 | 25 |
| Germany | 0.78 | | 19 | | 19 | | 19 |
| Estonia | | 6.28 | 20 | 84.67 | 20 | 84.67 | 20 |
| Greece | 2.60 | | 23 | | 23 | | 23 |
| Spain | 0.91 | | 21 | | 21 | | 21 |
| France | | 7.33 | 20 | 3.72 | 20 | 3.72 | 20 |
| Croatia | | 5.25 | 25 | | 25 | | 25 |
| Ireland | | 22.55 | 23 | 424–616 | 23 | 309–424 | 23 |
| Italy | 2.70 | | 22 | | 22 | | 22 |
| Cyprus | | 6.00 | 19 | | 19 | | 19 |
| Lithuania | | 3.10 | 21 | 64.03 | 21 | 64.03 | 21 |
| Latvia | | 2.71 | 21 | 65.16 | 21 | 65.16 | 21 |
| Luxembourg | 0.79 | | 15 | | 15 | | 15 |
| Hungary | | 5.47 | 27 | | 27 | 33.34 | 27 |
| Malta | 1.73 | | 18 | | 18 | | 18 |
| Holland | 7.59–47.48 | | 21 | 88.36 | 21 | 88.36 | 21 |
| Austria | 2.00 | | 20 | 0.00 | 20 | 0.00 | 20 |
| Poland | 1.84 | | 23 | 37.35 | 23 | 23–37 | 23 |
| Portugal | 9.43–26.45 | | 23 | | 13 | | 23 |
| Romania | 0.87 | | 24 | | 24 | 10.65 | 24 |
| Slovenia | | 12.10 | 22 | | 22 | | 22 |
| Slovakia | | 3.58 | 20 | | 20 | | 20 |
| Finland | | 32.05 | 24 | 339.00 | 24 | 339.00 | 24 |
| Sweden | | 20.62 | 25 | 267.47 | 25 | 267.00 | 25 |
| UK | | 23.95 | 20 | 334.11 | 20 | 334.11 | 20 |

*Source*: Author's calculations using data from WHO.

possible to apply reduced or different rates for wine, beer, and spirits. For years Portugal has violated EC provisions by imposing lower VAT on wine (13 percent) than on beer and spirits (both at 23 percent), thus favoring this locally produced beverage that is consumed most by the strata of society at a lower risk of road accidents.[12]

Member states, however, have some autonomy in determining excise taxes. In fact, unlike beer, the European Union has not established mandatory minimum excise duties for wine (which are zero in Italy), and governments can set different amounts for different types of beverages, as shown in table 7.3. Community legislation does insist, however, that the treatment is not discriminatory toward imported drinks either directly (similar drinks produced by competing countries, such as two wines produced by Italy and France) or indirectly (competing drinks produced predominantly by other countries—e.g., Swedish beer and Italian wine). Therefore, excise duties on beer and spirits cannot be increased to generate a substitution effect in favor of wine (Georgopoulos, 2009). This is the price to pay for compromises in the community between northern European countries' breweries and Mediterranean countries' wine producers and yet another instrument of economic policy which national governments have surrendered into the hands of the European Union.

Leaving to one side any favoritism toward one or other type of product, a desirable solution would, in any case, be to convince EU legislators to include all alcoholic beverages in a reduced VAT scheme (e.g., 10 percent) and to increase excise taxes to stimulate the purchase of products of superior quality that cause less damage to the human body, leaving unchanged the tax burden and therefore total consumption.

### 7.4.4   Limits on the Availability of Alcohol

Another measure authorities can adopt to reduce alcohol consumption is to limit availability. The restrictions may concern the following issues.

- *The minimum legal drinking age (MLDA)*: Although a minimum age has not been set in about twenty of the 147 countries registered by the WHO (2011), in the other countries it varies between fifteen-years-old and twenty-five-years-old, but is usually eighteen. Clearly a minimum age reduces alcohol consumption and abuse in the excluded age group, but the long-term effect is less obvious. A series of recent US studies have addressed this question by trying to measure the impact of lowering the minimum age from twenty-one to eighteen on alcohol consumption and abuse in adulthood. Cook and Moore (2001) showed that if people had resided in a state with MLDA of eighteen when they were fourteen, they had a 7 percent higher chance of alcohol abuse (binge drinking) four times a month at the age of about twenty-four. Norberg, Beirut, and Grucza (2009) analyzed alcohol use disorders and found a 32 percent lower incidence among twenty-one- to fifty-three-year-olds residing in states with a twenty-one-year-old MLDA compared with those residing in states with a lower MLDA. Lastly, Kaestner and Yarnoff (2011) showed that men who grew up in states where drinking is allowed at the age of eighteen through twenty recorded higher levels of alcohol consumption and road deaths by 20–33 percent and 10 percent respectively while the effect on women was null.

- *Opening hours and days of shops and bars*: They vary from country to country for on-premise and off-premise consumption. In England pubs had to close by law at 11 p.m. until 2005 while in Italy the sale of alcohol in nightclubs was allowed only until 2 a.m. for a short period in 2007. (For an international comparison, see WHO, 2011.)
- *Places for purchase or consumption*: In the United States each state decides if or which drinks can be sold in food stores (see Rickard, Costanigro, and Garg, 2013) while in Sweden drinks with an alcohol content of more than 3.5 degrees can only be purchased in public monopoly shops (Systembolaget). The sale of alcoholic beverages may also be prohibited for reasons of public safety in some places, such as stadiums or service stations on motorways.

### 7.4.5   Prohibition

The most restrictive policy of all is, of course, a total prohibition on the production, sale, and consumption of any alcoholic beverage. Currently absolute prohibition is imposed in several countries, all of which are either Muslim and/or monarchical or dictatorial regimes, such as Afghanistan, Saudi Arabia, Bangladesh, Brunei, some states of India, Iran, Kuwait, Libya, Mauritania, Pakistan, United Arab Emirates, Sudan, and Yemen. In the past, however, many other countries, including some Western countries, have adopted the same policy, such as Australia (Capital Territory, 1910–1928), Canada (1901–1924), Faroe Islands (1907–1992), Iceland (1915–1935), Norway (1916–1927), the Russian Empire/Soviet Union (1914–1923), Finland (1919–1932), and Hungary (March–August 1919). The United States imposed prohibition from 1920 to 1933, but the growing financial commitment to counter alcohol smuggling that was in any case only partially successful led both public opinion and the US Congress to change their minds. During this period of time the amount of illegal activity connected with alcohol production and commercialization grew dramatically (see Okrent, 2010, pp. 267–288). However, nowadays in the United States there are still many "dry" counties where the production, distribution, and sale of alcoholic beverages is forbidden (see chapter 8).[13]

There are various channels through which prohibition can influence alcohol prices (Miron and Zwiebel, 1991): (1) a fall in supply and increased production costs related to illegality and the risk of arrest; (2) a fall in demand because of the cost of finding the goods, the risk of being discovered, or finding poor quality products; (3) a fall in demand due to the growing sense of morality that condemns the consumption of alcohol; and (4) a fall in demand "in compliance with the law." Miron and Zwiebel's results showed that in the early years of American prohibition consumption contracted consistently but then rose to about 60–70 percent of the initial values to stay there even in the years following repeal. The authors concluded that the decrease in consumption was modest when compared with the increase in prices that more than tripled in just a little over a decade. The only channel that effectively reduced consumption was rising prices while fear of having problems with the law, a sense of duty, and social pressure played a negligible role.

# 8

## The Regulation of Supply

If ships generally function better than states, this happens for the simple reason that everyone accepts the role he is expected to play, while in states, generally, the less someone knows, the more eager he is to command.
—A phrase attributed to Massimo d'Azeglio (1798–1866)

Market regulation refers to all policies discussed in this second section of the book, ranging from the protection of health to the reduction of asymmetries, from the improvement of product quality by imposing strict minimum standards to the restoring of the market equilibrium. The regulation of supply, therefore, is a subset and consists in the array of instruments and policies adopted in different countries to influence the production (directly or indirectly through sales) of wine.

As already mentioned in the introduction, over time Europe and the United States have adopted completely different policies. In Mediterranean Catholic countries there has always been a strong culture of moderate daily consumption of wine during meals. In northern Protestant countries, instead, beer and also spirits are the prevalent beverages, overall consumption rates are lower, and public concern about the negative consequences of alcohol abuse is higher. In their empirical study of alcohol consumption patterns in the United States, Holt et al. (2006) found the same differences between these two major religious groups, with Catholics drinking more than Protestants.

These cultural and historical backgrounds have strongly affected the alcohol policies in the two continents. As the European Union was gradually being established after World War II, France had a dominant impact on the rules of the Common Market Organization (CMO). In this country wine is part of the national identity, and there is a very large number of people involved in its production, distribution, and sale (Banerjee et al., 2010). Banning alcohol would have been extremely unpopular among voters because of opposition from both consumers and producers. Wine makers have shown their determination and political strength during massive revolts (for example in Languedoc 1907 and in 2005, and in Bordeaux in 1974–1975; see Colman, 2008, pp. 18, 51–52, 55). Prohibition and policies to contain alcohol consumption were politically unthinkable and the target of policy makers was to reduce oversupply to ensure positive returns for wineries. This dirigisme was acquired by the European Union; they applied the same regulatory instruments adopted by France in the previous decades (incentives to uproot the vineyards, subsidies to distill wine excess, etc.). The main focus was neither public health nor consumer welfare but rather producers' economic wealth.

However, in the United States, religious Protestant temperance movements gained political influence over the second half of the nineteenth century until they managed to outlaw alcohol production and consumption. The focus of policy makers here was public health, not producers' wealth. When Prohibition was repealed in 1933, the new law provided single states and even counties the power to regulate every aspect of the production and sale of alcohol in order not to disappoint these movements. The legacy of Prohibition was "a chaotic patchwork of state regulations for producers and consumers to navigate" (Colman, 2008, p. 34). It was the temperance movements, not those of alcohol producers, that politicians did not want to disappoint—a completely different perspective.

This chapter starts by discussing the theoretical reasons which justify public intervention to regulate the market. It goes on to describe the sources of law in Europe and the United States, underlining which are the most important for the wine sector in the two continents. The third and fourth sections explain how the socioeconomic context and politics have shaped the regulation of agricultural markets in the two areas, both with their own inefficiencies.

## 8.1    Market Regulation

### 8.1.1    Reasons for Market Regulation

The need to regulate a market arises from failures, as illustrated in the previous chapters. One type of failure, discussed in chapter 6, is linked to the presence of information asymmetries that determine an inefficient allocation of resources. A second one, analyzed in chapter 7, is connected to the externalities that generate suboptimal or super optimal levels of production or consumption. Finally, a third one derives from the need to ensure competition among firms when the existence of economies of scale lead to the creation of natural monopolies. This latter case has not been discussed in detail because the wine market is characterized by the presence of thousands of companies spread across five continents with a very high level of competitiveness (Milhau, 1953).[1]

However, market regulation sometimes aims to do the opposite—that is, reduce competition to avoid the erosion of firms' profits. This is the case with the European Union's Common Agricultural Policy (CAP), whose main target is to ensure reasonable prices and good standards of living to the community of farmers by subsidizing producers and restricting the supply (see appendix 8.1). This is not a market failure;

with falling business opportunities, some firms would leave the market. Rather, it is the response of politicians to the pressures of organized defiant groups of interest. In the United States the temperance movement has been politically stronger than that of alcohol producers; therefore public policies have regulated the supply by prohibiting the production or hindering alcohol demand.

### 8.1.2    Measures to Prevent Market Failures

To counter abuse of a dominant position, information asymmetry, or negative externalities, public authorities can adopt a series of economic incentives or regulatory measures. The latter range from guaranteeing competition in the market (antitrust activity) to imposing labeling rules and minimum safety standards (at work, in public and private transport, the wholesomeness of food, the quality of teaching in schools, and universities, etc.). To resolve or at least mitigate these failures, four different—but not mutually exclusive—strategies may be adopted that give increasing power to the state at the expense of private individuals: self-discipline, private resolution of disputes, regulation, and finally, public ownership.

Self-discipline requires people to behave in a correct way and not to abuse others. This solution is obviously purely theoretical as the world is full of individuals and businesses that pursue their own interests and savagely trample on others. The private resolution of disputes means that individuals settle judicial controversies by relying on current regulations. The third solution requires regulation through the introduction of new laws and the possible establishment of a special authority with the task of overseeing the market to prevent the emergence of conflicts. The most restrictive policy consists in the state taking over total control of the market through nationalization.[2] These last two solutions have often been adopted in Western Europe, though there has been a progressive privatization of state monopolies since the 1980s resulting in the liberalization of markets and the establishment of specific guarantor authorities.[3]

### 8.1.3    Debate About Market Regulation

In the last hundred years, regulation by public authorities has grown steadily, but the debate about whether it has contributed or, on the contrary, hindered economic growth is still alive. There are three main theories on regulation (for an excellent review of the literature, see Shleifer, 2005). The first is about public interest and refers to Pigou (1920), the second is the contract theory associated with Coase (1960), and the third is Stigler's (1971) capture theory. The theory of public interest supports the need for state intervention in the economy, starting from the two assumptions that markets often fail and that public authorities are able to solve a problem through regulation. This theory was widely used during the twentieth century, and above all by socialist governments, to justify public intervention in the economy. It also indicates the measures to be taken—for example, price control in the case of a natural monopoly—or environmental, safety, and quality standards to avoid negative externalities of production or consumption.

These ideas were strongly criticized by the Chicago school of economics that claims that markets are able to solve almost all problems spontaneously through incentives or bargaining mechanisms among parties. If a product does not meet certain quality or safety standards, the consumer turns to a competing company, just as an employee changes company when economic or working conditions are unsatisfactory. When private negotiations are ineffective, efficiency can be restored by the courts (Coase, 1960), provided that they are able to enforce the laws and compensate for any injustice suffered.

State intervention is considered harmful because its employees are often incompetent or corrupt. Stigler (1971), in fact, goes even further in his criticism, going down to the very roots of the theory of public interest, and questions whether the state is actually a benevolent, competent, and impartial planner. In actual fact, governments often favor, rather than hinder, the interests of power groups. Even when they really want to pursue public interest, civil servants are often ill-prepared and end up producing effects that are very different from those intended.

While the theory of public interest is excessively pessimistic about the self-regulatory ability of the market, the same holds for the optimism shown by the Chicago school. Self-regulation works up to a certain point, beyond which it becomes anarchy in which the strongest, and not the righteous, often dominates. Judges can be corrupt or politicized, and the courts as well as governments are made up of men with their own weaknesses (Shleifer, 2005), so self-regulation suffers from very similar problems to regulation.

From a practical point of view, there is no single solution to all problems in the management of an economy. The choice of the best strategy must be carefully weighed and contextualized. Disagreements in condominiums are resolved most of the time amicably for the sake of peace and quiet, but there are things that the market often does not do spontaneously that have important consequences on the well-being of society. One example is the use of active and passive safety systems installed in cars that spread significantly after they were introduced by law. Moreover, people often are not fully aware of the negative consequences that can result from their actions, and compensation only partially repays the harm suffered. Now, who decides the rules in Europe and in the United States?

## 8.2   The Sources of Law

Modern legal systems have various sources of law. In the most important wine-producing countries the relationship between sources are regulated by hierarchical and chronological criteria. The hierarchical criterion provides that, when two conflicting laws come from different sources, those of the lower rank are invalid. The chronological criterion requires that the most recent source in time prevails over sources of the same level in accordance with the principle of *lex posterior derogat legi priori* (more recent laws modify earlier ones). The hierarchy of sources is listed below.

1. *The constitution*: It expresses broad principles and does not deal directly with agriculture.
2. *International treaties and agreements*: They regulate trade among countries, subsidies, and import duties and tend to be reciprocal.
3. *European laws (in Europe) and federal laws (in the United States)*: They apply to all countries/states within the union/federation and are passed by the European Parliament and by the US Congress, respectively. They are particularly important in Europe[4] because the European Union has been strongly influenced by French dirigisme and has regulated almost every aspect (e.g., the wine classification system, the maximum number of hectares and wine that countries can produce, the amount of subsidies to uproot vineyards and distill wine, etc.).
4. *National laws (in Europe) and state laws (in the United States)*: In this case they are more relevant in the United States because of the US Constitution's Twenty-First Amendment that repealed Prohibition in 1933. In order not to disappoint the temperance movement which was politically very strong—especially in religious areas—the same amendment gave local authorities (states and even counties) the power to regulate the production and distribution of alcoholic beverages (see section 8.4).
5. *Local laws*: As mentioned before, they are more important in the United States. In Europe, local authorities have some degree of autonomy in the regulation of the opening hours of shops, restaurants, and clubs. However, in the United States the Twenty-First Amendment gives much more power so that states and counties can even decide to stay "dry" and forbid the production and sale of alcohol.
6. *Judgments of courts*: They matter in the absence of written laws, which in the wine sector is very unlikely since it is heavily regulated.
7. *Customs*: Behavior repeated by people in the belief that they are following a law or that other individuals do the same. There must, therefore, be repetition over time and the belief that it is right or done by everyone.

This brief outline clearly shows the prominent role of the European legislature within the European Union and of state and local laws in the United States. The regulation of the wine classification has already been discussed in chapter 6. In the next two sections we will focus on the regulation of supply in the European Union and in the United States respectively.

## 8.3    Regulation of Supply in the European Union

### 8.3.1    Tools, Origin, and Development of the European Wine Policy

Of the four solutions mentioned at the beginning of the chapter (self-discipline, private dispute resolution, regulation and public ownership) the third is the most frequently adopted in the wine sector in line with the entire agri-food sector. It is implemented through the enactment of strict laws on the wholesomeness of drinks and on the wine-making practices allowed as well as through the creation of appellations of origin and rules on labeling. The EU wine system is undoubtedly the most regulated in the world. The legislature establishes practically everything—from which vines are allowed to the borders of each appellation or geographical indication and from wine-making practices to labeling (Marks, 2015, pp. 127–128).

Some measures have been introduced to protect both producers and consumers. Others, however, such as the attempt to influence supply to correct the disequilibrium in the market, reflect the interests of the winery lobbies and are aimed at redistributing income from potential newcomers (outsiders) or from consumers to existing producers (insiders).[5] Since the 1960s, in fact, the European Union has guaranteed a minimum price and the purchase of accumulated or distilled surpluses. In doing so, however, the subsidies system has accentuated structurally[6] rather than solved the problem of surpluses and has tried to remedy the situation by introducing planting rights and incentives for the grubbing up of vineyards.

From the very beginning, the European wine policy has mirrored French policy regarding both the wine classification system and supply regulation policies (Meloni and Swinnen, 2013). The system of appellations was, in fact, established in France in 1935 (the Bordeaux areas had already been classified in 1855) and was later extended to all EU countries through national laws (such as in Italy with law 930/1963), a year after the approval of the Council Regulation No. 24/62 establishing the first wine CMO.

Policies for restricting supply also originated in France where producers suffered fierce competition both from Italian and Spanish viticulturists as well as active French producers in the Algerian colony in the 1920s. Italian and Spanish competition was countered by raising customs duties while the Algerian threat led to the introduction of the Appellation d'Origine Contrôlée (AOC) system in 1935 to prevent the Algerian wines from being labeled as French (Colman, 2008, pp. 20–22). To restore the balance between the production and consumption of wine, between 1931 and 1935 the French authorities passed the *Statut Viticole* that contained various measures to restrict supply (such as the obligation to either store or distill part of excess production), the taxation of companies with productivity per hectare above a certain threshold, a ban on planting new vineyards for companies with more than ten hectares, and incentives for grubbing up existing vineyards. These measures were designed on the whole to hit primarily the Algerian producers who had, on average, large plots of intensively cultivated land and marketed low-quality wines, passing them off as French (Birebent, 2007; Simpson, 2011).

The incentives for grubbing up vineyards, though considerable (7,000 francs per hectare), proved to be ineffective. Only the worst producers, in fact, abandoned the cultivation of some land, and the overall impact on the quantities produced and the average quality was negligible (Milhau, 1953). World War II and the German occupation led to serious damage to the French vineyards so the *Statut Viticole* was set aside. In the years of the immediate postwar period, the focus was on the reconstruction of the production potential. However, in less than a decade, the age-old problem of overproduction reemerged, and the *Code du vin* was introduced with objectives and instruments similar to those of the *Statut Viticole*: incentives for the grubbing up of vineyards, penalties for high yields, and the management of overproduction through stock accumulation and crisis distillation (Meloni and Swinnen, 2013). Once again incentives for grubbing up did not have the desired effect since it was mainly the producers in regions where there had already been a spontaneous decline in viticulture that responded to economic incentives (Bartoli, 1986).

The establishment of the first wine CMO in 1962 led to a gradual harmonization of legislation in the six member countries, among which France and Italy had dominant positions with over 90 percent of production. In 1970 a compromise was reached between the more interventionist French and the more liberal Italian positions. In response to French requests, a guaranteed minimum price was established with the Council Regulation (EEC) Nos. 816/70 and 817/70 through support given to the accumulation of stocks and the distillation of table wines. However, in line with the Italian model, no regulation of planting rights or incentives for grubbing up were introduced (Arnaud, 1991).[7]

In the early 1970s excess production, partly favored by European support for the production of low-quality wines, increased and absorbed a growing amount of EC resources. French wine makers felt threatened by the competitiveness of Italian wine prices, which was heightened by the devaluation of the Italian currency lira. To put a stop to this "wine war" and reduce the size of the "European Wine Lake," the Council Regulation (EEC) No. 1162/76 introducing incentives for grubbing up vineyards and planting rights was passed in 1976. This last measure was similar to the quota system for milk and sugar.[8] So, in a short time, the French model became dominant, shaping EC legislation in its own image. This interventionist policy aimed at restricting supply has continued over the decades. Attempts by legislators to restrict production have proved, however, to be ineffective because of an increase in the productivity of the land. As a result, a new production peak was recorded in the 1980s (Corsi, Pomarici, and Sardone, 2004). Council Regulation (EC) No. 1493/1999 contains a series of measures including the provision of funds for the promotion of Community wine abroad.

Council Regulation (EC) No. 479/2008 on the common organization of the wine market pursued the ambitious goal of reducing waste and standardizing the European wine market to make it more efficient, transparent, and competitive. It included numerous measures. Some of them were praiseworthy and included the elimination of subsidies for the destruction of surpluses,[9] support for company investments, checks on compliance with production regulations that were no longer made by consortia but by third parties, and measures to ensure product traceability.[10] Others, however, were much more questionable, as for example the new Vino da Tavola (VdT), protected geographical indication (PGI), and protected designation of origin (PDO) classification system, which has been discussed before, as well as legislation on the planting, replanting, and grubbing up of vineyards that severely limited the planting of new vineyards and provided rewards for uprooting existing ones.[11] Green harvesting also aimed to restrict production by providing subsidies to producers. It consisted in "the total destruction or removal of grape bunches while still in their immature stage, thereby reducing the yield of the relevant area to zero" (Council of the European Union, 2008, Article 12), which should not be confused with the thinning out of bunches that is part of winter pruning. The whole regulation moved toward restrictions on production to increase the average price level and support farmers' incomes, as clearly shown in Recital 5:

> Increasing the competitiveness of the Community's wine producers; strengthening the reputation of Community quality wine as the best in the world; recovering old markets and gaining new ones in the Community and worldwide; creating a wine regime that operates through clear, simple and effective rules that balance supply and demand; creating a wine regime that preserves the best traditions of Community wine production, reinforcing the social fabric of many rural areas, and ensuring that all production respects the environment. (Council of the European Union, 2008)

It is also evident in Recital 2, summarizing the problem to be solved that the attention of European legislators was directed toward producers, not consumers:

> Wine consumption in the Community has been steadily diminishing and the volume of wine exported from the Community since 1996 has been increasing at a much slower rate than the respective imports. This has led to a deterioration of the balance between supply and demand which in turn puts producers' prices and incomes under pressure. (Council of the European Union, 2008)

The only (small) consolation remains the mea culpa, in the third recital, about the failure of past EC policy:

> Not all the instruments currently included in Regulation (EC) No 1493/1999 have proved effective in steering the wine sector towards a competitive and sustainable development. The market mechanism measures have often proved mediocre in terms of cost effectiveness to the extent that they have encouraged structural surpluses without requiring structural improvements. Moreover, some of the existing regulatory measures have unduly constrained the activities of competitive producers. (Council of the European Union, 2008)

Even if we did agree with the objectives of the expensive measures that aim to favor the grubbing up of vineyards but totally fail to consider the wellness of consumers, the problem of this policy lies in the fact that the European Union is not a closed economy and is no longer the exclusive producer of wine. "New" producers (primarily Australia, New Zealand, Chile, South Africa, Argentina, and the United States) are invading world markets with their products, and this tends to nullify, or at least strongly weaken, the effects of the EC strategy to restrict production in the European area. All this is happening in spite of the considerable cost of more than €1 billion incurred by the European Union for the three-year period 2009–2011 to incentivize the eradication of often unprofitable vineyards.

Above all, one wonders whether it would not have been better to have left the market to itself. In time, the worst producers would have left the market spontaneously, and there would have been more resources to allocate to the innovation of the most competitive companies wishing to focus on quality. The EC legislators' attempts to rebalance supply and demand, which have been going on for decades, appear costly and useless in the now globalized economy. The level of incentives for grubbing up (a one-off payment of between €1,740 and €14,760 per hectare depending on the yield) does not seem enough to encourage an entrepreneur to leave his business, unless it is running at a bad loss, in which case the grubbing up would happen without any subsidy. Some observers believe that we must offer a way out for entrepreneurs who are no longer competitive and help them to convert production. It would be interesting to see the costs and benefits of this policy in the future with the hope of not reading in the next wine CMO another mea culpa like those in Article 5 quoted above.

### 8.3.2   Recurring Cycles in the European Wine Policy

Even though the economic system is changing rapidly, the problems and their solutions do not seem to have changed substantially in the course of time. In an article that appeared a number of years ago in the *Economic Journal*, Charles Gide analyzed the causes of the crisis in the French wine market in a lucid and precise manner. The author noted how, according to many economists, the main cause of the collapse in wine prices was excess production, in which case there was no better solution than to rely on the ancient law of supply and demand. A fall in prices would induce some farmers to abandon the cultivation of vines, and this would lead to a rebalancing between supply and demand. According to the author, however, the true root of the problem was not excess production but rather a lack of demand, a much more serious question. The production of wine can be limited by law, but individuals cannot be forced to increase their consumption of alcoholic drinks. Gide also commented that wine makers should have restricted their production spontaneously to restore a balance between supply and demand and focused on quality, not quantity. Interestingly enough, Gide's article was published in 1907![12]

In looking at today's Europe, there is no point in deluding ourselves. With constantly falling domestic consumption and increasing international competition, restrictions on supply and grubbing-up premiums will certainly not solve the long-standing problem of the imbalance between supply and demand. After more than a hundred years, the European Union has finally taken note and decreed, again with Council Regulation (EC) No. 479/2008, the liberalization of the sector starting from January 1, 2016, through the abolition of planting rights, incentives for grubbing up vineyards, and subsidies for concentrated and corrected musts as well as the abolition of distillation measures.

Before proceeding with total liberalization, a final grubbing-up program was planned for a total of 175,000 hectares between 2008 and 2011. It was designed to encourage the exit of the less competitive wine makers from the market and the restructuring and/or conversion of vineyards to improve the competitiveness of those who intended to remain in business. The system of planting rights had to cease on December 31, 2015, unless national governments decided to postpone this date by three years. In the three years from 2007–2008 to 2010–2011, 272.528 hectares were removed in Europe (table 8.1), of which 161,164 received EC incentives (59 percent of the total uprooted, 5 percent of the area planted in 2007–2008).

Regulation (EU) No. 1308/2013 establishing a common organization of agricultural product markets approved the abolition of all the previous restrictions on supply: the milk quota system from April 1, 2015, sugar quotas from October 1, 2017, and vineyard planting rights from January 1, 2016. Producers' lobbies, however, have fiercely disputed what they define as the "wild liberalization" of the market and have succeeded in winning a postponement leading to "controlled" liberalization. Up to now the rights to plant a vineyard had to be bought from another producer, but during the transitional period from 2016 to 2030 free permits will have to be requested based on the availability of single states. National authorities may issue new authorizations for an annual amount not exceeding 1 percent of the national vineyards, with the possibility of reducing this level and concentrating the emission in the most valuable areas, taking into account the recommendations of the protection consortia. Rights and permits present a number of differences. Rights, in fact, last eight years and can be transferred or purchased while permits have a three-year term, are not transferable, and are free of charge. Despite these differences, however, the two instruments pursue the same objective of restricting production. Two steps forward and one step backward, therefore, in an overall picture that in half a century has been made up of high and low points.

## 8.4   Regulation of Supply in the United States

### 8.4.1   The Temperance Movement and Prohibition (Volstead Act)

The history of market regulation in the United States is completely different from Europe and is closely linked with the temperance movement, whose ultimate goal was abstinence from alcohol (see Colman, 2008, pp. 29–36). In the nineteenth century its

**Table 8.1**
Change in the rooted surface and uprooting with EU subsidies, 2007/2008–2010/2011

| Country | Rooted surface | | | Change 2007/2008–2010/2011 | | |
|---|---|---|---|---|---|---|
| | 2007/2008 | 2010/2011 | % Change | Change 2007/2008–2010/2011 | Of which uprooted with subsidies | % of the total uprooted with subsidies of the total planted with subsidies in 2007/2008 |
| Spain | 1,098 | 968 | –12% | –130 | 94 | 9% |
| France | 848 | 806 | –5% | –42 | 23 | 3% |
| Italy | 700 | 66 | –5% | –36 | 28 | 4% |
| Portugal | 240 | 236 | –1% | –3 | 4 | 2% |
| Germany | 102 | 102 | 0% | 0 | 0 | 0% |
| Greece | 71 | 67 | –5% | –4 | 2 | 3% |
| Austria | 50 | 46 | –8% | –4 | 1 | 2% |
| Luxembourg | 1 | 1 | –1% | 0 | 0 | 0% |
| Hungary | 81 | 72 | –12% | –10 | 6 | 7% |
| Slovak Republic | 20 | 19 | –9% | –2 | 1 | 3% |
| Czech Republic | 17 | 17 | 0% | 0 | 0 | 0% |
| Slovenia | 17 | 16 | –8% | –1 | 0 | 1% |
| Cyprus | 13 | 9 | –33% | –4 | 2 | 13% |
| Malta | 1 | 1 | –24% | 0 | 0 | 0% |
| Romania | 181 | 182 | 0% | 0 | 1 | 0% |
| Bulgaria | 104 | 69 | –34% | –35 | 0 | 0% |
| Total EU | 3,549 | 3,277 | –8% | –272 | –161 | 5% |

*Source:* Author's calculations using data from Tables 1 and 2, Annex II of the EU (2012).

influence spread across the United States; the first city banning alcohol and declaring itself "dry" was Portland (Maine) in 1843, followed by the state of Oregon in 1844, which forbade the sale of spirits, and then by Evanston (Illinois). The fight against alcohol slowed down during the American Civil War because Congress needed to raise taxes to finance the army. In fact, the alcohol industry became an important source of state revenues and was thus legitimated. The business was large, and it was easy to make producers and consumers pay taxes.

After the end of the Civil War the temperance movement gained political strength again, and by 1919 they had managed to pass Prohibition regulations in thirty-three of the forty-eight states. After a first attempt in 1875, Congress finally approved a national ban (the Volstead Act) on alcohol production, sale, and consumption in 1919 (the law came into effect in January 1920). The Association Against the Prohibition Amendment (AAPA) managed to convince the United States that alcohol had harmful consequences not only on society but also on the economy by reducing workers' efficiency, especially when working with the technologically advanced machines that recent innovations had introduced.

This was a terrible blow for the infant American wine industry. During what President Herbert Hoover called "the noble experiment," the number of wineries fell from more than one thousand to around 150 in California. Wine production survived thanks to three loopholes: wine for sacramental use in churches, wine for medicinal purposes, and grapes sold for home production. In fact, each household was allowed to produce a small amount of "nonintoxicating" cider and fruit juice.[13] Therefore, the paradox of Prohibition is that, during the 1920s, due to the ban of alcohol, the acres under vines increased, even though the quality of produced wine obviously fell.

The illegal alcohol industry also flourished thanks to the weak enforcement of the Volstead Act, which was carried out by officers from the US Department of the Treasury instead of the US Department of Justice. In 1920 the Treasury Department established the Bureau of Prohibition for this purpose, though it did not have much interest in fighting illegal alcohol trade because there were no excise taxes to collect. Only in 1930 did this bureau become part of the Department of Justice.

The Volstead Act produced a series of negative effects on the US economy. First of all, a number of economic activities involved in the production and distribution of alcoholic beverages had to close down. Second, many states had to face a significant reduction in tax revenues. Third, the balance between benefits and costs became uncertain over time. In fact, after a first period of a marked fall, per capita consumption of alcohol returned to around 70 percent of pre-Prohibition levels while corruption among public officers was spreading and the costs to enforce the law were booming.

The Volstead Act was also increasingly criticized because mass violations of Prohibition engendered widespread disrespect for the law, a condition usually termed "lawlessness." In his inaugural speech in 1929 President Hoover declared: "Our whole system of self-government will crumble either if officials elect what laws they will enforce or if citizens elect what laws they will support. The worst evil of disregard for some law is that it destroys respect for all law" (Hoover, 1929). Just as the Anti-Saloon League had been the main promoter of Prohibition, the AAPA dominated the making of repeal. This association was led and financed by some of the richest US entrepreneurs, such as Pierre du Pont of du Pont Chemicals and John Raskob of General Motors. The main reason for their involvement was economic: by restoring alcohol taxes they hoped to reduce the fiscal pressure on their firms.

The Great Depression, which started in 1929, provided valuable support for the repeal for two reasons. First, "it destroyed any credibility for the long-standing prohibitionist claim that Prohibition brought prosperity, and it fueled the new fantasy that repeal would end the depression by putting men back to work by stimulating the economy" (Levine, 1985, p. 72). Second, public authorities were more and more concerned about the discontent of people caused by both the economic situation and the alcohol ban that had given rise to protests and riots in major US cities. The 1917 Bolshevik Russian Revolution promoted a widespread "Red Scare," especially after the 1919 series of bombings by the Italian anarchist Luigi Galleani and the 1920 terrorist attack in Wall Street, which killed thirty-eight people and injured hundreds more. The idea of leaving young, unemployed people, with limited social and marriage opportunities because of the economic difficulties and without the chance to drink any alcoholic beverage, was risky in a period when Communism was gaining consensus all over the world. This problem became even more severe after the 1929 crisis, which fed people's anger toward entrepreneurs and capitalists. Over time the consensus around Prohibition shrank until the Congress approved the Twenty-First Amendment in 1933, voiding the Eighteenth Amendment.

### 8.4.2   The Politics of Repeal and the Three-Tier System

When Prohibition was repealed, every state and even every county was allowed to decide how to regulate the alcohol market. First, they had to decide if they wanted to remain "dry"—thereby forbidding the production and sale of alcoholic beverages—or become "wet." Then, if they opted for the second alternative, every local government had to decide what, when, and where to sell the different beverages, the amount of taxes to impose, the conditions for shipment, and so on. The repeal was made optional rather than mandated by federal law. This was meant to allow dry states to continue their fight against alcohol (ab)use. US states enacted about four thousand different laws to regulate the sector. Some maintained the ban for many years: Utah remained dry until 1959 and Mississippi—the last one—until 1966. Eighteen states imposed a monopoly on distribution and/or sales.

All over the country, the distribution and sale of alcohol must remain separate. Most states do not allow the direct sale to consumers by imposing a number of constraints. Direct sales to private buyers or retailers (e.g., restaurants, bars, wine shops) can be completely forbidden[14] or strongly limited—for example, by prohibiting shipments, allowing purchases only at the producers' facilities, imposing a maximum number of bottles a person can buy (which affects the average shipping cost), and demanding a number of complex and time-consuming bureaucratic tasks to be completed by firms.[15] Most states require an additional license to sell alcohol directly to consumers; some even require consumers to purchase a license to order alcoholic beverages. Periodic (annual or even quarterly) reports on shipments and taxes may have to be produced for each state where the producer sells directly to consumers or retailers. This means submitting up to three hundred reports per year. Producers also have to create records for each buyer to ensure they do not exceed the amounts allowed in a certain time window decided by law. Additional restrictions can include checking the buyers' identity by sending a picture of their identity card, imposing some packaging requirements, and so forth. It turns out that in many US states, it is easier for a consumer to buy a gun than to ship a bottle of good Californian wine (Colman, 2008, p. 2).

This is the so called "three-tier system" where around 90 percent of wine is sold to retailers and consumers through distributors, 50 percent of which are sold by the five largest ones. Producers sell to distributors, who sell to retailers, who sell to consumers. Obviously, the longer the chain, the higher the final price paid by consumers due to the markups added in each ring. The economic rationale for this is to ensure economic competition and to prevent monopolies caused by vertical integration, as in the eighteenth century when many bottlers imposed their products on the saloons they owned. Since the production, distribution, and sale of alcohol was in the hands of bootleggers and criminal organizations during Prohibition, a license system would have cleaned the sector up. Another reason for the adoption of regulatory controls was to enhance socioeconomic welfare by increasing controls, quality, and prices, thereby reducing abuse and its negative consequences, especially on underage drinking.

Apart from being extremely complex, with thousands of different laws in the US states, the three-tier system has a number of important drawbacks. From the point of view of consumers, it implies higher prices that can be good if it reduces alcohol abuse, but it can be bad if it pushes people to reduce quality to leave the overall budget and total quantities unchanged.[16] This can be a serious problem if low-quality products generate more harmful effects on health.[17]

From the point of view of producers, direct sales to consumers and retailers can be an important channel, especially for small wineries that do not have large portfolios of products and large economies of scale (Colman, 2008, pp. 92–93; Thornton, 2013, p. 126). Distributors often privilege large companies or even conglomerates because they can supply large quantities of cheaper products and offer all they need, from beer to wine and spirits. This is more efficient from a logistic and economic point of view, but small producers tend to be excluded, especially in those states where the distribution of alcohol is concentrated. Indeed and unfortunately, while the number of wine firms has been constantly growing, the number of distributors has been shrinking. Small producers also have a limited scale and higher production costs, so they have to charge a higher price. Bypassing the distributor can provide a substantial contribution to the firms' budget. Further, relying on a distributor does not allow full control of the marketing strategy. In fact, distributors autonomously decide the market positioning (e.g., luxury or cheap restaurants) and the promotion effort and policy, and once again, they normally put more effort into marketing for the large producers.

Whether these state and county laws pursue the social welfare of the community as a whole or rather the specific economic interests of some groups (large producers and distributors) remains an open question. However, over time the US Supreme Court has invalidated a number of local laws which regulate alcohol distribution and sale because they were not serving the public interest.

Why has such a complex and questionable system been put in place? The reason is simple: politics! Politicians want to be reelected and therefore tend to support the groups of voters which are more organized and contribute the most to parties during electoral campaigns. On the one hand, as already mentioned, the market of alcohol distribution is very profitable and concentrated. It is easier to form a powerful lobby when there are few subjects with strong interests than when there are many with weak incentives. The Wine and Spirits Wholesalers Association and large distributors like Southern Glazer's carry on their political lobbying activity by financing congressional elections on a regular basis. Yet, the anti-alcohol religious movements continue their battle to forbid—or at least minimize—alcohol consumption and abuse. An odd alliance emerged between these two groups that have very different purposes—the first reducing competition and increasing profits and the second minimizing the negative externalities of alcohol abuse—and it has managed to file a number of motions against free distribution. (See Thornton, 2013, pp. 3, 130–147, and Colman, 2008, pp. 89–99 for a detailed description of the three-tier system.)

### Appendix 8.1: The Common Agricultural Policy

#### From the ECSC to the CAP

In the immediate postwar period the governments of European countries were aware that, to safeguard peace, economic prosperity had to be guaranteed by encouraging cooperation. The first step of the long journey undertaken by the future European Union was the creation of the European Coal and Steel Community (ECSC) in 1951. The ECSC had been proposed a year earlier in Paris in the Schuman Declaration. The treaty clearly states that the reason for its existence was to guarantee the peace of peoples through well-being and economic stability (Preamble).[18]

After more than half a century of peace it seems difficult to conceive, but the founding fathers considered the economy more as a means to guarantee peace than as the ultimate goal when the treaty was signed. The ECSC was invested with several functions aimed at ensuring the economic prosperity of member states (Article 2) by ensuring to all consumers an orderly supply at the lowest possible prices (Article 3). The intention, therefore, was that prices be as low as possible. The same treaty also established a common market by prohibiting import and export duties, quantitative restrictions on the movement of products, public subsidies and special charges, and restrictive practices (Article 4).

It was the first step toward a free and united market, though limited only to coal and steel for the time being. The proposal was so positive that the principles of a future united Europe and the intention to create a common market for goods and atomic energy were announced in an official declaration at the end of the Messina Conference in 1955. The declaration of intent was followed by the signing of the Treaty of Rome in 1957 establishing the European Economic Community (EEC; later to become the European Union, EU) and the European Atomic Energy Community (EURATOM).

The treaty establishing the EEC reaffirmed the reasons for economic progress aimed at ensuring peace and the objective of creating a single market free of customs duties and barriers to goods or persons (Article 3). This article also provided for the establishment of a common policy in the agricultural sector, which was described in detail in Title II. In the aftermath of the war, in fact, Europe was not self-sufficient in its food supply, and despite the high proportion of the population engaged in agriculture, it depended heavily on imports from the Americas. Farmers experienced hard times, and food security was uncertain.

The objectives of the future agricultural policy were (Article 39) to increase agricultural productivity by promoting technical progress, to ensure a fair standard of living for the agricultural community, to stabilize markets, to assure the availability of supplies, and to ensure that supplies reached consumers at reasonable prices. In this list of targets, producers' needs were at the top and those of consumers at the bottom. To attain the objectives, the European Community created a common organization of agricultural markets with common rules on competition (Article 40) and established a system of guaranteed minimum prices as the pivotal tool for supporting agriculture (Article 44).

The Stresa Conference of 1958 reaffirmed and outlined in more detail the principles of the CAP set out in the treaty that found the EEC. It provided for the setting of agricultural prices at an average level compared to those in force in the European Union, the creation of a single market with free circulation of agricultural products, the protection of the internal market, and the creation of an EC budget to bear all the costs associated with the application of the common policy.

### First Phase: Support of the Supply

In 1962 the first Common Market Organizations (CMOs) were created for cereals, pork, poultry, eggs, fruit, vegetables, and wine. A minimum price was guaranteed to the manufacturer in exchange for which the product was withdrawn from the market at the expense of the community. At the same time the European Agricultural Guarantee and Guidance Fund was established to finance CAP expenditure, with the "Guarantee" section acting on the prices and markets of agricultural products and the "Guidance" section providing EC funding to implement structural policies.

In the 1960s, agricultural production expanded enormously, and the Europe of Six (Germany, Italy, France, Netherlands, Belgium, and Luxembourg) achieved food self-sufficiency. In 1972 measures to encourage the modernization of businesses were introduced, including early retirement for older farmers and help and training for younger ones. In 1975 a special program to help less favored and mountainous areas was launched.

### Second Phase: Restrictions on the Supply

However, from the 1970s onward production surpluses grew, and public expenditure that was necessary to guarantee minimum prices, destroy surpluses, or subsidize exports to non-EU countries mushroomed. Public opinion did not tolerate this huge waste of public money aimed at supporting an ever-smaller share of the working population, especially in light of globalization that required the use of resources to face new needs and challenges.

The EC approach changed rapidly. Whereas previously it had favored the growth of the agricultural sector using a range of tools, now it was trying to restrict production to limit surpluses. The measures adopted ranged from the allocation of quotas (e.g., by country in the sugar sector in 1979 and milk in 1984 or the planting rights of vineyards) to the reduction of guaranteed minimum prices. In 1986 a co-responsibility levy for big surpluses of cereals was imposed while in 1988 a ceiling on CAP spending was fixed through "stabilizers" that set a limit on the quantities that could receive support.

In 1992, with Commissioner Ray MacSharry's reform, efforts to restrict production and to open markets intensified. Guaranteed minimum prices were reduced by 30 percent (15 percent for beef), which brought the prices of European foodstuffs close to those of the rest of the world and led to a fall in surpluses. The decrease in price support was completely compensated by the introduction of direct aid to farmers. Support was "decoupled" from production levels through the introduction of payments per hectare, thus reducing support for producers with higher yields per hectare. Therefore, support was no longer given for production but rather for producers to promote quality rather than quantity.

In its initial stages the CAP absorbed up to 80 percent of the EC budget, but over time its impact has progressively fallen to less than 50 percent to favor investments in other sectors considered to be more strategic, such as infrastructure and scientific research.

### Third Phase: Quality and Liberalization (with Caps)

In the 1990s the EC responded to the growing demand for quality food and environmental protection. With the Council Regulation (EEC) No. 2092/91, the production of raw materials and organic food was regulated for the first time, while the Council Regulation (EEC) Nos. 2081/92 and 2082/92 established the protected designations of origin (PDO; protected geographical indication, or PGI; and traditional specialities guaranteed, or TSG).

The "Santer Package" (named after then-Commissioner Jacques Santer), better known as "Agenda 2000," continued and intensified the stabilization of the CAP budget, reducing further the guaranteed minimum price as direct support for farmers and thus forcing the sector toward greater market orientation. The most important innovation, however, was the creation of a second pillar—namely, rural development—to which 20 percent of the budget was allocated. While the first pillar deals with agricultural development and the market measures to support farmers' incomes as discussed above, the second pillar aims to make farms more competitive through production differentiation and through the development of new sources of income and employment while at the same time as protecting the culture, environment, and heritage of rural areas.

The 2003 CAP reform reflects the increased willingness to invest in food quality and environmental protection by introducing cross-compliance: direct payments become conditional on compliance with certain EC (eighteen regulations) and national (good agricultural practices established by each member state) standards in matters of food safety, animal welfare, plant health, environmental protection, etc.

Lastly, the CAP Reform 2014–2020 has frozen the agricultural budget until 2020 at the nominal values of 2013, with a consequent contraction of real values. To increase the competitiveness of European agriculture, greater reliance has been placed on market mechanisms starting from 2016 (apart from national exceptions) with the abolition of quotas (e.g., in the sugar, wine, and milk sectors), thus allowing European supply to adapt to trends in world demand. A further impulse is expected from measures to promote the modernization of existing companies and the creation of start-ups by young farmers.

Funds and instruments are being set up to favor insurance against damage to crops, plants, and animals while a reserve of €400 million (at 2011 prices) has been created for unpredictable and exceptional damage. If resources are going to shrink, they will have to be used more efficiently. For this reason, a part of the direct payments is conditional on following three good practices: greening practices, the diversification of production, and measures for environmental protection. Member states will have the right to adopt schemes that encourage young farmers and small- to medium-sized businesses. Rural development objectives continue to play an important role with the first and second pillars operating in closer association.

## Schizophrenic Approach to Supply Regulation

The CAP is a complex and constantly evolving policy, but what is of most interest here in the context of this volume is identifying the key elements. According to theory, public intervention in the economy should aim to (1) guarantee competition between companies, (2) reduce information asymmetries and protect public health, and (3) combat externalities of production and consumption. The second objective has been pursued through a series of regulations that have introduced ever stricter quality standards and stringent classification and labeling systems for products so that from this point of view the European Union today is the safest area in the world. The creation and strengthening of PGI and PDO quality collective brands have played a fundamental role in supporting small producers[19] and finest food. Many of the measures of the second pillar aim to protect the environment by making agriculture more sustainable, now leading us to the third objective.

The policy that first favored (from 1962 to the end of the 1970s), then limited (from the introduction of quotas on sugar in 1979 to 2015), and finally liberalized (from 2016 onward) production seems schizophrenic. The production incentive policies were conceived in the 1950s when Europe was not yet self-sufficient in agriculture, even though rapid technological progress could have suggested that food security could be reached even without any kind of public subsidy. It is hard to distinguish the growth of production as a result of technological development from the contribution of the CAP, but with an enormous use of resources, it exacerbated a trend that already existed, forcing the authorities to make a sea change less than twenty years later and put a limit on surpluses and a brake on public spending. The guaranteed minimum price (higher than in the rest of the world) and the destruction of surpluses proved to be distortive policies.[20] These two instruments caused a structural excess of supply and a waste of public money and ended up subsidizing primarily the lower-level products that would probably not have been bought by consumers. It was decided, therefore, more than fifty years after the creation of the CAP to substantially reduce support for the quantities produced and to liberalize the market abolishing supply restrictions.

# Economic Policy Conclusions

Beer is made by men, and wine by God.
—Sentence attributed to Martin Luther (1483–1546)

Over the last few decades the global wine market has experienced enormous upheavals both on the demand and the supply side. Two convergence processes have been affecting the absolute quantity of alcohol consumed and the type of preferred beverages while the market has experienced the aggressive entry of the so-called "New World" countries. These two phenomena have been a serious challenge to Mediterranean Europe, a traditional producer and consumer of wine, but have been a great opportunity for New World countries.

Markets are becoming more integrated—which increases the variety of products available—and quality has been steadily growing, together with buyers' skills. Consumers are benefiting from this situation, but the point of view of producers is more problematic. The fall in consumption in Mediterranean Europe has generated a structural oversupply of wine in the world, which has been continuing for decades. The economic strategy of firms from New World countries relies on the adoption of technologies aimed at producing standardized products of good quality and on the large economies of scale, which allow for lower prices. Instead, firms from Old World countries invest in the promotion of tradition and terroir to differentiate themselves and avoid price wars that would erode profit margins. Further, since domestic consumptions of wine have been falling for decades, "exporting" has become a necessary condition for Mediterranean countries.

Government policies are very different as well. Europe has been heavily influenced by the French dirigisme, and the EU Common Agricultural Policy has regulated almost every aspect of wine production, distribution, and sale. Most importantly, the EU has tried to eliminate the world oversupply of wine by providing a number of incentives whose effects have been controversial. The United States, instead, adopted the three-tier system after the repeal of Prohibition in 1933, where the producer has to sell to the distributor, who has to sell to the retailer, who can finally sell to the buyer. However, every state or even every county can decide how to regulate the distribution and sale of alcohol so that nowadays there are thousands of laws in the United States and several barriers to direct-to-consumers sales and shipments (see Riekhof and Sykuta, 2005, for more details). These impediments give considerable bargaining power to distributors and represent a serious obstacle to (especially small) producers as this distribution channel has been shown to positively affect firm growth and gross profit margins (see Newton, Gilinsky, and Jordan, 2015, for a study on US wineries).

In such a difficult environment, surviving this "wine war" depends on six key variables.

1. *Quality of products.*

   In recent decades there has been a constant increase in the quality of wine produced in every part of the world and a parallel increase in the expectations of consumers who have become very well informed and ever more demanding. Therefore, producing shoddy wines is no longer sustainable. The consortia for the protection of appellations must work constantly to improve production specifications without giving in to the temptation to lower minimum quality standards to achieve short-term benefits. At the same time the adoption of best-practice and innovative cultivation techniques needs to be encouraged through research and development, and it should not be deprecated but rather fully endorsed, although excesses and adulterations should be avoided. The role of universities and winemakers' and agronomists' associations is fundamental, and they must also receive strong and adequate public support.

2. *Changes to the tax system.*

   Alcoholic beverages are subject to two types of taxation: excise duties on alcohol content and ad valorem tax. Taxes on alcohol content favor an increase in quality because, based on the assumption that the best products are more expensive, their relative weight decreases for higher-range goods. Taxes ad valorem do not change the relative prices between products of different qualities. Therefore, if we intend to raise the quality of products, then the tax burden should be moved from value-added taxes to excise taxes, leaving the total tax burden unchanged.

3. *Marketing and a clear wine classification system.*

   Perceived quality is, however, more important than actual quality (Cardebat, 2017, p. 44) since it directly influences purchase choices and consumers' willingness to pay. Marketing campaigns are essential to impose the company brand and influence sales, but they remain the prerogative of large groups that are just a small minority in the Old World. For this reason, well-managed EC appellations and funds provided by the European Union for protection consortia to promote wine abroad become important instruments of industrial policy. The same holds true for the funds provided by the Market Access Program of the US Department of Agriculture for marketing, promotion, and research.

The appellation system, if regulated intelligently, can mitigate the information asymmetries between producer and consumer so that small businesses, without the necessary financial resources to build a solid individual reputation, are able to benefit from the reputation of a prestigious collective brand at a low cost. Over the years, the Appellation d'Origine Contrôlée (AOC) and Denominazione di Origine Controllata (DOC) systems have been criticized because of the excessive number of appellations (Colman, 2008, pp. 60–62),[1] which sometimes even have similar names and overlap geographically in countries like Italy.[2] Yet other criticisms report a weak correlation between hierarchical categories (e.g., Denominazione di Origine Controllata e Garantita [DOCG], DOC, IGT) and product quality as well as a growing share of wines classified as excellent, a term that risks losing all significance. Appellations that have a small number of producers should therefore be eliminated unless they are "pearls in the national winescape."

In addition, a system should be devised that can, on the one hand, stop an ever-increasing number of wines from being classified as excellent and, on the other, avoid having membership in the top segment being a poor reflection of the real hierarchy of quality, thus perpetuating the status quo forever, rather like an aristocratic title. Downgrading, therefore, should not be a theoretical and remote possibility but rather a concrete one and as frequent as relegation in soccer/football leagues. It would allow other wines to level up and prevent overcrowding and the emptying of the information content on the label once a cry of *todos cabelleros* ("everyone's a winner") has been made. The choice of who decides which appellations are upgraded and downgraded is, obviously, complex from both a technical and a political point of view.

4. *Competition and support to small wineries.*
Market concentration usually increases when there is no booming consumer demand or radical product innovations, and this has been the case in the wine sector (see Thornton, 2013, p. 289). The problem is that large companies and conglomerates exert their market power at the expense of consumers and reduce product differentiation. In addition to preventing abuse of a dominant position, public support—especially in terms of services and promotion—can be crucial for the survival of small firms and are often those producing some of the best and nonstandardized wines.

Further, in the United States the problem of growing market concentration involves not only wine producers but also distributors. Here it is necessary to end the three-tier system that generates higher prices to consumers and discourages the purchase of quality wines. In addition, distributors tend to favor large companies and corporations because they are able to offer a diversified portfolio of products at competitive prices and are well known, and therefore logistic and marketing costs are minimized. However, this is detrimental to small, emerging, quality firms, especially since most states discourage—through costly administrative tasks—the direct-to-consumer sales of alcohol. Without having the chance to fully rely either on distributors or on direct sales, small producers have a hard life, and this is bad for both competition and qualitative excellence.

5. *Economies of scale and competitive prices.*
The high-end wines that can be sold at high prices represent a small part of global wine production (Thornton, 2013, p. 218). Wines of general consumption are generally from the medium or low levels and are aimed at a clientele who are very sensitive to the quality/price relationship or even just price. Price largely reflects the structure of average production costs, which in turn are an inverse function of economies of scale: large companies have greater bargaining power in the purchase of production factors and can achieve better organizational efficiency leading to lower average costs.

The average size of European companies is much smaller than New World companies (see Thornton, 2013, p. 289). This guarantees, on the one hand, a great variety in production, but on the other hand it involves a competitive disadvantage on the cost side and sometimes even for quality since the adoption of some machinery and special wine-making techniques can be too burdensome for small businesses. Therefore, there is a need for consolidation in many countries (Colman, 2008, p. 108).

From this point of view cooperatives play a fundamental role in compacting production potential spread over hundreds or thousands of firms with vineyards that are too small to survive in the market. Their task is to engage in the constant improvement of quality, identifying internal management rules that can minimize the risk of opportunistic behavior by members. But public authorities should promote a process of aggregation among smaller private companies through appropriate tax incentives to make them competitive and capable of surviving.

In most US states direct-to-consumer sales are either forbidden or discouraged. However, they increase competition and reduce prices (Ellig and Wiseman, 2007) and therefore can increase the demand for quality wine and the firms' investments to achieve excellence.

6. *Promotion of the wine culture among consumers.*
The imbalance between demand and supply of wine in the countries of Mediterranean Europe is caused not by the growth of production—which has been declining for decades—but by the collapse of domestic consumption. The answer

of the EC legislators has consisted in a succession of policies aimed at "bureau-cratically" restoring equilibrium and favoring restrictions on production while encouraging exports to non-European countries. These attempts have proven to be useless for the most part because they have often generated the opposite effect to that desired. Above all, production could not be restricted in non-European countries, and consumption, the other side of the coin, has been completely ignored. With an almost stationary population and a 50–70 percent decline in per capita consumption in the countries of Mediterranean Europe in less than half century, the authorities have tried to act only on the supply side.

Instead, a national wine education plan actively involving sommelier associa-tions could bring young people closer to this drink. While keeping in strict accor-dance with the protection of public health with the aim of not increasing total consumption of alcohol, the objective would be to encourage consumers' willing-ness to pay. It would also hope to reverse the process of substituting wine with beer and spirits that has been taking place for years in many countries. Wine is often perceived as a complex product, especially by consumers with a low level of wine education (Thornton, 2013, p. 237), so this type of training could succeed in influ-encing their preferences. Indeed, in his experiment with Canadian students Sagala (2013) showed that attending a wine appreciation course increased their monthly wine budget and promoted wine consumption in a socially responsible manner. It is very strange and totally unacceptable that beer is becoming "trendier" than wine in countries such as Italy and Spain that have an ancient wine-making tradition.

Courses introducing wine should be given at the beginning of adulthood and include lessons on health protection given by specialized medical personnel that illustrate the benefits of moderate consumption and warn about damage from abuse. Tasting knowingly—and not swigging—alcohol must be the model of con-sumption for young people from the time of their coming of age because bad habits drag on through life with very serious and irreparable consequences for health. Awareness campaigns such as "Wine in Moderation" are not only ineffec-tive from the point of view of the results achieved but also conceptually wrong. Limiting the slogan to wine alone, in fact, puts this drink in a bad light when it actually plays a very marginal role in the phenomenon of juvenile binge drinking and road deaths. It would, therefore, be more appropriate to change the slogan to "Alcohol in Moderation" and divert resources toward training programs set up in the way mentioned above.

Courses about approaching wine held by sommelier associations should also be supported abroad by cultural institutes to increase the reputation of national wines, to enhance the loyalty of buyers, and to further the export of quality wines. All these measures should be accompanied by the promotion of "wine roads," with adequate economic incentives for the renovation of wineries by famous architects. Wine should be considered as a cultural good producing a number of positive externalities—especially in the tourism sector where it is likely to attract people with higher levels of income and education, thereby promoting economic development (Towse, 2010, pp. 530–533). Indeed, food and wine tourism has an enormous potential but is only minimally exploited, and for this reason it should be supported with public grants (Marks, 2015, pp. 186–187, 193). It is also essential to develop a serious public transport policy (subways, buses, taxis) to allow people to leave their cars at home in the evening.[3]

*Richness of vineyard heritage.*
New World countries are very competitive on the cost side and often also on the quality side, but as they do not have native vines, they are forced to plant international varieties that have been successfully cultivated in all the other continents. Therefore, they are not distinctive, unlike the countries of the Old World that have hundreds of native vines in addition to those adopted internationally. Some New World countries are trying to create new "native" grapes artificially (McKee, 2016), but it is difficult to say whether they will be good and whether they will be easy to sell internationally.

The richness of the vineyard heritage is a strong point when it comes to a sophisticated and curious clientele, but it can be an obstacle for less experienced buyers who find themselves faced with a jungle of vines and appellations that they have never heard of. Argentina is the perfect example of a country which has built its success on just a few international vines. Indeed, many consumers immediately associate it with Malbec, a French vine that has found its ideal terroir overseas. In general, vineyards in the New World are largely dominated by about ten varieties.

For this reason, this debated point was mentioned but not numbered as a seventh key variable. Even though many observers consider it as a potentially strategic variable, there is no evidence that it positively affects sales and prices. Actually, the anecdotal evidence collected among wine producers and oenologists suggests that promoting niche grape varieties is difficult because people do not know them. Further, once tasted, these vines are barely remembered. Thus, the much-vaunted importance of the immense vineyard heritage of the Old World deserves to be further investigated.

# Notes

## Chapter 1

1. See https://data.worldbank.org/indicator/sp.pop.totl.

2. The Russian Federation has been arbitrarily classified as a country of the New World because of its historical preference for vodka, even though geographically and historically (also as far as wine production is concerned) it could be considered, to all intents and purposes, as belonging to the Old World.

3. This figure has been omitted but is available on request.

4. The averages have not been weighted for the resident population.

5. In Argentina, Chile, and Uruguay, 85 percent of the population are considered to be of European descent as opposed to 55 percent in Brazil and less than 15 percent in Mexico and Peru (Aizenman and Brooks, 2008).

6. It is important to distinguish consumption expressed in money terms (euro, dollars, etc.) from that expressed in quantity. The latter, in turn, can be expressed in liters of drink or in liters of pure alcohol consumed. The results can change significantly, because beer is on average the least alcoholic of the three types of drinks mentioned (generally between 4.5 and 6 degrees, or percentage alcohol by volume) and the least expensive, while distillates are generally the most expensive with an alcohol content of around 40 degrees.

7. Australia, Canada, New Zealand, and the United States signed a trade agreement to reduce customs duties and barriers and to recognize each other's wine-making techniques, practices, and labeling and bottling systems (Castaldi, Cholette, and Hussain, 2006).

8. The lack of comparable data on the costs of transporting wine by road, rail, and sea means it is not possible to verify Bartlett's (2009, p. 7) statement that the decline in shipping costs and the simultaneous increase in costs on the road have made the cost of trading wine from Australia or the south of France to Great Britain very similar.

9. Data downloaded in September 2013 from www.fao.org.

10. Author's calculations using data downloaded in September 2013 from www.fao.org.

11. A similar shift in consumer preferences from white wine to red wine as a result of medical information was recorded in Ontario in the 1990s (Dyack and Goddard, 2001).

12. Author's calculations using data downloaded in September 2013 from www.fao.org.

13. Even though in many states it is theoretically possible to buy alcohol from a producer either at its facility or by arranging a shipment, the majority of US states impose a number of

restrictions which make it very difficult to do so, and therefore around 90 percent of wine is sold through retailers who buy from distributors. See chapter 8 for a detailed explanation of the US distribution system.

14. Freedom of entry is largely limited by EU community law on planting rights and limits on the entry of new producers within an appellation.

15. Vines native to other countries and those used for the production of table grapes have been omitted from the list of vines entered in the national registers.

16. To draw an analogy with finance, it conceptually recalls the capital asset pricing model.

17. For a literature review of the causes for the introduction of the geographic appellations and for changes in their delineation, see Meloni and Swinnen (2018a).

18. For a more in-depth discussion of individual and collective reputation, see chapter 6.

## Chapter 2

1. For example, in Muslim countries (see chapter 1).

2. Other possible physical reasons reported in French et al. (2010, p. 86) are metabolism, the pharmacokinetics of alcoholism, and the effect of alcoholism on the volume of the brain.

3. See Castriota (2015, p. 57) for Italian data on alcohol consumption by age.

4. Spirits, on the other hand, are mainly made up of water and alcohol while the contribution of other nutrients is minimal.

5. These variables are not considered as separate determinants in this discussion since they affect wine consumption by way of age.

6. For a broad overview of psychology studies, see the National Institute on Alcohol Abuse and Alcoholism (2005).

7. Psychoactive substances act on mechanisms and processes of the brain. This category includes alcohol, hallucinogens, amphetamines, hemp (hashish and marijuana), cocaine, ecstasy, opioids (heroin and morphine), and tobacco.

8. The influence of friends through imitation also applies to smoking, the use of cigarettes and drugs, church attendance, and the dropping out of school (Gaviria and Raphael, 2001; Lundborg, 2006) but may also work in the opposite way. Christakis and Fowler (2008), in fact, found imitation phenomena among individuals belonging to groups or networks where other people had stopped, rather than started, smoking.

9. A series of scientific studies has shown that level of education influences cigarette consumption, food quality, and frequency of physical activity in a decisive way (Huerta and Borgonovi, 2010).

10. See Tiziano Terzani's accounts (2008) from Japan in the 1980s about the army of "salary men" that were working for multinationals and that were forced to attend frequent and exhausting alcoholic evenings with managers, colleagues, and customers.

11. Di Tella, MacCulloch, and Oswald (2001) and Becchetti, Castriota, and Giuntella (2010) used Eurobarometer data to measure, net of income, the social costs of unemployment, distinguishing between unemployed status (a binary variable that assumes value one if the respondent is unemployed and zero otherwise) and the unemployment rate (the percentage of the workforce that claims to be looking for a job).

12. The symptoms of a high correlation between regressors are (1) small changes in the data cause enormous changes in parameter estimates; (2) the coefficients present very high standard errors and low significance levels, even though they are collectively significant and $R^2$ of the regression is high; and (3) the coefficients have the "wrong" sign or an unlikely magnitude. See Greene (2000), pp. 255–256.

13. If two groups of individuals—one educated, rich, and in employment; the other illiterate, poor, and unemployed—present different characteristics (e.g., the family context) that are not observable but influence the regressors, then a positive or negative correlation between socioeconomic status and alcohol consumption may not be due to the variables considered but to those omitted. In other words, unexpected ex ante differences (e.g., coming from lower socioeconomic families, with problems, etc.) can affect both alcohol consumption and socioeconomic status (education, income, and employment status), making identification and the exact quantification of the parameters difficult.

14. Economics textbooks often give the classic example of a reduction in the purchase of potatoes for meat.

15. For another review of 132 studies on the elasticity of demand for alcoholic beverages and the influence of the choice of data and methodologies on results, see Gallet (2007).

16. In his extensive meta-analysis Nelson (2014) calculated the price elasticity of beer of −0.20.

17. Scientific literature has also analyzed the cross-elasticities of beverages sold in off-trade and on-trade businesses as well as drinks belonging to the same category but with different quality segments. In fact, if the price of alcoholic drinks sold in bars and restaurants increases, people can decide to consume at home. The same applies to the consumption of average or top-of-the-line products that can be replaced with others of the same type but that is of a lower quality.

18. Price ranges should be reviewed periodically to take account of inflationary pressures and adapted to the economic context (e.g., per capita income, disposable income, tax system).

19. Robert Tinlot (2001, p. 9), former general manager of the International Organization of the Vineyard and Wine (OIV), states that "there is no wine region in our world that does not try to value its vineyards and their output without reference to the character that they inherit from the place where the wine is produced. Consumers who visit producers are particularly sensitive to the beauty of the landscape, to the architecture of the villages and to any other element that belongs to the region of production."

20. For a quick review, see AIS (2005a), pp. 31–33.

21. For a study on the role of terroir on the price of vineyards, see Cross, Plantinga, and Stavins (2017).

22. Oak barrels increase wine quality but are expensive. Whether they increase or decrease profits is not clear. However, Sims and Quintanar (2017) showed that if over the past fifteen years US winemakers had purchased French oak barrels early (in April, with a discount) rather than in September (with no discount), even accounting for lost interest they would have decreased the costs of the winery by over $60,000 per year.

23. For example, the use of wood chips in the fermentation phase of the must to save on the cost of barrels.

24. However, given that technological choices depend partly on the terrain, the effect of technology on quality can only be accounted for by an approach that uses instrumental variables.

25. Gergaud and Ginsburgh (2008) used wine prices as an indicator of quality implicitly assuming that there is an almost perfect correlation between quality and price. This approach, however, is questionable, since the price of wine can be influenced by many other variables in addition to quality, such as belonging to famous appellations, advertising, and sales strategies. Ginsburgh, Monzak, and Monzak (2013), however, did not discuss the determinants of quality in their econometric analysis of wine prices even though they adopted a similar approach to Gergaud and Ginsburgh (2008).

26. See Jonathan Nossiter's 2004 documentary film *Mondovino*.

27. Analysis of the chemical profile of wine, reverse osmosis, micro-oxygenation, co-pigmentation, and analysis of the olfactory profile of wine are just some of the new technologies on which the biggest and most innovative companies are working, especially in the New World, given the regulatory and cultural constraints existing in Europe.

28. For a review of the most important conglomerates producing alcoholic beverages around the world and the growing market concentration, see Thornton (2013), pp. 293–297.

29. Fraser (2005) examined the supply contracts for grapes used in the main Australian regions and found that in the areas that produce lower quality grapes more attention is given to the evaluation of grape quality to establish penalties and premiums. Instead, those that produce higher quality grapes tend to involve the buyer more in the definition of the rules to regulate the various phases of the production process in a more meticulous way and to carry out stricter controls on suppliers. Further, grape prices are often determined according to the selling price of the bottles of wine, thus binding the economic result of the raw material supplier to that of the wine producer.

30. New York diamond merchants, for example, informally conduct economic transactions within the community making significant savings in the fees to be paid to lawyers: failure to pay the amount due would cause irreparable damage to their reputation, which would affect any future business (Guiso, Sapienza, and Zingales, 2004).

31. See also appendix 2.1 for an application of Veronelli's data used by Castriota, Delmastro, and Curzi (2013).

32. From a more macroeconomic point of view Bukenya (2008) showed that prices are an inverse function of accumulated stocks. The author was able to reconstruct reliable historical series for Argentina, Australia, France, Germany, Italy, Spain, and the United States. In Oczkowski (2006) prices are also influenced by market imbalances. Bentzen and Smith (2002), however, analyzed the price of a sample of Californian and Australian wines sold in the country of origin and in Scandinavian countries (Sweden, Norway, Denmark, and Finland). These countries are far from the production regions and are characterized by high taxes and excise duties on alcohol to discourage alcohol abuse. Nevertheless, the price of the wine was not necessarily higher in the Scandinavian countries, and indeed, sometimes it was lower.

33. En primeur wines are those sold but not yet available in the market.

34. This topic will be dealt with in detail in chapter 6.

35. For a comparison of the role of certifications of origin in the olive oil and wine markets, see Cacchiarelli et al. (2016).

36. Organic wines are produced from organic grapes, but in some countries (e.g., Europe and Canada) they can contain sulfites while in others (e.g., the United States) they cannot. Biodynamic wines are made with organic grapes and in addition follow Rudolf Steiner's rules in the preparation of the soil and respect of the land.

37. The effect of organic production on sensory quality is, however, uncertain and limited to anecdotal evidence since there is limited empirical literature.

## Chapter 3

1. Nowadays, with falling transport costs, spatial differentiation à la Hotelling (1929) is irrelevant.

2. A similar situation arises if we consider a model of oligopolistic competition à la J.L.F. Bertrand without product differentiation. Competition becomes so fierce that profits disappear.

3. Cuttings are the vine seedlings aged between one and two years.

4. This problem, of course, does not exist for the less prestigious areas where wine production is in freefall as it is not profitable.

5. See chapter 8.

6. According to WorldAtlas, "there are about 18 million people living in the dry area of the US, which is about 10 percent of the total area of the US." https://www.worldatlas.com /articles/dry-counties-of-the-united-states.html.

7. In the wine sector there are no companies that dominate the industry like Microsoft, Danone, or Coca-Cola. Nevertheless, there are some multinationals (in Australia, Foster's; in the United States, American Constellation Brands, Gallo, and Mondavi; in France, LVMH; etc.) that have a significant share of some markets and many other companies that act as regional operators.

8. The law prohibits the production of wine with table grapes. The sale of wine grapes as a fruit to eat is allowed but is very rare as table grapes are preferred for this purpose.

9. These companies hold a monopoly in the purchase and distribution of beverages with an alcohol content above a certain percentage that varies according to the country (e.g., 3.5 percent in Sweden, 4.7 percent in Norway, etc.). The companies in question are Systembolaget in Sweden, Vinmonopolet in Norway, Alko in Finland, Vínbúð in Iceland, Rúsdrekkasøla Landsins in the Faroe Islands, SAQ in Québec, and LCBO in Ontario.

10. This operation is very expensive and is, in fact, funded by the European Union.

11. "Return on invested capital (ROIC) is the appropriate measure of profitability for strategy formulation, not to mention for equity investors. Return on sales or the growth rate of profits fail to account for the capital required to compete in the industry. Here, we utilize earnings before interest and taxes divided by average invested capital less excess cash as the measure of ROIC. This measure controls for idiosyncratic differences in capital structure and tax rates across companies and industries," Porter (2008).

12. The ROIC of private firms and that of cooperatives cannot be compared.

13. "Intangible assets, such as a particular technology, accumulated consumer information, brand name, reputation and corporate culture, are invaluable to the firm's competitive power. In fact, these invisible assets are often the only real source of competitive edge that can be sustained over time," Itami (1987).

14. For a more systematic exposition, see basic microeconomics texts such as Krugman and Wells (2006) and Becchetti, Bruni, and Zamagni (2011). Oligopoly is not relevant since there are many companies in the wine sector.

15. The figure assumes that market price is higher than minimum average total costs and therefore the firm makes profits. If, however, the price never manages to cover even the average variable costs, then the company should stop production and exit the market. Finally, if the price manages to cover average variable—but not fixed—costs, then the firm will continue production in the short term.

## Chapter 4

1. Companies can be classified in various ways—for example, according to legal form (sole proprietorship, limited liability company, joint-stock company, cooperative, etc.)—but here it has been arbitrarily decided to proceed following the criterion of vertical integration because this influences the quality of the product and, in turn, the reputation of the company.

2. The "donative-labor hypothesis" has been studied by, among others, Hansmann (1980), Preston (1989), Frank (1996), and Rose-Ackerman (1996).

3. "That the entrepreneur aims at maximizing his profits is one of the most fundamental assumptions of economic theory. So much so that it has almost come to be regarded as equivalent to rational behavior, and as an axiom, which is self-evident and needs no proofs or justifications. Doubts have been raised by several writers whether maximising his profits is always the entrepreneur's best policy. But such doubts were few and have died away without reverberation; mainly, I think, because it has never been made clear what exactly profit maximization implies; and perhaps also because we have a vested interest in maintaining this assumption—it makes economic analysis so much simpler" (Scitovszky, 1943, p. 57).

4. "First of all, there is the dream and the will to found a private kingdom, usually, though not necessarily, also a dynasty. ... Then there is the will to conquer: the impulse to fight, to prove oneself superior to others, to succeed for the sake, not of the fruit of success, but of success itself. From this aspect, economic action becomes akin to sport—there are financial races, or rather boxing-matches. The financial result is a secondary consideration, or, at all events, mainly valued as an index of success and as a symptom of victory, the displaying of which very often is more important as a motive of large expenditure than the wish for the consumers' goods themselves ... Finally, there is the joy of creating, of getting things done, or simply of exercising, one's energy and ingenuity" (Schumpeter, 1911, p. 93).

5. The commitment made by an owner who directly manages a company can also vary greatly. In fact, the objective of an entrepreneur is twofold: to maximize profits and balance work and free time. Scitovszky (1943) developed a model in which the entrepreneur's indifference curves are a function of monetary income (given by the company's profit and the manager's "salary") and leisure time (the inverse function of time dedicated to work). The commitment of an owner-manager to work in the company is influenced by his preferences regarding the two inputs of the utility function.

6. Investments in quality require, of course, availability of capital and a not too high intertemporal discount rate. In other words, the "impatient" enterprise favors short-term results, risking the chances of improving reputation and future profitability.

7. The wholesomeness of products is taken for granted.

8. As underlined by Bénabou and Tirole (2003), one of the foundations of economic science is that individuals, be they workers or children, react to incentives while psychologists and sociologists consider rewards and punishments as counterproductive because they reduce intrinsic motivation.

9. The owner's passion, however, should be kept quite distinct from his skills, which is why Delmastro (2007) considers separately—and evaluates positively—the contribution given by external consultants who bring new skills and experiences acquired in other contexts. "The fact that the owner is involved in the company by following the production phase directly appears to be a prerequisite for the inclusion of the quality of wine in the utility function. The owner-winemaker therefore represents a proxy of the owner's attitudes towards his product, and does not appear to incorporate the effect of his skills (in fact, the owner often takes care of the enological part without having particular skills and/or qualifications)" (Delmastro, 2007, p. 69).

10. Pennerstorfer and Weiss (2006, p. 12): "Assuming that the members of the cooperative are paid according to the quantity they deliver and that the quality of the inputs is non-contractible between independent actors, there is a strong incentive to free-ride and deliver low quality. This free rider problem among members of cooperatives is a well-recognized problem in the literature."

11. In a study on the reputation of Italian wineries, Castriota and Delmastro (2012) found a null result for the cooperative type.

12. This information was found on the ICA website, last accessed in 2013 (the page is no longer available).

13. "Clearly, in most introductory textbooks, co-operative economic organizations either are entirely ignored or receive only a passing mention" (Hill, 2000, p. 283).

14. Similarly, worker cooperatives have proven to dismiss fewer people compared with competitors during periods of economic contraction, thus protecting both jobs and working conditions (Craig and Pencavel, 1992, 1994; Bonin, Jones, and Putterman, 1993; Burdin and Dean, 2009).

15. "However, as this research has shown, there is a considerable share of co-operatives that allows loose contractual relations—soft or shadow membership. Policy makers should therefore be aware that the beneficial aspects attributed to co-operatives in rural development programs may be overestimated" Pascucci et al. (2012, p. 71).

16. "Access policies therefore must strike the right balance between the protection of investment and openness" (Rey and Tirole, 2007, p. 1063). "New members free ride on the investment of established members (had we introduced uncertainty, free riding might have been even more of an issue as potential members could join the joint venture only if it turns successful). This induces underinvestment (the horizon problem) or even prevents the cooperative from getting off the ground" (p. 1084).

17. As cited in Becattini (2002), p. 84.

18. A calculation of the number of districts gives different results depending on the definition used. ISTAT (2001), in the 8° Censimento Generale dell'Industria e dei Servizi (Eighth General Census of Industry and Services), counted 156; the Osservatorio Nazionale Distretti Italiani (the Italian National Districts Observatory) (2013) in its Fourth Report counted 101; and Intesa Sanpaolo Bank (2013) in its Sixth Annual Report counted 144.

19. The importance of districts extends to many areas of the world as well as China. As reported in the Fourth Report of the Italian National Districts Observatory (2013), "the Ministry of Foreign Trade in China noted that 50 percent of production in the more industrialized areas of the country is organized in specialized districts, consequently the government decided to continue investing in those areas" (pp. 27–28).

20. Porter (2000, p. 27) argued that all clusters are desirable independently of what they produce: "All clusters can be desirable, and all offer the potential to contribute to prosperity. What matters is not what a nation (location) competes in but rather how it does so. Instead of targeting, therefore, all existing and emerging clusters deserve attention."

21. This methodology can easily be applied to other regions such as Chianti or Bordeaux. For an in-depth analysis of the input-output methodology applied to the wine sector and its mathematical properties, see Ciaschini and Socci (2005).

## Chapter 5

1. "Given the space, there is every argument for buying wine young, at its opening price, and 'laying it down' in cellar or cupboard until it reaches perfect maturity. Wine merchants are not slow to point out that it appreciates in monetary, as well as gastronomic, value out of all proportion to the outlay" (Johnson, 1971, p. 38).

2. The term "stylized facts" was introduced by Kaldor (1961) and refers to representations simplified by recurrent and agreed empirical cases.

3. Italy is not included because very few firms are listed on the stock exchange. The largest companies are cooperatives, and the others are closely controlled by families or are small.

4. For a more in-depth discussion of the theoretical foundations of derivative instruments, see Björk (1998).

5. Aghion et al. (2009) demonstrated the importance of the development of financial markets in neutralizing the negative effects of exchange rate volatility on the growth of productivity. But, the link between derivatives and growth is subject to some debate, especially in light of their immense growth that is not justified by the size of the real economy and the financial crisis that hit global markets in 2008. As highlighted by Blundell-Wignall and Atkinson (2011), derivatives are largely used not to cover against risk but for short-term speculation and regulatory and fiscal arbitrage. Therefore, it was the misuse of derivatives that was the real cause of the worsening of the crisis (Fink, Haiss, and Hristoforova, 2006) while the usefulness of these tools should not be put in doubt. Positive, but weak, effects of the size of the OTC derivatives markets on economic growth were found by Becchetti and Ciampoli (2012).

6. There are two approaches to identifying insurance conditions. The first is the actuarial one in which the goal is to identify the conditions that must be met so that an event is, at least in theory, insurable. The second is an approach of general equilibrium whose purpose is to identify the conditions that lead to equilibrium with Pareto-efficient solutions in which every type of risk is insured. The goals of the two approaches are different, but most of the conditions coincide. Berliner (1982), Williams (1997), Stahel (2003), and Dorfman (2004) belong to the first; Borch (1962), Arrow (1965), and Shiller (1993) to the second; and Gollier (2005) provides a general description of the two approaches.

7. This last problem is particularly relevant in many developing countries and smaller towns in advanced countries.

8. The countries are Brazil, Costa Rica, India, Japan, Mexico, the Philippines, and the United States.

9. For a detailed discussion of the weather index insurance, see IFAD (2011).

10. The country that has seen the greatest development of index insurances of both types is India. In 2006 there were over eighteen million farmers with insurance linked to the average productivity of the region (see Skees, Barnett, and Collier, 2008).

11. Futures and options, open interest, December 2019, see https://stats.bis.org/statx/srs/table /d1?f=pdf.

12. For data on OTC derivatives, see https://stats.bis.org/statx/srs/tseries/OTC_DERIV/H :A:A:A:5J:A:5J:A:TO1:TO1:A:A:3:C?t=D5.1&p=20172&x=DER_RISK.3.CL_MARKET _RISK.T:B:D:A&o=w:19981.,s:line.nn,t:Derivatives%20risk%20category.

13. For a more in-depth discussion of the theoretical foundations, see Bodie, Kane, and Marcus (2005, chapters 6, 8–10, 13).

14. Empirical studies (e.g., Friend and Blume, 1975; Grossman and Shiller, 1981) have found risk aversion between two and four, which corresponds to the profile of a risk-averse person.

15. The two securities could also be one portfolio of bonds (D) and the other a portfolio of stocks (E).

16. A critique of this model is that SMB and HML are not necessarily specific risk factors—that is, the securities of these companies do not show greater yields by virtue of their greater riskiness. A possible alternative explanation may be the exaggerated reaction of investors to previous successes and failures of a security leading to excessive sales or purchases (Lakonishok, Shleifer, and Vishny, 1994).

## Chapter 6

1. "Common examples [of asymmetric information contexts] include mundane transactions in which a person buys a bottle of wine with unknown quality" (Bar-Isaac and Tadelis, 2008, p. 275).

2. For a digression on the different definitions and meanings of cultural goods, see Towse (2010, pp. 151–152).

3. "Such information and beliefs about the seller's skill and behavior, which we refer to as the seller's "reputation," are a consequence of many things. These include direct observations on past performance, experience with other sellers, reports from third parties, actions that the seller may undertake outside of the transaction, and numerous other factors" (Bar-Isaac and Tadelis, 2008, p. 277). The definition provided by Cabral (2005, p. 4) is much more concise, but similar: reputation is the situation "when agents believe a particular agent *to be* something."

4. The concept of reputation invests all fields of the economy and goes beyond its boundaries influencing the outcome of economic and noneconomic transactions between agents, which may be companies (Kreps, 1990), banks (Gorton, 1996), central banks (Barro and Gordon, 1983), public debt managers (Drudi and Prati, 2000), minority shareholders (Gomes, 2000), managers (Yermack, 2004), internal controllers (Sridhar, 1994), participants in auctions (Houser and Wooders, 2006), criminals (Lott, 1996), and governments committed to countering requests for independence in some regions (Walter, 2009). The need to defend the reputation of a state or its prime minister can even be the (con)cause of armed conflicts (Dafoe, Renshon, and Huth, 2014).

5. For a study on the effect of word of mouth on sales of books, see Chevalier and Mayzlin (2006).

6. The theoretical benefits of reputation have been debated in the literature. Ely and Välimäki (2003) build a theoretical model where a sequence of short-lived players interacts with the long-run agent. Results show that the reputational concern of the long-run player to look good in the current period results in the loss of all surplus. That is, the observability of past actions might actually lower the long-run player's payoff. In a laboratory experiment, Grosskopf and Sarin (2010) find that reputation is rarely harmful and its beneficial effects are not as strong as theory suggests.

7. Yu, Bouamra-Mechemache, and Zago (2018) developed a model to explain the rationale of nested names where collective labels are effective in reaching uninformed buyers while individual brands help firms to reach informed buyers.

8. See also Masset, Weisskopf, and Cossutta (2015) for a study examining the ratings of twelve influential wine critics on the Bordeaux en primeur market. For an application to the gastronomic market, see Gergaud, Smeets, and Warzynski (2010): "For most chefs, having his restaurant being awarded one or more stars in the famous Michelin Guide Rouge represents a major achievement, a recognition of their work, and also increased notoriety generating a significant stream of future revenues. In this specific industry, experts play a decisive role, and reputation of restaurants and chefs are basically established according to their opinion" (p. 1).

9. "We found a great deal of evidence that (past or present) expert scores have been found to be positively correlated with wine prices independently of the specific countries, wine magazines, or experts (e.g., Landon and Smith, 1998; Angulo et al., 2000; Schamel and Anderson, 2003; Costanigro, McCluskey, and Mittelhammer, 2007)" (Costanigro et al., 2010, p. 1344).

10. Of the sixty-seven companies that had acquired an international reputation in 2006, twenty-eight had one star, twenty-nine had two stars, and only ten had three stars. Reputation, both national and international, is measured with an ordinal scale ranging from zero to three, with the difference that zero always indicates no stars, but in the case of national reputation it means presence in the *Espresso* guide, whereas in the case of international reputation it means there is no mention in Hugh Johnson's guide.

11. For a detailed review of the theoretical literature, see Bar-Isaac and Tadelis (2008).

12. It should, however, be remembered that this positive correlation refers to a sample of companies selected on the basis of quality and, therefore, is not representative of the whole sector. If the sample were made up of both companies reviewed in wine guides (which are generally smaller and sell through the Horeca channel) and nonreviewed companies (which are generally larger and sell through mass market retailing), the correlation would, in all probability, be negative.

13. These results are in contrast with the studies on the quality of wine that have shown the negative effects of cooperatives (Frick, 2004; Dilger, 2004; Delmastro, 2007) and the relevance of a winemaker as a consultant (Delmastro, 2007).

14. Gallo Nero is the historic trademark of the Consorzio del Chianti Classico DOCG and appears on the bottle foil (capsule) or on the label.

15. Only if at least 85 percent of the grapes come from the same vintage. The indication of the vintage can have an advantage in terms of image ("vintage" wine), but it means products of different vintages (blend) cannot be mixed. Since generic wine does not lend itself to aging, the

indication of the vintage can be counterproductive if quite a long time has elapsed since the grape harvest.

16. According to the Italian Ministry Decree No. 381 of March 19, 2010, varietal wines are "wines without designations of origin or geographical indication, which show, on the label, the vintage and/or the name of one or more varieties of grapes from which they were produced, without any link to a production area. The certification is based on documentation ascertaining that the optional indications that are intended to be included on the label are truthful." Only varieties of grapes specified by each member state are allowed. For Italy, the varieties are Cabernet, Cabernet Franc, Cabernet Sauvignon, Chardonnay, Merlot, Sauvignon, and Syrah for wine, and Moscato, Malvasia, Trebbiano, Pinot Bianco, Pinot Grigio, and Pinot Nero for sparkling wines.

17. Data downloaded from www.istat.it on March 23, 2020.

18. Alternatively, it has been suggested to aim for the opposite strategy by conferring the recognition of DOCG for the largest number of wines possible because in this way the origin, traceability, and quality of the products is guaranteed, whereas other instruments should be found to indicate quality as happens with the cru or the grand cru within the same French appellation (AIS, 2005b, p. 10). In doing so, however, the proven system of pyramid classification that goes from common wines to DOCG would come to an end.

19. Defrancesco et al. (2012) found a positive effect for the geographical indication of the Argentine Malbec on consumers' willingness to pay in the New World but not on those in the Old World.

20. "A delimited grape-growing region having distinguishing features as described in part 9 of the TTB regulations and a name and a delineated boundary as established in part 9 of the TTB regulations (27 CFR part 9)" (US Alcohol and Tobacco Tax and Trade Bureau, 2012, p. 3).

21. The Alcohol and Tobacco Tax and Trade Bureau manual speaks of "distinguishing features." "A petition must explain, and provide substantive evidence of, the distinguishing features of the proposed AVA that differentiate the area from what surrounds it in all directions. Distinguishing features are also referred to or characterized as 'geographical features.' The regulations mention climate, geology, soils, and physical features as distinguishing features; these examples reflect the types of features most often mentioned in AVA petitions. They are intended to be illustrative only, and other relevant features may be relied on in AVA petitions. When comparing the distinguishing features inside the proposed AVA boundary to the different features outside that boundary, the petition should explain how the features in question affect viticulture both within and outside the proposed AVA" (US Alcohol and Tobacco Tax and Trade Bureau, 2012, p. 14).

22. The website of the US Government Publishing Office reports the data referring to November 21, 2014 (US Government Publishing Office, 2014).

23. However, empirical evidence has provided conflicting results about the impact of family management on company profitability, with Anderson and Reeb (2003) and Lee (2006) finding a positive effect; Filatotchev et al. (2005) and Westhead and Howorth (2006), a negative effect; and Daily and Dollinger (1992) and Villalonga and Amit (2006), a null effect. The only empirical evidence of the effect of ownership on the quality of products is, instead, the study by Frick (2004) on German data showing a superior quality of products of cellars managed by external managers.

24. "The availability of information may benefit large firms disproportionally by inflating audiences' familiarity with their activities" (Fombrun and Shanley, 1990, p. 224).

25. An external oenologist is a person who provides his consulting services (e.g., by suggesting the best cuts) for a number of companies, unlike the internal oenologist "Cantiniere" (wine maker) who works exclusively for one company and actually produces the wine. The first work is purely intellectual while in the second there is a great component of manual skills.

26. "If too many firms are admitted to the brand, the incentive to free ride necessarily overrides the reputation effect and reduces the incentive to invest, relative to stand-alone firms. This is because once the brand is sufficiently large, the marginal contribution of an individual member's investment to the brand's visibility and reputation becomes negligible, in comparison to the payoff from free riding" (Fishman et al., 2008, p. 4).

27. The lack of producer traceability is one of the causes of excessive "extraction" of collective reputation by the individual company that deviates from virtuous behavior in Winfree and McClucskey's (2005) model.

## Chapter 7

1. For this and other contributions, Ronald Coase was awarded the Nobel Prize for economics in 1991.

2. While wine production favors tourism, the opposite is also true. Fischer and Gil-Alana (2009) showed that German tourism to Spain influences the export flows of Spanish wines to Germany. As a result, tourism produces not only direct short-term effects on the economy but also indirect effects protracted over time.

3. The cost disease argument claims that in a typical cultural performance, the labor share of the total costs rises over time, thereby increasing the price of performances more than the overall inflation rate. Since rising prices discourage consumers and cultural goods are important for the identity of countries, public support can be a solution to avoid an "artistic deficit." Whether the share of labor costs has been rising in the wine sector is, however, an open question. The artistic deficit can be even conceived in terms of diversity, and subsidies could encourage less popular works and products, as shown by Pierce (2000) and Heilbrun (2001) using data on US opera companies.

4. Daily consumption does not mean an average of glasses per week or month but the actual consumption on the day of reference as consumption of the same quantities of alcohol in a limited period (binge drinking) or over several days generates very different consequences.

5. "Current research and public-health perspectives on alcohol emphasize harms disproportionately relative to benefits. The major exception is research establishing beneficial effects of moderate drinking on cardiovascular health and overall mortality. In addition, much observational and experiential data suggest the widespread prevalence of positive drinking experiences" (Peele and Brodsky, 2000).

6. Sacks et al. (2015) estimate the costs of alcohol abuse in the United States and find that 76.7 percent are due to binge drinking and 9.7 percent, to underage drinking.

7. "Diseases and injuries attributed to alcohol kill millions and harm tens of millions of people each year worldwide. But the death and injury that strike at all strata of society can be reduced through prevention and treatment policies that are shown to work—if governments will adopt and enforce them (Box 15). Indeed, it is a significant shortcoming in all countries that alcohol-attributable death, disease and injury receive so little attention in public health and safety policy" (WHO, 2011, p. 40).

8. For a detailed analysis of the cost of road deaths and safety policies, see WHO's (2013) *Global Status Report on Road Safety*. For some analyses of the negative effect of alcohol on accidents, see Levitt and Porter (2001), Baughman et al. (2001), and Carpenter and Dobkin (2011). The role of speed limits as a deterrent was analyzed by Brown et al. (1990) and Baum et al. (1991) while Lave (1985) demonstrated with US data that speed variability is even more relevant than the maximum speed because it increases the number of times cars overtake one another. The importance of passive safety has been studied by, among others, Cohen and Einav (2003) and French et al. (2009).

9. As emphasized by Young and Bielinska-Kwapisz (2006), however, laws carrying an increased tax burden on alcoholic beverages may not be considered as completely exogenous since the authorities can introduce these changes because of widespread abuse.

10. Excise duty contributes to forming the value of products; hence, VAT on products subject to excise duty also weighs on excise duty itself.

11. "Generally, studies have found that wine drinkers tend to have a healthier lifestyle profile than beer or spirits drinkers, but generally, wine drinkers have shown better socioeconomic levels that can positively influence the health indicators" (Baltieri et al., 2009). People who drink wine seem, therefore, to be different (better) than those who drink beer and spirits.

12. Portugal, like all other countries in Mediterranean Europe, has been witnessing the substitution of wine with beer, and young people today tend to drink more of the second than the first (see chapters 1 and 2).

13. According to WorldAtlas, "there are about 18 million people living in the dry area of the US, which is about 10% of the total area of the US. After the repeal of the prohibition in 1933, a huge proportion of the population persistently supported the prohibition. While some states chose to maintain their prohibition, others allowed local counties to decide if they wanted to continue with prohibition within their borders." https://www.worldatlas.com/articles/dry -counties-of-the-united-states.html.

## Chapter 8

1. "Marché viticole est un marché atomistique, assez semblable au marché idéal de la théorie classique car les producteurs et les consommateurs sont extrêmement nombreux et aucun d'eux pris individuellement ne peut par sa volonté ou par son action modifier sensiblement le marché Même en négligeant les petits producteurs here ne commercialisent souvent q'une fraction infime de leur récolte les vendeurs importants se comptent par centaines de milliers qui exclut toute tentative de cartel" (Milhau, 1953, pp. 701–702). However, over the last decades market concentration has been increasing, and antitrust authorities have intervened to authorize mergers and acquisitions, provide opinions to governments, and evaluate anticompetitive agreements (see Minutorizzo, 2019).

2. As seen in chapter 7, total state control in the alcoholic beverages sector can take the form of a monopoly of production or sales, but the market may also disappear completely with the introduction of prohibition.

3. In some cases, the two concepts have been confused in public opinion. In fact, by "privatization," we mean the sale of a public company to private subjects while "liberalization" means the opening of the market to competition from new operators. Privatization, therefore, does not automatically entail liberalization. The privatization of infrastructure, such as motorways and airports, for example, is unlikely to lead to increased competition since there is usually

only one highway that joins two metropolitan areas and only one airport (especially if international) in a city. In these cases, there is usually a transition from a public monopoly to a private monopoly with little or no benefits for the consumer (or even a worsening of conditions). The case of Italian and British motorways and airports are, from this point of view, perfect examples.

4. While EU regulations are directly applicable, directives bind states to objectives that the countries will pursue by enacting laws and specifying the means by which the objectives will be achieved. Unlike regulations and directives, which apply generally, decisions concern a single country. Finally, recommendations are not binding.

5. For a detailed analysis of the role and mechanisms of lobbying in the wine sector, see chapter 3 of Gaeta and Corsinovi (2014).

6. "The EU tries to cope with the situation by siphoning wine out of the lake for distillation (for example, into vinegar) and by grubbing up vines from the vineyards on the hills around the lake. [However] the problem is that EU-financed distillation is a positive stimulant of overproduction of largely undrinkable wine, since it maintains less efficient growers of poor quality wine which would have given up long since if it were not for the EU support system. ... The EU is losing ground in the expanding middle sector of the market [to New World wines]. ... The EU thus finds itself running a wine support policy that costs around 1.5 billion [euros] a year, involving the annual destruction of an average of 2–3 billion litres of substandard and undrinkable wine" (Grant, 1997, pp. 137–138).

7. "L'organisation commune du marché vitivinicole pour les vins de table s'est avérée une des plus délicates à mettre en place dès le départ, le règlement n° 24 du 4/04/1962 en jette les bases toujours d'actualité. Il s'agissait, en effet, de fusionner deux marchés, le français et l'italien, que tout séparait, entre lesquels n'existait alors pas d'échanges commerciaux réguliers et qui représentaient déjà 50 percent de la production mondiale de vin. Le faudra huit ans et les accords d'Evian pour parvenir, en 1970, à un fragile compromis. Les crises passés avaient installé en France une organisation dirigiste et centralisée du secteur. Schématiquement tout était sévèrement contrôlé par l'Etat: cadastre viticole, surface plantée, classement des cépages, déclaration de récolte, prestations viniques, quantum de commercialisation, jusqu'aux mises en marché échelonnées dans le temps. En contrepartie, l'Etat intervenait sistématiquement pour soutenir les cours du vin, qui étaient en moyenne de 25 percent supérieurs aux cours italiens, en octroyant des facilités de financement des stocks et en prenant en charge la distillation des excédents par le monopole des alcools. La fraude, très sévèrement réprimée, demeurait quasi impossible. En Italie où, bien au contraire, régnait le plus grand libéralisme, ce qui était interdit s'avérait souvent possible. En 1970, la doctrine italienne, plus libérale, prévalut. La plantation et la replantation de vignobles ne furent plus soumises qu'à des règles qualitatives, la commercialisation des vins ne fit l'objet d'aucune disposition obligatoire, les règles de production entérinaient même des distorsions de concurrence entre les différents pays. Brussels estimait qu'il suffirait de prévoir quelques interventions conjoncturelles de soutien du marché des vins de table (distillation, aide au stockage) and qu'il n'y avait pas interdépendance entre ce marché et celui des vins de qualité, seule une protection efficace aux frontières avec le «prix de référence», et une aide à l'exportation sous forme de restitutions complétaient l'édifice" (Arnaud, 1991, p. 6).

8. Planting rights were initially conceived as a temporary measure, but were constantly renewed (ten times between 1976 and 2008). "The planting rights regime was introduced at EU level in 1976 with Council Regulation (EEC) No. 1162/76 of 17 May 1976. The context

in the years before 1976 was of an excessive and growing production (especially of low quality table wines) in relation to the available outlets. Following the Commission's proposal, the
Council decided to introduce a ban on any new plantings, in order to limit the production of
table wines and prevent structural surpluses. This ban was initially set for the period between
1 December, 1976 and 30 November, 1978. In this first regulation three exceptions to the general ban were established: 1. new plantings aimed at the production of quality wines produced
in a specified region (qwpsr) in the Member States where the respective production in recent
years represented less than 50 percent of total wine production; 2. new plantings established
in the context of the execution of farm development plans (Directive 72/159/EEC); 3. new
plantings in Member States with an annual wine production below 5,000 hl. In the period
between 1976 and 2008 the expiring date of the planting rights regime was prolonged ten
times on the basis of Council regulations. The justifications were most frequently the permanent risk of 'structural surpluses affecting the sector,' 'the situation on the wine sector market'
or 'tendency in the next few years for production to exceed foreseeable needs'" (European
Union, 2012, p. 5).

9. Over 75 percent of the subsidies reserved for the European wine sector were earmarked for
the destruction of surpluses, often produced specifically for this purpose. Every year between
12 percent and 22 percent was destroyed by distillation (European Commission, 2009).

10. "In order to provide for a satisfactory level of traceability of the products concerned, in
particular in the interest of consumer protection, provision should be made for all the products covered by this Regulation to have an accompanying document when circulating within
the Community" (Council of the European Union, 2008, preamble, Recital 78).

11. Articles 91 and 92 established the criteria for planting and replanting vineyards while Article 100 established criteria for eligibility for grubbing premiums. The award was divided into
eight segments depending on the production ascertained. It ranged from €1,740 for one hectare with a yield of 20 hectoliters per hectare (hl/ha) to €14,760 for one hectare with a yield of
160 hl/ha. The amounts gradually decreased in the two campaigns following the 2008/2009
campaign (see Commission of the European Communities, 2008, Annex XV). The maximum
limit for grubbing is equal to 10 percent of the total area under vines in the region.

12. See the article by Castriota and Delmastro (2010).

13. "The Volstead Act, the federal law that provided for the enforcement of Prohibition, also
left enough loopholes and quirks that it opened the door to myriad schemes to evade the
dry mandate. One of the legal exceptions to the Prohibition law was that pharmacists were
allowed to dispense whiskey by prescription for any number of ailments, ranging from anxiety
to influenza. Bootleggers quickly discovered that running a pharmacy was a perfect front for
their trade. As a result, the number of registered pharmacists in New York State tripled during
the Prohibition era. As Americans were also allowed to obtain wine for religious purposes,
attendance rose at churches and synagogues, and cities saw a large increase in the number
of self-professed rabbis who could obtain wine for their congregations. The law was unclear
when it came to Americans making wine at home. With a wink and a nod, the American
grape industry began selling kits of juice concentrate with warnings not to leave them sitting
too long or else they could ferment and turn into wine. Home stills were technically illegal,
but Americans found they could purchase them at many hardware stores, while instructions
for distilling could be found in public libraries in pamphlets issued by the U.S. Department
of Agriculture. The law that was meant to stop Americans from drinking was instead turning
many of them into experts on how to make it" (Lerner, n.d.).

14. See Wine Institute and Avalara (n.d.) for an updated map of laws on direct-to-consumer shipping.

15. A 2005 Supreme Court ruling found that states permitting direct-to-consumer shipping must give the same right to both in-state and out-of-state producers. Nowadays, thirty-nine states allow interstate shipping of alcoholic beverages. For a study on how various economic and public interest factors affect the likelihood that a state adopts a change in its direct shipment regulation and the nature of that change, see Reikhof and Sykuta (2005).

16. Gruenewald et al. (2006), using Swedish data from 1984 to 1994, find that "consumers respond to price increases by altering their total consumption and by varying their brand choices. Significant reductions in sales were observed in response to price increases, but these effects were mitigated by significant substitutions between quality classes."

17. Surprisingly, while anecdotal evidence indicates that cheap, low-quality alcoholic beverages are bad for health and can increase the severity of hangover, there is no scientific evidence of this, apart from studies on unrecorded alcohol, which is more toxic. For a systematic, computer-assisted review of the literature, see Rehm, Kanteres, and Lachenmeier (2010).

18. For the original treaties, see https://eur-lex.europa.eu/collection/eu-law/treaties/treaties-founding.html.

19. For a study on the positive effects of names being reported on labels for small French dairies, see Bontemps, Bouamra-Mechemache, and Simioni (2013).

20. Meloni and Swinnen (2013) concluded that "One of the most striking conclusions of economic studies on the EU's wine markets is that the policies have caused—rather than resolved—some major distortions in the wine sector" and referred to the text of the report by the European Commission (2004): "Distillation of wine measures are neither effective nor efficient in eliminating structural surpluses. Distillation measures involve fairly high EU expenditure. The short-term income support through buying-in of wines for distillation stabilizes surplus production in the long-term. … Additionally, continuous implementation of distillation measures producing industrial alcohol out of wine might be an incentive for higher yields."

## Economic Policy Conclusions

1. "At some point, the AOC system was questioned, because the high number of wines with this name caused more confusion than clarity for the consumer" (AIS, 2012, p.10). When Vaseth (2011) was discussing the European system of wine classification, he spoke of *Lost in Translation* (taking up the title of the film by Sofia Coppola) and the "Da Vino Code" (jokingly changing the title of the book by Dan Brown).

2. For example, Chianti DOCG/Chianti Classico DOCG, Prosecco DOC/Prosecco di Conegliano, and Valdobbiadene DOCG. Stallcup (2005) quite rightly notes that "the traditional approach to wine education has been and continues to be too complex for non-experts. Most consumers only want to be able, from time to time, to buy a good bottle of wine without having to follow a stochastic calculation course or theoretical physics in French."

3. The solution to Saturday night accidents caused by drunk driving cannot be the early closing time of nightclubs or restrictions on the sale of alcoholic beverages. In doing so, the state tries to cover up its failure to provide services and solutions for citizens with prohibitionist policies. It is like a doctor giving crutches to a patient who is limping because he is unable to cure the leg: it is best to change the doctor!

# References

Abbey, A. (2002). "Alcohol-Related Sexual Assault: A Common Problem among College Students." *Journal of Studies on Alcohol*, Suppl. No. 14, pp. 118–128.

Abel, A.B. (1990). "Asset Prices under Habit Formation and Catching Up with the Joneses." *American Economic Review Papers and Proceedings*, Vol. 80, pp. 38–42.

Abrams, B.A., and Lewis, K.A. (1995). "Cultural and Institutional Determinants of Economic Growth: A Cross-Sectional Analysis." *Public Choice*, Vol. 83, No. 3–4, pp. 273–289.

Aghion, P., Bacchetta, P., Rancière, R., and Rogoff, K. (2009). "Exchange Rate Volatility and Productivity Growth: The Role of Financial Development." *Journal of Monetary Economics*, Vol. 56, pp. 494–513.

Agrawal, A., Dick, D.M., Bucholz, K.K., Madden, P.A.F., Cooper, M.L., Sher, K.J., and Heath, A.C. (2008). "Drinking Expectancies and Motives: A Genetic Study of Young Adult Women." *Addiction*, Vol. 103, pp. 194–204.

Aizenman, J., and Brooks, E. (2008). "Globalization and Taste Convergence: The Cases of Wine and Beer." *Review of International Economics*, Vol. 6, pp. 217–233.

Akerlof, G.A. (1970). "The Market for 'Lemons': Quality Uncertainty and the Market Mechanism." *The Quarterly Journal of Economics*, Vol. 84, No. 3, pp. 488–500.

Albæk, S., and Schultz, C. (1998). "On the Relative Advantage of Cooperatives." *Economics Letters*, Vol. 59, pp. 397–401.

Ali, H.H., Lecocq, S., and Visser, M. (2008). "The Impact of Gurus: Parker Grades and 'En Primeur' Wine Prices." *Economic Journal*, Vol. 118, No. 529, pp. 158–173.

Ali, H.H., and Nauges, C. (2007). "The Pricing of Experience Goods: The Example of en Primeur Wine." *American Journal of Agricultural Economics*, Vol. 89, No. 1, pp. 91–103.

Ali, M., and Dwyer, D. (2010). "Social Network Effects in Alcohol Consumption among Adolescents." *Addictive Behaviors*, Vol. 35, pp. 337–342.

Allen, D.W., and Lueck, D. (2002). *The Nature of the Farm: Contracts, Risk, and Organization in Agriculture*. Cambridge, MA: The MIT Press.

Allen, F. (1984). "Reputation and Product Quality." *RAND Journal of Economics*, Vol. 15, No. 3, pp. 311–327.

Allingham, M.G., and Sandmo, A. (1972). "Income Tax Evasion: A Theoretical Analysis." *Journal of Public Economics*, Vol. 1, pp. 323–338.

Almenberg, J., and Dreber, A. (2011). "When Does the Price Affect the Taste? Results from a Wine Experiment." *Journal of Wine Economics*, Vol. 6, No. 1, pp. 111–121.

Alston, J.M., Fuller, K.B., Lapsey, J.T., Soleas, G., and Tumber, K.B. (2015). "Splendide Mendax: Fale Label Claims about High and Rising Alcohol Content of Wine." *Journal of Wine Economics*, Vol. 10, No. 3, pp. 275–313.

Amadieu, P., and Viviani, J.L. (2011). "Intangible Expenses: A Solution to Increase the French Wine Industry Performance?" *European Review of Agricultural Economics*, Vol. 38, No. 2, pp. 237–258.

Anderson, K. (2013). *Which Winegrape Varieties Are Grown Where?* Adelaide, AUS: University of Adelaide Press.

Anderson, K., Norman, D., and Wittwer, G. (2004). "The Global Picture." In *The World's Wine Markets. Globalization at Work*, ed. K. Anderson. Cheltenham, UK: Edward Elgar.

Anderson, R.C., and Reeb, D.M. (2003). "Founding-Family Ownership and Firm Performance: Evidence from the S&P 500." *Journal of Finance*, Vol. 58, No. 3, pp. 1301–1327.

Andersson, F. (2002). "Pooling Reputations." *International Journal of Industrial Organization*, Vol. 20, No. 5, pp. 715–730.

Angulo, A.M., Gil, J.M., and Gracia, A. (2001). "The Demand for Alcoholic Beverages in Spain." *Agricultural Economics*, Vol. 26, pp. 71–83.

Angulo, A.M., Gil, J.M., Gracia, A., and Sanchez, M. (2000). "Hedonic Prices for Spanish Red Quality Wine." *British Food Journal*, Vol. 102, No. 7, pp. 481–493.

Aristei, D., Perali, F., and Pieroni, L. (2008). "Cohort, Age and Time Effects in Alcohol Consumption by Italian Households: A Double-Hurdle Approach." *Empirical Economics*, Vol. 35, No. 1, pp. 29–61.

Arnaud, C. (1991). "Le Vin et l'Organisation Commune de Marché Entre Paris et Bruxelles: un Dialogue Quelquefois Difficile." *Économie Rurale*, Vol. 204, No. 204, pp. 3–10.

Arrow, K.J. (1965). *Aspects of the Theory of Risk Bearing.* Yrjo Jahnsson Lectures, Helsinki. (Reprinted in *Essays in the Theory of Risk Bearing*, Chicago, IL: Markham Publishing, 1971).

Arrow, K.J. (1968). "The Economics of Moral Hazard: Further Comment." *American Economic Review*, Vol. 58, No. 3, pp. 537–539.

Arrow, K. (1972). "Gifts and Exchanges." *Philosophy & Public Affairs*, Vol. 1, No. 4, pp. 343–362.

Arthur, B. (1990). "Positive Feedbacks in the Economy." *Scientific American*, Vol. 262, No. 2, pp. 92–99.

Asero, V., and Patti, S. (2009). *From Wine Production to Wine Tourism Experience: The Case of Italy.* AAWE Working Paper No. 52.

Ashenfelter, O. (2008). "Predicting the Quality and Prices of Bordeaux Wine." *Economic Journal*, Vol. 118, pp. 174–184.

Ashenfelter, O., and Storchmann, K.H. (2010). "Using Hedonic Models of Solar Radiation and Weather to Assess the Economic Effect of Climate Change: The Case of Mosel Valley Vineyards." *Review of Economics and Statistics*, Vol. 92, No. 2, pp. 333–349.

Ashenfelter, O., and Storchmann, K.H. (2016). "Wine and Climate Change: A Review of the Economic Implications." *Journal of Wine Economics*, Vol. 11, No. 1, pp. 105–138.

Associazione Italiana Sommelier (AIS) (2005a). *Il Mondo del Sommelier.* Milan, Italy.

Associazione Italiana Sommelier (AIS) (2005b). *L'Italia dei Vini DOCG*. Milan, Italy.

Associazione Italiana Sommelier (AIS) (2012). *Il Vino nel Mondo*. Milan, Italy.

Aylward, D. (2008). "Towards a Cultural Economy Paradigm for the Australian Wine Industry." *Prometheus*, Vol. 26, No. 4, pp. 373–385.

Baan, R., Straif, K., Grosse, Y., Secretan, B., El Ghissassi, F., Bouvard, V., Altieri, A., and Cogliano, V. (2007). "Carcinogenicity of Alcoholic Beverages." *The Lancet Oncology*, Vol. 8, No. 4, pp. 292–293.

Baldi, L., Vandone, D., and Peri, M. (2010). "Is Wine a Financial Parachute?" In *System Dynamics and Innovation in Food Networks*, ed. M. Fritz, U. Rickert, and G. Schiefer, pp. 472–487. Bonn, Germany: University of Bonn–ILB Press.

Balogh, P., Békési, D., Gorton, M., Popp, J., and Lengyel, P. (2016). "Consumer Willingness to Pay for Traditional Food Products." *Food Policy*, Vol. 61, pp. 176–184.

Balsa, A.I., Giuliano, L.M., and French, M.T. (2011). "The Effects of Alcohol Use on Academic Achievement in High School." *Economics of Education Review*, Vol. 30, No. 1, pp. 1–15.

Baltieri, D.A., Daró, F.R., Ribeiro, P.R., and De Andrade, A.G. (2009). "The Role of Alcoholic Beverage Preference in the Severity of Alcohol Dependence and Adherence to the Treatment." *Alcohol*, Vol. 43, pp. 185–195.

Banerjee, A., Duflo, E., Postel-Vinay, G., and Watts, T. (2010). "Long-Run Health Impacts of Income Shocks: Wine and Phylloxera in Nineteenth-Century France," *The Review of Economics and Statistics*, Vol. 92, No. 4, pp. 714–728.

Banfield, E.C. (1958). *The Moral Basis of a Backward Society*. New York: Free Press.

Bang, H. (1998). "Analysing the Impact of the Liquor Industry's Lifting on the Ban on Broadcast Advertising." *Journal of Public Policy and Marketing*, Vol. 17, No. 1, pp. 132–138.

Barber, N., Taylor, C., and Strick, S. (2009). "Wine Consumers' Environmental Knowledge and Attitudes: Influence on Willingness to Purchase." *International Journal of Wine Research*, Vol. 1, pp. 59–72.

Bar-Isaac, H., and Tadelis, S. (2008). "Seller Reputation." *Foundations and Trends in Microeconomics*, Vol. 4, No. 4, pp. 273–351.

Barnes, L.B., and Hershon, S.A. (1976). "Transferring Power in the Family Business." *Harvard Business Review*, Vol. 54, pp. 105–114.

Barnett, B.J. (2000). "The U.S. Federal Crop Insurance Program." *Canadian Journal of Agricultural Economics*, Vol. 48, pp. 539–551.

Barnett, B.J., and Mahul, O. (2007). "Weather Index Insurance for Agriculture and Rural Areas in Lower-Income Countries." *American Journal of Agricultural Economics*, Vol. 89, No. 5, pp. 1241–1247.

Barney, J. (1991). "Firm Resources and Sustained Competitive Advantage." *Journal of Management*, Vol. 17, pp. 99–120.

Barro, R., and Gordon, D. (1983). "Rules, Discretion, and Reputation in a Model of Monetary Policy." *Journal of Monetary Economics*, Vol. 12, pp. 101–121.

Bartlett, C.A. (2009). *Global Wine War 2009: New World versus Old*. Harvard Business School Case Study No. 303–056.

Bartoli, P. (1986). "Les Primes d'Arrachage et la Régression du Vignoble. Une Analyse d'Impact de la Politique Viticole." *Économie Rurale*, Vol. 175, pp. 3–19.

Barzel, Y. (1976). "An Alternative Approach to the Analysis of Taxation." *Journal of Political Economy*, Vol. 84, No. 6, pp. 1177–1197.

Batty, D., Lewars, H., Emslie, C., Benzeval, M., and Hunt, K. (2008). "Problem Drinking and Exceeding Guidelines for 'Sensible' Alcohol Consumption in Scottish Men: Associations with Life Course Socioeconomic Disadvantage in a Population-Based Cohort Study." *BMC Public Health*, Vol. 8, pp. 1–7.

Baughman, R., Conlin, M., Dickert-Conlin, S., and Pepper, J. (2001). "Slippery When Wet: The Effects of Local Alcohol Access on Highway Safety." *Journal of Health Economics*, Vol. 20, No. 6, pp. 1089–1096.

Baum, H.M., Wells, J.K., and Lund, A.K. (1991). "The Fatality Consequences of the 65 mph Speed Limits, 1989." *Journal of Safety Research*, Vol. 22, pp. 171–177.

Baum-Baicker, C. (1985). "The Psychological Benefits of Moderate Alcohol Consumption: A Review of the Literature." *Drug and Alcohol Dependence*, Vol. 15, No. 4, pp. 305–322.

Baumol, W.J. (1986). "Unnatural Value: Or Art Investment as Floating Crap Game." *American Economic Review—Papers and Proceedings*, Vol. 76, No. 2, pp. 10–14.

Baumol, W.J., and Bowen, W.G. (1966). *Performing Arts: The Economic Dilemma.* Hartford, CT: Twentieth Century Fund.

Becattini, G. (1990). "The Marshallian Industrial District as a Socio-Economic Notion." In *Industrial Districts and Interfirm Cooperation in Italy*, ed. F. Pyke, G. Becattini, and W. Sengenberger, pp. 38–51. Geneva, Switz.: International Institute for Labour Studies.

Becattini, G. (2002). "From Marshall's to the Italian 'Industrial Districts.' A Brief Critical Reconstruction." In *Complexity and Industrial Clusters*, ed. A. Quadrio Curzio and M. Fortis, pp. 83–106. Heidelberg, Germany: Physica-Verlag.

Becchetti, L., Bruni, L., and Zamagni, S. (2011). *Microeconomia.* Bologna, Italy: Il Mulino.

Becchetti, L., and Castriota, S. (2011). "Wage Differentials in Italian Social Enterprises." *Economia Politica*, Year 28, No. 3, pp. 323–367.

Becchetti, L., Castriota, S., and Conzo, P. (2015). "Quantitative Analysis of the Impacts of Fair Trade." In *Handbook of Research on Fair Trade*, ed. L. Raynolds and E. Bennett. Cheltenham, UK: Edward Elgar.

Becchetti, L., Castriota, S., and Depedri, S. (2014). "Working in the For-Profit versus Not-For-Profit Sector: What Difference Does It Make? An Inquiry on Preferences of Voluntary and Involuntary Movers." *Industrial and Corporate Change*, Vol. 23, No. 4, pp. 1087–1120.

Becchetti, L., Castriota, S., and Giuntella, O. (2010). "The Effects of Age and Job Protection on the Welfare Costs of Inflation and Unemployment." *European Journal of Political Economy*, Vol. 26, No. 1, pp. 137–146.

Becchetti, L., Castriota, S., and Tortia, E. (2013). "Productivity, Wages and Intrinsic Motivations." *Small Business Economics*, Vol. 41, No. 2, pp. 379–399.

Becchetti, L., and Ciampoli, N. (2012). *What Is New in the Finance-Growth Nexus: OTC Derivatives, Bank Assets and Growth.* AICCON Working Paper No. 110.

Becker, G.S. (1962). "Investment in Human Capital: A Theoretical Analysis." *The Journal of Political Economy*, Vol. 70, No. 5, pp. 9–49.

Becker, G.S. (1968). "Crime and Punishment: An Economic Approach." *The Journal of Political Economy*, Vol. 76, pp.169–217.

Becker, G.S., and Murphy, K.M. (1988). "A Theory of Rational Addiction." *The Journal of Political Economy*, Vol. 96, No. 4, pp. 675–700.

Beckman, L.J., and Ackerman, K.T. (1995). "Women, Alcohol, and Sexuality." *Recent Developments in Alcoholism*, Vol. 12, pp. 267–285.

Bénabou, R., and Tirole, J. (2003). "Intrinsic and Extrinsic Motivation." *Review of Economic Studies*, Vol. 70, pp. 489–520.

Benfratello, L., Piacenza, M., and Sacchetto, S. (2009). "Taste or Reputation: What Drives Market Prices in the Wine Industry? Estimation of a Hedonic Model for Italian Premium Wines." *Applied Economics*, Vol. 41, pp. 2197–2209.

Ben-Ner, A., and Ellman, M. (2013). "The Contributions of Behavioural Economics to Understanding and Advancing the Sustainability of Worker Cooperatives." *Journal of Entrepreneurial and Organizational Diversity*, Vol. 2, No. 1, pp. 75–100.

Ben-Ner, A., and Putterman, L. (1998). *Economics, Values, and Organization*. Cambridge, MA: Cambridge University Press.

Bentzen, J., Leth-Sorensen, S., and Smith, V. (2002). *Prices of French Icon Wines and the Business Cycle: Empirical Evidence from Danish Wine Auctions*. CIES Discussion Paper No. 0224.

Bentzen, J., and Smith, V. (2002). *Wine Prices in the Nordic Countries: Are They Lower Than in The Region of Origin?* CIES Discussion Paper No. 0223.

Berger, M.C., and Leigh, J.P. (1988). "The Effect of Alcohol Use on Wages." *Applied Economics*, Vol. 20, No. 10, pp. 1343–1351.

Berle, A.A., and Means, G.C. (1932). *The Modern Corporation and Private Property*. New York: Macmillan.

Berliner, B. (1982). *Limits of Insurability of Risks*. Englewood Cliffs, NJ: Prentice-Hall.

Bernoulli, J. (1713). Chapter 4. In *Ars Conjectandi: Usum & Applicationem Praecedentis Doctrinae in Civilibus, Moralibus & Oeconomicis*, trans. Oscar Sheynin. http://www.sheynin.de/download/bernoulli.pdf.

Berthomeau, J. (2001). "Comment Mieux Positionner les Vins Français sur les Marchés d'Exportation?" Paris: Ministère de l'Agriculture, ONIVINS.

Berument, H., Akdi, Y., and Atakan, C. (2005). "An Empirical Analysis of Istanbul Stock Exchange Sub-Indexes." *Studies in Nonlinear Dynamics and Econometrics*, Vol. 9, No. 3, Art. 5.

Bijman, J., Iliopoulos, C., Poppe, K.J., Gijselinckx, C., Hagedorn, K., Hanisch, M., Hendrikse, G.W.J., Kühl, R., Ollila, P., Pyykkönen and P van der Sangen, G. (2012). *Support for Farmers' Cooperatives*. European Commission Final Report, Brussels, Belgium.

Birebent, P. (2007). *Hommes, Vignes et Vins de l'Algérie Française: 1830–1962*. Nice, France: Jacques Gandini Editions.

Björk, T. (1998). *Arbitrage Theory in Continuous Time*. Oxford, UK: Oxford University Press.

Blundell-Wignall, A., and Atkinson, P. (2011). "Global SIFIs, Derivatives, and Financial Stability." *OECD Journal: Financial Market Trends*, Vol. 2011, No. 1, pp. 1–34.

Bodie, Z., Kane, A., and Marcus, A.J. (2005). *Investments* (6th ed.). New York: McGraw-Hill.

Bogetoft, P. (2005). "An Information Economic Rationale for Cooperatives." *European Review of Agricultural Economics*, Vol. 32, No. 2, pp. 191–217.

Bombrun, H., and Sumner, D.A. (2003). *What Determines the Price of Wine? The Value of Grape Characteristics and Wine Quality Assessments.* University of California Agricultural Issues Center No. 18.

Bonin, J.P., Jones, D.C., and Putterman, L. (1993). "Theoretical and Empirical Studies of Producer Cooperatives: Will Ever the Twain Meet?" *Journal of Economic Literature*, Vol. 31, No. 3, pp. 1290–1320.

Bonomo, Y.A., Bowes, G., Coffey, C., Carlin, J.B., and Patton, G.C. (2004). "Teenage Drinking and the Onset of Alcohol Dependence: A Cohort Study over Seven Years." *Addiction*, Vol. 99, pp. 1520–1528.

Bonroy, O., Garapin, A., Hamilton, S., and Souza Monteiro, D.M. (2018). "Free-Riding on Product Quality in Cooperatives: Lessons from an Experiment." *American Journal of Agricultural Economics*, Vol. 101, No. 1, pp. 89–108.

Bontemps, C., Bouamra-Mechemache, Z., and Simioni, M. (2013). "Quality Labels and Firm Survival: Some First Empirical Evidence." *European Review of Agricultural Economics*, Vol. 40, No. 3, pp. 413–439.

Borch, K. (1962). "Equilibrium in a Reinsurance Market." *Econometrica*, Vol. 30, No. 3, pp. 424–444.

Bordieu, P. (1985). "The Forms of Capital." In *Handbook of Theory and Research for the Sociology of Education*, ed. J.C. Richardson. New York: Greenwood.

Borges, G., Cherpitelb, C., and Mittlemanc, M. (2004). "Risk of Injury After Alcohol Consumption: A Case-Crossover Study in the Emergency Department." *Social Science & Medicine*, Vol. 58, pp. 1191–1200.

Borzaga, C., and Tortia, E. (2006). "Worker Motivations, Job Satisfaction, and Loyalty in Public and Nonprofit Social Services." *Nonprofit and Voluntary Sector Quarterly*, Vol. 35, No. 2, pp. 225–248.

Box, G.E.P., and Cox, D.R. (1964). "An Analysis of Transformations." *Journal of the Royal Statistical Society, Series B*, Vol. 26, pp. 211–252.

Brazil, N., and Kirk, D.S. (2016). "Uber and Metropolitan Traffic Fatalities in the United States." *American Journal of Epidemiology*, Vol. 184, No. 3, pp. 192–198.

Brecklin, L.R. (2002). "The Role of Perpetrator Alcohol Use in the Injury Outcomes of Intimate Assaults." *Journal of Family Violence*, Vol. 17, No. 3, pp 185–197.

Brown, D.B., Maghsoodloo, S., and McArdle, M.E. (1990). "The Safety Impact of the 65 mph Speed Limit: A Case Study Using Alabama Accident Records." *Journal of Safety Research*, Vol. 21, pp. 125–139.

Bruwer, J. (2003). "South African Wine Routes: Some Perspectives on the Wine Tourism." *Tourism Management*, Vol. 24, pp. 423–435.

Bukenya, O. (2008). "Do Fluctuations in Wine Stocks Affect Wine Prices?" In *Commodity Modeling and Pricing: Methods for Analyzing Resource Market Behavior*, ed. P.V. Schaeffer, pp. 136–166. Hoboken, NJ: Wiley.

Buono, R., and Vallariello, G. (2002). "Introduzione e Diffusione della Vite (Vitis vinifera L.) in Italia." *Delpinoa*, Vol. 44, pp. 39–51.

Burdin, G., and Dean, A. (2009). "New Evidence on Wages and Employment in Worker Cooperatives Compared with Capitalist Firms." *Journal of Comparative Economics*, Vol. 37, No. 4, pp. 517–533.

Burgdorf, J., Lennon, C., and Teltser, K. (2019). *Do Ridesharing Services Increase Alcohol Consumption?* Andrew Young School of Policy Studies Research Paper Series No. 19-23.

Burkart, M., Panunzi, F., and Shleifer, A. (2003). "Family Firms." *Journal of Finance*, Vol. 58, pp. 2167–2202.

Burke, S., Schmied, V., and Montrose, M. (2006). *Parental Alcohol Misuse and the Impact on Children.* Western Sydney University Working Paper.

Burton, B.J., and Jakobsen, J.P. (1999). "Measuring Returns on Investments in Collectibles." *Journal of Economic Perspectives*, Vol. 13, No. 4, pp. 193–212.

Burton, B.J., and Jakobsen, J.P. (2001). "The Rate of Return on Investment in Wine." *Economic Inquiry*, Vol. 39, No. 3, pp. 337–350.

Cabral, L. (2005). "The Economics of Trust and Reputation: A Primer." New York University, Mimeo.

Cabral, L., and Hortaçsu, A. (2010). "The Dynamics of Seller Reputation: Evidence from eBay." *Journal of Industrial Economics*, Vol. 58, No. 1, pp. 54–78.

Cacchiarelli, L., Carbone, A., Laureti, T., and Sorrentino, A. (2016). "The Value of the Certifications of Origin: A Comparison Between the Italian Olive Oil and Wine Markets." *British Food Journal*, Vol. 118, No. 4, pp. 824–839.

Cadbury, A. (2000). *Family Firms and Their Governance: Creating Tomorrow's Company from Today's.* London: Egon Zehnder International.

Caesar, J. (1869). *The Gallic Wars*, trans. W. A. McDevitte and W. S. Bohn. n.p. http://classics.mit.edu/Caesar/gallic.html.

Cakir, M., and Balagtas J.V. (2012). "Estimating Market Power of U.S. Dairy Cooperatives in the Fluid Milk Market." *American Journal Agricultural Economics*, Vol. 94, No. 3, pp. 647–658.

Caldwell, T., Rodgers, B., Clark, C., Jefferis, B., Stansfeld, S.A., and Power, C. (2008). "Life-course Socioeconomic Predictors of Midlife Drinking Patterns, Problems and Abstention: Findings from the 1958 British Birth Cohort Study." *Drug and Alcohol Dependence*, Vol. 95, pp. 269–278.

Calfee, J.E., and Schergata, C. (1994). "The Influence of Advertising on Alcohol Consumption: A Literature Review and an Econometric Analysis of Four European Nations." *International Journal of Advertising*, Vol. 13, No. 4, pp. 287–310.

Camargo, C.A., Hennekens, C.H., Gaziano, J.M., Glynn, R.J., Manson, J.E., and Stampfer, M.J. (1997a). "Prospective Study of Moderate Alcohol Consumption and Mortality in US Male Physicians." *Archives of Internal Medicine*, Vol. 157, pp. 79–85.

Camargo, C.A., Stampfer, M.J., Glynn, R.J., Gaziano, J.M., Manson, J.E., Goldhaber, S.Z., and Henneken, C.H. (1997b). "Prospective Study of Moderate Alcohol Consumption and Risk of Peripheral Arterial Disease in US Male Physicians." *Circulation*, Vol. 95, pp. 577–580.

Campbell, J., and Cochrane, J. (1999). "By Force of Habit: A Consumption-Based Explanation of Aggregate Stock Market Behavior." *Journal of Political Economy*, Vol. 107, No. 2, pp. 205–251.

Capitello, R., and Agnoli, L. (2009). *Development of Strategic Options for Italian Wine Cooperatives through a New Membership Integration Pattern.* Working Paper.

Cardebat, J.M. (2017). *Economie du Vin.* Paris: La Decouverte.

Cardebat, J.M., Faye, B., Le Fur, E., and Storchmann, K. (2017). "The Law of One Price? Price Dispersion on the Auction Market for Fine Wine." *Journal of Wine Economics*, Vol. 12, No. 3, pp. 302–331.

Cardebat, J.M., and Figuet, J.M. (2004). "What Explains Bordeaux Wine Prices?" *Applied Economics Letters*, Vol. 11, pp. 293–296.

Cardebat, J.M., and Figuet, J.M. (2010). "Is Bordeaux Wine an Alternative Financial Asset?" Presented at the Fifth International Academy of Wine Business Research Conference, February 8–10, Auckland, New Zealand.

Cardebat. J.M., and Jiao, L. (2018). "The Long-Term Financial Drivers of Fine Wine Prices: The Role of Emerging Markets." *The Quarterly Review of Economics and Finance*, Vol. 67, pp. 347–361.

Carpenter, C., and Dobkin, C. (2011). "The Minimum Legal Drinking Age and Public Health." *Journal of Economic Perspectives*, Vol. 25, No. 2, pp. 133–156.

Carrell, S.E., Hoekstra, M., and West, J.E. (2011). "Does Drinking Impair College Performance? Evidence from a Regression Discontinuity Approach." *Journal of Public Economics*, Vol. 95, No. 1, pp. 54–62.

Casswell, S., Pledger, M., and Hooper, R., (2003). "Socioeconomic Status and Drinking Patterns in Young Adults." *Addiction*, Vol. 98, No. 5, pp. 601–610.

Casta, J.F., Ramond, O.J., and Escaffre, L. (2008). *Economic Properties of Intangibles Recognized Under Domestic Accounting Standards: Evidence from Pre-Ifrs Major European Markets*. Working paper.

Castaldi, R., Cholette, S., and Hussain, M. (2006). *A Country-Level Analysis of Competitive Advantage in the Wine Industry*. DEIAgra Working Paper No. 6002.

Castriota, S. (2015). *Economia del Vino*. Milan, Italy: EGEA-Bocconi University Press.

Castriota, S. (2018). *Does Excellence Pay Off? Quality, Reputation and Vertical Integration in the Wine Market*. AAWE Working Paper N. 227.

Castriota, S., and Delmastro, M. (2009). "L'Europa non Apprezza il Buon Vino." *La Voce*, July 30.

Castriota, S., and Delmastro, M. (2010). "E Bruxelles Mette il Tappo al Vino." *La Voce*, July 30.

Castriota, S., and Delmastro, M. (2012). "Seller Reputation: Individual, Collective, and Institutional Factors." *Journal of Wine Economics*, Vol. 7, No. 1, pp. 49–69.

Castriota, S., and Delmastro, M. (2015). "The Economics of Collective Reputation: Evidence from the Wine Industry." *American Journal of Agricultural Economics*, Vol. 97, No. 2, pp. 469–489.

Castriota, S., Delmastro, M., and Curzi, D. (2013). "Tasters' Bias in Wine Guides' Quality Evaluations." *Applied Economics Letters*, Vol. 20, No. 12, pp. 1174–1177.

Castriota-Scanderbeg, A., Hagberg, G.E., Cerasa, A., Committeri, G., Galati, G., Patria, F., Pitzalis, S., Caltagirone, C., and Frackowiak, R. (2005). "The Appreciation of Wine by Sommeliers: A Functional Magnetic Resonance Study of Sensory Integration." *Neuroimage*, Vol. 25, No. 2, pp. 570–578.

Catalano, R., Dooley, D., Wilson, G., and Hough, R. (1993). "Job Loss and Alcohol Abuse: A Test Using Data from the Epidemiologic Catchment Area Project." *Journal of Health and Social Behavior*, Vol. 34, No. 3, pp. 215–225.

Chang, H.S., and Bettington, N. (2001). *Demand for Wine in Australia: Systems versus Single Equation Approach*. University of New England Working Paper No. 2001–5.

Charles, K.K., and DeCicca, P. (2008). "Local Labor Market Fluctuations and Health: Is There a Connection and for Whom?" *Journal of Health Economics*, Vol. 27, pp. 1532–1550.

Charters, S., and Ali-Knight, J. (2002). "Who Is the Wine Tourist?" *Tourism Management*, Vol. 23, pp. 311–319.

Checkoway, H., Powers, K., Smith-Weller, T., Franklin, G.M., Longstreth, W.T., and Swanson, P.D. (2002). "Parkinson's Disease Risks Associated with Cigarette Smoking, Alcohol Consumption, and Caffeine Intake." *American Journal of Epidemiology*, Vol. 155, No. 8, pp. 732–738.

Chevalier, J., and Mayzlin, D. (2006). "The Effect of Word of Mouth on Sales: Online Book Reviews." *Journal of Marketing Research*, Vol. 43, pp. 345–354.

Chew, K.-K., Bremner, A., Stuckey, B., Earle, C., and Jamrozik, K. (2009). "Alcohol Consumption and Male Erectile Dysfunction: An Unfounded Reputation for Risk?" *Journal of Sexual Medicine*, Vol. 6, No. 5, pp. 1386–1394.

Chi, G., Zhoua, X., McClurea, T.E., Gilbert, P.A., Cosby, A.G., Zhang, L., Robertson, A.A., and Levinson, D. (2011). "Gasoline Prices and Their Relationship to Drunk-Driving Crashes." *Accident Analysis and Prevention*, Vol. 43, pp. 194–203.

Chiva-Blanch, G., Arranz, S., Lamuela-Raventos, R.M., and Estruc, R. (2013). "Effects of Wine, Alcohol and Polyphenols on Cardiovascular Disease Risk Factors: Evidences from Human Studies." *Alcohol Alcohol*, Vol. 48, No. 3, pp. 270–277.

Choi, I.G., Son, H.G., Yang, B.H., Kim, S.H., Lee, J.S., Chai, Y.G., Son, B.K., Kee, B.S., Park, B.L., Kim, L.H., Choi, Y.H., and Shin, H.D. (2005). "Scanning of Genetic Effects of Alcohol Metabolism Gene (ADH1B and ADH1C) Polymorphisms on the Risk of Alcoholism." *Human Mutation*, Vol. 26, pp. 224–234.

Christakis, N.A., and Fowler, J.H. (2008). "The Collective Dynamics of Smoking in a Large Social Network." *New England Journal of Medicine*, Vol. 358, No. 21, pp. 2249–2258.

Christiansen, C. (1953). *Management Succession in Small and Growing Enterprises*. Boston, MA: Division of Research, Graduate School of Business Administration, Harvard University.

Ciaschini, M., and Socci, C. (2005). "Multiplier Impact of Wine Activity on Inter-Industry Interactions." *Agricultural Economics Review*, Vol. 6, No. 2, pp. 61–74.

Clark, D.B., and Bukstein, O.G. (1998). "Psychopathology in Adolescent Alcohol Abuse and Dependence." *Alcohol Health and Research World*, Vol. 22, No. 2, pp. 117–121.

Cleary, M., Hunt, G.E., Matheson, S., and Walter, G. (2009). "Psychosocial Treatments for People with Co-occurring Severe Mental Illness and Substance Misuse: Systematic Review." *Journal of Advanced Nursing*, Vol. 65, pp. 238–258.

Clements, K.W., and Johnson, L.W. (1983). "The Demand for Beer, Wine, and Spirits: A Systemwide Analysis." *The Journal of Business*, Vol. 56, No. 3, pp. 273–304.

Cloninger, C.R. (1987). "A Systematic Method for Clinical Description and Classification of Personality Variants: A Proposal." *Archives of General Psychiatric*, Vol. 44, No. 6, pp. 573–588.

Coase, R.H. (1960). "The Problem of Social Cost." *Journal of Law and Economics*, Vol. 3, pp. 1–44.

Cohen, A., and Einav, L. (2003). "The Effects of Mandatory Seat Belt Laws on Driving Behavior and Traffic Fatalities." *Review of Economics and Statistics*, Vol. 85, No. 4, pp. 828–843.

Cohen, S., Tyrrell, D.A., Russell, M.A., Jarvis, M.J., and Smith, A.P. (1993). "Smoking, Alcohol Consumption, and Susceptibility to the Common Cold." *American Journal of Public Health*, Vol. 83, No. 9, pp. 1277–1283.

Coker, A.L., Hanks, J.S., Eggleston, K.S., Risser, J., Tee, P.G., Chronister, K.J., Troisi, C.L., Arafat, R., and Franzini, L. (2006). "Social and Mental Health Needs Assessment of Katrina Evacuees." *Disaster Management & Response*, Vol. 4, No. 3, pp. 88–94.

Coleman, J.S. (1990). *Foundations of Social Theory*. Cambridge, MA: Harvard University Press.

Coleman, L., and Cater, S. (2003). "What Do We know About Young People's Use of Alcohol?" *Education and Health*, Vol. 21, pp. 50–55.

Colen, L., and Swinnen, J. (2010). *Beer Drinking Nations: The Determinants of Global Beer Consumption*. LICOS Discussion Paper 270/2010. (Reprinted in *The Economics of Beer*, ed. J. Swinnen, Oxford, UK: Oxford University Press, 2011.)

Colman, T. (2008). *Wine Politics*. Berkeley: University of California Press.

Colombo, M.G., and Delmastro, M. (2004). "Delegation of Authority in Business Organizations: An Empirical Test." *Journal of Industrial Economics*, Vol. 52, No. 1, pp. 53–80.

Combris, P., Lecocq, S., and Visser, M. (1997). "Estimation of a Hedonic Price Equation for Bordeaux Wine: Does Quality Matter?" *Economic Journal*, Vol. 107, No. 441, pp. 390–402.

Commission of the European Communities (2008). "Commission Regulation (EC) No. 555/2008 of 27 June 2008 laying down detailed rules for implementing Council Regulation (EC) No. 479/2008 on the common organisation of the market in wine as regards support programmes, trade with third countries, production potential and on controls in the wine sector." *Official Journal of the European Union*, June 30. https://eur-lex.europa.eu/legal-content/EN/TXT/HTML/?uri=CELEX:32008R0555&from=en.

Conigrave, K.M., Hu, B.F., Camargo, C.A. Jr., Stampfer, M.J., Willett, W.C., and Rimm, E.B. (2001). "A Prospective Study of Drinking Patterns in Relation to Risk of Type 2 Diabetes Among Men." *Diabetes*, Vol. 50, pp. 2390–2395.

Cook, M.L. (1994). "The Role of Management Behaviour in Agricultural Co-operatives." *Journal of Agricultural Co-operation*, Vol. 9, pp. 42–58.

Cook, M.L (1995). "The Future of U.S. Agricultural Cooperatives: A Neo-Institutional Approach." *American Journal of Agricultural Economics*, Vol. 77, pp. 1153–1159.

Cook, P.J., and Durrance, C.P. (2013). "The Virtuous Tax: Lifesaving and Crime-Prevention Effects of the 1991 Federal Alcohol-Tax Increase." *Journal of Health Economics*, Vol. 32, No. 1, pp. 261–267.

Cook, P.J., and Moore, M.J. (1993). "Economic Perspectives on Alcohol-Related Violence." In *Alcohol-Related Violence: Interdisciplinary Perspectives and Research Directions*, ed. S.E. Martin, pp. 193–212. Rockville, MD: National Institute on Alcohol Abuse and Alcoholism, No. 93–3496.

Cook, P.J., and Moore, M.J. (2000). "Alcohol." In *Handbook of Health Economics* (Vol. 1, Part B), ed. A.J. Culyer and J.P. Newhouse, pp. 1629–1673, Amsterdam, Netherlands: Elsevier.

Cook, P.J., and Moore, M.J. (2001). "Environment and Persistence in Youthful Drinking Patterns." In *Risky Behavior among Youths: An Economic Analysis*, ed. Jonathan Gruber, pp. 375–438. Chicago, IL: University of Chicago Press.

Corrado, R., and Oderici, V. (2008). *Tradition or Control? Change of Product Categories in the Italian Wine Industry*. University of Bologna Working Paper.

Corsi, A., and Ashenfelter, O. (2000). *Predicting Italian Wines Quality from Weather Data and Experts' Ratings*. Working Paper.

Corsi, A., Pomarici, E., and Sardone, R. (2004). "Italy." In *The World's Wine Markets: Globalization at Work*, ed. K. Anderson. Cheltenham, UK: Edward Elgar.

Corsi, A., and Strøm, S. (2013). "The Price Premium for Organic Wines: Estimating a Hedonic Farm-Gate Price Equation." *Journal of Wine Economics*, Vol. 8, No. 1, pp. 29–48.

Costanigro, M., McCluskey, J.J., and Goemans, C. (2010). "The Economics of Nested Names: Name Specificity, Reputations, and Price Premia." *American Journal of Agricultural Economics*, Vol. 92, No. 5, pp. 1339–1350.

Costanigro, M., McCluskey, J.J., and Mittelhammer, R.C. (2007). "Segmenting the Wine Market Based on Price: Hedonic Regression When Different Prices Mean Different Products." *Journal of Agricultural Economics*, Vol. 58, No. 3, pp. 454–466.

Cottino, A. (1995). "Italy." In *International Handbook on Alcohol and Culture*, ed. D.B. Heath, pp. 156–167. Westport, CT: Greenwood Press.

Council of the European Union (2008). "Council Regulation (EC) No. 479/2008 of 29 April 2008 on the common organisation of the market in wine, amending Regulations (EC) No. 1493/1999, (EC) No. 1782/2003, (EC) No. 1290/2005, (EC) No. 3/2008 and repealing Regulations (EEC) No. 2392/86 and (EC) No. 1493/1999." *Official Journal of the European Union*, June 6. https://eur-lex.europa.eu/LexUriServ/LexUriServ.do?uri=OJ:L:2008:148:0001:0061:EN:PDF.

Court, A.T. (1939). *Hedonic Price Indexes with Automobile Examples in the Dynamics of Automobile Demand*. New York: The General Motors Corporation, pp. 99–117.

Craig, B., and Pencavel, J. (1992). "The Behaviour of Worker Cooperatives, the Plywood Companies of the Pacific North-East." *American Economic Review*, Vol. 82, No. 5, pp. 1083–1105.

Craig, B., and Pencavel, J. (1994). "The Empirical Performance of Orthodox Models of the Firm: Conventional Firms and Worker Cooperatives." *Journal of Political Economy*, Vol. 102, No. 4, pp. 718–744.

Cross, R., Färe, R., Grosskopf, S., and Weber, W.L. (2013). "Valuing Vineyards: A Directional Distance Function Approach." *Journal of Wine Economics*, Vol. 8, No. 1, pp. 69–82.

Cross, R., Plantinga, A.J., and Stavins, R.N. (2011). "What Is the Value of Terroir?" *American Economic Review: Papers & Proceedings*, Vol. 101, No. 3, pp. 152–156.

Cross, R., Plantinga, A.J., and Stavins, R.N. (2017). "Terroir in the New World: Hedonic Estimation of Vineyard Sale Prices in California." *Journal of Wine Economics*, Vol. 12, No. 3, pp. 282–301.

Crozet, M., Head, K., and Mayer, T. (2012). "Quality Sorting and Trade: Firm-level Evidence for French Wine." *Review of Economic Studies*, Vol. 79, No. 2, pp. 609–644.

Crum, R.M, Helzer, J.E., and Anthony, J.C. (1993). "Level of Education and Alcohol Abuse and Dependence in Adulthood: A Further Inquiry." *American Journal of Public Health*, Vol. 83, No. 6, pp. 830–837.

Cuellar, S.S., Karnowsky, D., and Acosta, F. (2009). "The Sideways Effect: A Test for Changes in the Demand for Merlot and Pinot Noir Wines." *Journal of Wine Economics*, Vol. 4, No. 2, pp. 1–14.

Cupples, L.A. (2000). "Effects of Smoking, Alcohol and APOE Genotype on Alzheimer Disease: The MIRAGE Study." *Alzheimer Report*, Vol. 3, pp. 105–114.

Cyr, D., and Kusy, M. (2007). "Canadian Ice Wine Production: A Case for the Use of Weather Derivatives." *Journal of Wine Economics*, Vol. 2, No. 2, pp. 145–167.

Dafoe, A., Renshon, J., and Huth, P. (2014). "Reputation and Status as Motives for War." *Annual Review of Political Science*, Vol. 17, pp. 371–393.

Daily, C.M., and Dollinger, M.J. (1992). "An Empirical Examination of Ownership Structure in Family and Professionally Managed Firms." *Family Business Review*, Vol. 5, No. 2, pp. 117–136.

Dave, D., and Saffer, H. (2008). "Alcohol Demand and Risk Preference." *Journal of Economic Psychology*, Vol. 29, pp. 810–831.

Dawson, D.A., Grant, B.F., Chou, S.P., and Pickering, R.P. (1995). "Subgroup Variation in U.S. Drinking Patterns: Results of the 1992 National Longitudinal Alcohol Epidemiologic Study." *Journal of Substance Abuse*, Vol. 7, pp. 331–344.

De Castro, J.M. (2000). "Eating Behavior: Lessons from the Real World of Humans." *Nutrition*, Vol. 16, pp. 800–813.

Dee, T.S. (2001). "Alcohol Abuse and Economic Conditions: Evidence from Repeated Cross-Sections of Individual Level Data." *Health Economics*, Vol. 10, pp. 257–270.

Defourney, J., Estrin, S., and Jones, D.C. (1985). "The Effects of Workers' Participation on Enterprise Performance." *International Journal of Industrial Organisation*, Vol. 3, pp. 197–217.

Defrancesco, E., Orrego, J.E., and Gennari, A. (2012). "Would 'New World' Wines Benefit from Protected Geographical Indications in International Markets? The Case of Argentinean Malbec." *Wine Economics and Policy*, Vol. 1, No. 1, pp. 63–72.

Dekonink, K., and Swinnen, J. (2012). *Peer Effects in Alcohol Consumption: Evidence from Russia's Beer Boom*. LICOS Discussion Paper 316/2012, Katholieke Universiteit Leuven.

Delmastro, M. (2007). "Un'Analisi sulle Determinanti della Qualità dei Vini." *L'Industria*, Year 28, No. 1, pp. 57–79.

Delmond, A.R., and McCluskey, J.J. (2018). *Impact of Regional Designations on Returns from Collective Reputations*. Washington State University Working Paper.

Delmond, A.R., McCluskey, J.J., and Winfree, J.A. (2018). "Product Quality and Reputation in Food and Agriculture." In *The Routledge Handbook of Agricultural Economics*, ed. G.L. Cramer, K.P. Paudel, and A. Schmitz. London: Routledge.

Detemple, J.B., and Zapatero, F. (1991). "Asset Prices in an Exchange Economy with Habit Formation." *Econometrica*, Vol. 59, No. 6, pp. 1633–1657.

Devalos, M.E., Fang, H., and French, M.T. (2012). "Easing the Pain of an Economic Downturn: Macroeconomic Conditions and Excessive Alcohol Consumption." *Health Economics*, Vol. 21, pp. 1318–1335.

Devine, L., and Lucey, B. (2015). "Is Wine a Premier Cru Investment?" *Research in International Business and Finance*, Vol. 34, pp. 33–51.

di Garoglio, P. G., and Desmireanu, B. (1961). "Vino." In *Enciclopedia Italiana—III Appendice*. Rome: Istituto della Enciclopedia Italiana. http://www.treccani.it/enciclopedia/vino_res-0fb08a98-87e9-11dc-8e9d-0016357eee51_(Enciclopedia_Italiana)-/.

Dilger, A. (2004). *In Vino Veritas: The Effects of Different Management Configurations in German Viniculture*. Discussion Paper, GEBA.

Dills, A.K., and Mulholland, S.E. (2017). *Ride-Sharing, Fatal Crashes, and Crime*. https://ssrn.com/abstract=2783797.

Dimson, E., Rousseau, P.L., and Spaenjers, C. (2015). "The Price of Wine." *Journal of Financial Economics*, Vol. 118, No. 2, pp. 431–449.

Dimson, E., and Spaenjers, C. (2011). "Ex post: The Investment Performance of Collectible Stamps." *Journal of Financial Economics*, Vol. 100, No. 2, pp. 443–458.

Di Tella, R., MacCulloch, R., and Oswald, A. (2001). "Preferences over Inflation and Unemployment: Evidence from Surveys of Happiness." *American Economic Review*, Vol. 91, pp. 335–341.

Di Vittorio, A., and Ginsburgh, V. (1994). *Pricing Red Wines of Médoc Vintages from 1949 to 1989 at Christie's Auctions*. Working Paper. (Republished in French in the *Journal de la Société Statistique de Paris*, Vol. 137 [1996], pp. 19–49.)

Djousse, L., Biggs, M.L., Mukamal, K.J., and Siscovick, D.S. (2007). "Alcohol Consumption and Type 2 Diabetes Among Older Adults: The Cardiovascular Health Study." *Obesity*, Vol. 15, pp. 1758–1765.

Doll, R., Peto, R., Hall, E., Wheatley, K., and Gray, R. (1994). "Mortality in Relation to Consumption of Alcohol: 13 Years' Observations on Male British Doctors." *British Medical Journal*, Vol. 309, pp. 911–918.

Donnelley, R.G. (1964). "The Family Business." *Harvard Business Review*, Vol. 42, pp. 93–105.

Donovan, C., and McEwan, R. (1995). "A Review of the Literature Examining the Relationship Between Alcohol Use and HIV Related Sexual Risk-Taking in Young People." *Addiction*, Vol. 90, No. 3, pp. 319–328.

Dorfman, M.S. (2004). *Introduction to Risk Management and Insurance* (8th ed.). Upper Saddle River, NJ: Prentice Hall.

Doucet, C. (2002). "Activités viticoles et devéloppement regional." Tesi di Dottorato, Université de Bordeaux IV, Bordeaux.

Droomers, M., Schrijvers, C.T., Stronks, K., and Mackenbach, J.P. (1999). "Educational Differences in Excessive Alcohol Consumption: The Role of Psychosocial and Material Stressors." *Preventive Medicine*, Vol. 29, pp. 1–10.

Drudi, F., and Prati, A. (2000). "Signalling Fiscal Regime Sustainability." *European Economic Review*, No. 44, pp. 1897–1930.

Dubois, P., and Nauges, C. (2010). "Identifying the Effect of Unobserved Quality and Expert Reviews in the Pricing of Experience Goods: Empirical Application on Bordeaux Wine." *International Journal of Industrial Organization*, Vol. 28, pp. 205–212.

Dunkelberg, W., Moore, C., Scott, J., and Stull, W. (2013). "Do Entrepreneurial Goals Matter? Resource Allocation in New Owner-Managed Firms." *Journal of Business Venturing*, Vol. 28, No. 2, pp. 225–240.

Durrance, C.P., Golden, S., Perreira, K., and Cook, P. (2011). "Taxing Sin and Saving Lives: Can Alcohol Taxation Reduce Female Homicides?" *Social Science & Medicine*, Vol. 73, pp. 169–176.

Dyack, B., and Goddard, E.W. (2001). "The Rise of Red and the Wane of White: Wine Demand in Ontario Canada." Presented at the Forty-Fifth Annual Conference of the Australian Agricultural and Resource Economics Society, January 23–25, Adelaide, Australia.

Economides, N. (1999). "Quality Choice and Vertical Integration." *International Journal of Industrial Organization*, Vol. 17, pp. 903–914.

Edenberg, H.J., and Foroud, T. (2006). "The Genetics of Alcoholism: Identifying Specific Genes Through Family Studies." *Addiction Biology*, Vol. 11, pp. 386–396.

Elder, R.W., Lawrence, B., Ferguson, A., Naimi, T.S., Brewer, R.D., Chattopadhyay, S.K., Toomey, T.L., and Fielding, J.E. (2010). "The Effectiveness of Tax Policy Interventions for Reducing Excessive Alcohol Consumption and Related Harms." *American Journal of Preventive Medicine*, Vol. 38, No. 2, pp. 217–229.

Ellig, J., and Wiseman, A.E. (2007). "The Economics of Direct Wine Shipping." *Journal of Law, Economics and Policy*, Vol. 3, No. 2, pp. 255–270.

Ely, J.C., and Välimäki, J. (2003). "Bad Reputation." *Quarterly Journal of Economics*, Vol. 118, No. 3, pp. 785–814.

Ely, M., Hardy, R., Longford, N.T., and Wadsworth, M.E. (1999). "Gender Differences in the Relationship Between Alcohol Consumption and Drink Problems Are Largely Accounted for by Body Water." *Alcohol and Alcoholism*, Vol. 34, No. 6, pp. 894–902.

Ettner, S.L. (1997). "Measuring the Human Cost of a Weak Economy: Does Unemployment Lead to Alcohol Abuse?" *Social Science & Medicine*, Vol. 44, No. 2, pp. 251–260.

European Commission (2004). *Ex-post evaluation of the Common Market Organisation for Wine*. November. https://ec.europa.eu/info/food-farming-fisheries/key-policies/common-agricultural-policy/evaluation-policy-measures/products-and-markets/ex-post-evaluation-common-market-organisation-wine_en.

European Commission (2009). "Market Situation (History)." Accessed 2013. http://ec.europa.eu/agriculture/markets/wine/facts/winehist_fr.pdf.

European Union (2012). "The EU System of Planting Rights: Main Rules and Effectiveness." Working Document, Directorate General of Agriculture and Rural Development, Directorate C—Economics of agricultural markets and single CMO, Brussels, Belgium.

Fabbri, F. (2011). *L'Italia Cooperativa. Centocinquant'Anni di Storia e di Memoria. 1861–2011*. Ediesse, Rome, Italy.

Fagan, J. (1990). "Intoxication and Aggression." *Crime and Justice*, Vol. 13, pp. 241–320.

Fama, E.F., and French, K.R. (1992). "The Cross Section of Expected Returns." *Journal of Finance*, Vol. 47, No. 2, pp. 427–465.

Fama, E.F., and French, K.R. (1993). "Common Risk Factors in the Returns on Stocks and Bonds." *Journal of Financial Economics*, Vol. 33, pp. 3–56.

Fama, E.F., and French, K.R. (1995). "Size and Book-to-Market Factors in Earnings and Returns." *Journal of Finance*, Vol. 50, No. 1, pp. 131–155.

Fazel, S., Khosla, V., Doll, H., and Geddes, J. (2008). "The Prevalence of Mental Disorders Among the Homeless in Western Countries: Systematic Review and Meta-Regression Analysis." *PLOS Medicine*, Vol. 5, pp. 1670–1681.

FEANTSA (2009). *Homelessness and Alcohol*. Briefing Paper.

Fehr, E., Klein, A., and Schmidt, K.M. (2007). "Fairness and Contract Design." *Econometrica*, Vol. 75, No. 1, pp. 121–154.

Fehr, E., and Zych, P.K. (1998). "Do Addicts Behave Rationally?" *Scandinavian Journal of Economics*, Vol. 100, No. 3, pp. 643–662.

Fernández-Olmos, M., Rosell-Martínez, J., and Espitia-Escuer, M.A. (2009). "Vertical Integration in the Wine Industry: A Transaction Costs Analysis on the Rioja DOCa." *Agribusiness*, Vol. 25, No. 2, pp. 231–250.

Ferraro, L. (2014). "Investire in Grandi Rossi? Meglio dell'Oro e del Petrolio." *Corriere della Sera*, March 6.

Ferrier, G.D., and Porter, P.K. (1991). "The Productive Efficiency of US Milk Processing Co-Operatives." *Journal of Agricultural Economics*, Vol. 42, No. 2, pp. 161–173.

Filatotchev, I., Piesse, J., and Lien, Y.C. (2005). "Corporate Governance and Performance in Publicly Listed, Family-Controlled Firms: Evidence from Taiwan." *Asia Pacific Journal of Management*, Vol. 22, No. 3, pp. 257–283.

Fink, G., Haiss, P., and Hristoforova, S. (2006): "Credit, Bonds, Stocks and Growth in Seven Large Economies." EuropaInstitut WU-Wien Working Paper No. 70.

Fischer, C., and Gil-Alana, L.A. (2009). "The Nature of the Relationship Between International Tourism and International Trade: The Case of German Imports of Spanish Wine." *Applied Economics*, Vol. 41, pp. 1345–1359.

Fisher, I. (1926). *Prohibition at Its Worst*. New York: MacMillan.

Fisher, J.C. (1993). *Advertising, Alcohol Consumption, and Abuse: A Worldwide Survey*. Westport, CT: Greenwood Press.

Fishman, A., Finkelshtain, I., Yacouel, N., and Simhon, A. (2008). *The Economics of Collective Brands*. Discussion Paper, 14.08, The Hebrew University of Jerusalem.

Fleckinger, P. (2007). *Collective Reputation and Market Structure: Regulating the Quality vs. Quantity Trade-Off*. Discussion Paper, Ecole Polytechnique.

Fleming, M.F., Barry, K.L., Manwell, L.B., Johnson, K., and London, R. (1997). "Brief Physician Advice for Problem Alcohol Drinkers: A Randomized Controlled Trial in Community-Based Primary Care Practices." *Journal of the American Medical Association*, Vol. 277, No. 13, pp. 1039–1045.

Fogarty, J.J. (2006). "The Return to Australian Fine Wine." *European Review of Agricultural Economics*, Vol. 33, No. 4, pp. 542–561.

Fogarty, J.J. (2007). *Rethinking Wine Investment in the UK and Australia*. AAWE Working Paper No. 6.

Fogarty, J.J. (2010a). "The Demand for Beer, Wine and Spirits: A Survey of the Literature." *Journal of Economic Surveys*, Vol. 24, No. 3, pp. 428–478.

Fogarty, J.J. (2010b). "Wine Investment and Portfolio Diversification Gains." *Journal of Wine Economics*, Vol. 5, No. 1, pp. 119–131.

Fogarty, J.J., and Jones, C. (2011). "Return to Wine: A Comparison of the Hedonic, Repeat Sales and Hybrid Approaches." *Australian Economic Papers*, Vol. 50, No. 4, pp. 147–156.

Fogarty, J.J., and Sadler, R. (2016). "To Save or Savor: A Review of Approaches for Measuring Wine as an Investment." *Journal of Wine Economics*, Vol. 9, No. 3, pp. 225–248.

Fombrun, C., and Shanley, M. (1990). "What's in a Name? Reputation Building and Corporate Strategy." *Academy of Management Journal*, Vol. 33, No. 2, pp. 233–258.

Foran, H.M., and O'Leary, K.D. (2008). "Alcohol and Intimate Partner Violence: A Meta-Analytic Review." *Clinical Psychology Review*, Vol. 28, No. 7, pp. 1222–1234.

Francioni, B., Vissak, T., and Musso, F. (2017). "Small Italian Wine Producers' Internationalization: The Role of Network Relationships in the Emergence of Late Starters." *International Business Review*, Vol. 26, No. 1, pp. 12–22.

Frank, R.H. (1996). "What Price the High Moral Ground?" *Southern Economic Journal*, Vol. 63, pp. 1–17.

Fraser, I. (2003). "The Role of Contracts in Wine Grape Supply Coordination: An Overview." *Agribusiness Review*, Vol. 11, No. 5, pp. 1–16.

Fraser, I. (2005). "Microeconometric Analysis of Wine Grape Supply Contracts in Australia." *Australian Journal of Agricultural and Resource Economics*, Vol. 49, pp. 23–46.

Freeman, D.G. (1999). "A Note on 'Economic Conditions and Alcohol Problems.'" *Journal of Health Economics*, Vol. 18, No. 5, pp. 659–668.

Freeman, R. (1997). "Working for Nothing: The Supply of Volunteer Labour." *Journal of Labour Economics*, Vol. 15, pp. 140–166.

French, M.T., Gumus, G., and Homer, J.F. (2009). "Public Policies and Motorcycle Safety." *Journal of Health Economics*, Vol. 28, pp. 831–838.

French, M.T., and Maclean, J.C. (2006). "Underage Alcohol Use, Delinquency, and Criminal Activity." *Health Economics*, Vol. 15, pp. 1261–1281.

French, M.T., Norton, E.C., Fang, H., and MacLean, J.C. (2010). "Alcohol Consumption and Body Weight." *Health Economics*, Vol. 19, No. 7, pp. 814–832.

French, M.T., and Zarkin, G.A. (1995). "Is Moderate Alcohol Use Related to Wages? Evidence from Four Worksites." *Journal of Health Economics*, Vol. 14, No. 3, pp. 319–344.

French M.T., and Zavala, S.K. (2007). "The Health Benefits of Moderate Drinking Revisited: Alcohol Use and Self-Reported Health Status." *American Journal of Health Promotion*, Vol. 21, No. 6, pp. 484–491.

Frey, B.S. (1993). "Does Monitoring Increase Effort? The Rivalry with Trust and Loyalty." *Economic Inquiry*, Vol. 31, No. 4, pp. 663–670.

Frey, B.S., and Pommerehne, W.W. (1989). *Muses and Markets: Explorations in the Economics of the Arts*. Oxford, UK: Basil Blackwell.

Friberg, R., and Grönqvist, E. (2012). "Do Expert Reviews Affect the Demand for Wine?" *American Economic Journal: Applied Economics*, Vol. 4, No. 1, pp. 193–211.

Frick, B. (2004). "Does Ownership Matter? Empirical Evidence from the German Wine Industry." *Kyklos*, Vol. 57, No. 3, pp. 357–386.

Frick, B., and Simmons, R. (2013). "The Impact of Individual and Collective Reputation on Wine Prices: Empirical Evidence from the Mosel Valley." *Journal of Business Economics*, Vol. 83, No. 2, pp. 101–119.

Friend, I., and Blume, M.E. (1975). "The Demand for Risky Assets." *American Economic Review*, Vol. 65, No. 5, pp. 900–922.

Fueller, T.D. (2011). "Moderate Alcohol Consumption and the Risk of Mortality." *Demography*, Vol. 48, pp. 1105–1125.

Fukuyama, F. (1995). *Trust: The Social Virtues and the Creation of Prosperity*. New York: Free Press.

Fulton, M. (1999). "Cooperatives and Member Commitment." *Finnish Journal of Business Economics*, Vol. 4, pp. 418–437.

Furubotn, E., and Pejovich, S. (1970). "Property Rights and the Behaviour of the Firm in a Socialist State: The Example of Yugoslavia." *Zeitschrift fur Nationalokonomie*, Vol. 30, pp. 431–454.

Gaeta, D., and Corsinovi, P. (2014). *Economics, Governance, and Politics in the Wine Market*. New York: Palgrave Macmillan.

Galanis, D.J., Joseph, C., Masaki, K.H., Petrovitch, H., Ross, G.W., and White, L. (2000). "A Longitudinal Study of Drinking and Cognitive Performance in Elderly Japanese American Men: The Honolulu-Asia Aging Study." *American Journal of Public Health*, Vol. 90, No. 8, pp. 1254–1259.

Gallet, C.A. (2007). "The Demand for Alcohol: A Meta-Analysis of Elasticities." *Australian Journal of Agricultural and Resource Economics*, Vol. 51, pp. 121–135.

Gaviria, A., and Raphael, S. (2001). "School-Based Peer Effects and Juvenile Behavior." *Review of Economics and Statistics*, Vol. 83, No. 2, pp. 257–268.

Gayer, T. (2004). "The Fatality Risks of Sport-Utility Vehicles, Vans, and Pickups Relative to Cars." *Journal of Risk and Uncertainty*, Vol. 28, No. 2, pp. 103–133.

Georgopoulos, T. (2009). *Taxation of Alcohol and Consumer Attitude. Is the ECJ Sober?* AAWE Working Paper No. 37.

Gergaud, O., and Ginsburgh, V. (2008). "Natural Endowments, Production Technologies and the Quality of Wines in Bordeaux. Does Terroir Matter?" *Economic Journal*, Vol. 118, pp. 142–157.

Gergaud, O., and Livat, F. (2004). *Team Versus Individual Reputations: A Model of Interaction and Some Empirical Evidence*. Université de Reims Champagne-Ardenne Working Paper.

Gergaud, O., and Livat, F. (2007). *How Do Consumers Use Signals to Assess Quality?* American Association of Wine Economists AAWE Working Paper No.3.

Gergaud, O., Smeets, V., and Warzynski, F. (2010). *Stars War in French Gastronomy: Prestige of Restaurants and Chefs' Careers*. Working Paper, Aarhus School of Business.

Gide, C. (1907). "The Wine Crisis in South France." *Economic Journal*, Vol. 17, No. 67, pp. 370–375.

Ginsburgh, V. (1998). "Absentee Bidders and the Declining Price Anomaly in Wine Auctions." *Journal of Political Economy*, Vol. 106, No. 6, pp. 1302–1319.

Ginsburgh, V., Monzak, A., and Monzak, M. (2013). "Red Wines of Medoc: What Is Wine Tasting Worth?" *Journal of Wine Economics*, Vol. 8, No. 2, pp. 159–188.

Goldstein, R., Almenberg, J., Dreber, A., Emerson, J.W., Herschkowitsch, A., and Katz, J. (2008). "Do More Expensive Wines Taste Better? Evidence from a Large Sample of Blind Tastings." *Journal of Wine Economics*, Vol. 3, No. 1, pp. 1–9.

Gollier, C. (2005). "Some Aspects of the Economics and Catastrophe Risk Insurance." In *Catastrophic Risks and Insurance*, ed. OECD. Paris: OECD Publishing, Policy Issues in Insurance, No. 8.

Golub, S.S., and Tomasik, B. (2008). *Measures of International Transport Cost for OECD Countries*. OECD Economics Department Working Papers N.609, Paris.

Gomes, A. (2000). "Going Public Without Governance: Managerial Reputation Effects." *Journal of Finance*, Vol. 55, No. 2, pp. 615–646.

Goode, J. (2016). *I Taste Red: The Science of Tasting Wine*. Berkeley: University of California Press.

Goodhue, R.E., Heien, D.M., Lee, H., and Sumner, D.A. (2003). "Contracts and Quality in the California Winegrape Industry." *Review of Industrial Organization*, Vol. 23, pp. 267–282.

Gorton, G. (1996). "Reputation Formation in Early Bank Note Markets." *Journal of Political Economy*, Vol. 104, pp. 46–97.

Gottlieb, N.H., and Baker, J.A. (1986). "The Relative Influence of Health Beliefs, Parental and Peer Behaviors and Exercise Program Participation on Smoking, Alcohol Use and Physical Activity." *Social Science & Medicine*, Vol. 22, No. 9, pp. 915–927.

Grant, W. (1997). *The Common Agricultural Policy*. London: MacMillan.

Grebitus, C., Lusk, J.L., and Nayga, R.M. (2013). "Effect of Distance of Transportation on Willingness to Pay for Food." *Ecological Economics*, Vol. 88, pp. 67–75.

Greene, W. (2000). *Econometric Analysis* (4th ed.). Upper Saddle River, NJ: Prentice Hall International Editions.

Greenwood, B.N., and Wattal, S. (2016). *Show Me the Way to Go Home: An Empirical Investigation of Ride Sharing and Alcohol Related Motor Vehicle Homicide*. Fox School of Business Research Paper No. 15-054.

Greving, J.P., Lee, J.E., Wolk, A., Lukkien, C., Lindblad, P., and Bergström, A. (2007). "Alcoholic Beverages and Risk of Renal Cell Cancer." *British Journal of Cancer*, Vol. 97, pp. 429–433.

Grifoni, D., Mancini, M., Maracchi, G., Orlandini, S., and Zipoli, G. (2006). "Analysis of Italian Wine Quality Using Freely Available Meteorological Information." *American Journal of Enology and Viticulture*, Vol. 57, pp. 339–346.

Griliches, Z. (1961). "Hedonic Price Indexes for Automobiles." In *An Econometric Analysis for Quality Change: Studies in New Methods of Measurement*, ed. Z. Griliches, pp. 55–89. Cambridge, MA: Harvard University Press.

Grodstein, F., Colditz, G.A., Hunter, D.J., Manson, J.E., Willett, W.C., and Stampfer, M.J. (1994). "A Prospective Study of Symptomatic Gallstones in Women: Relation with Oral Contraceptives and Other Risk Factors." *Obstetrics & Gynecology*, Vol. 84, pp. 207–214.

Grosskopf, B., and Sarin, R. (2010). "Is Reputation Good or Bad? An Experiment." *American Economic Review*, Vol. 100, No. 5, pp. 2187–2204.

Grossman, M., Chaloupka, F.J., and Sirtalan, I. (1998). "An Empirical Analysis of Alcohol Addiction: Results from the Monitoring the Future Panels." *Economic Inquiry*, Vol. 36, No. 1, pp. 39–48.

Grossman, S.J., and Shiller, R.J. (1981). "The Determinants of the Variability of Stock Market Prices." *American Economic Review*, Vol. 71, No. 2, pp. 222–227.

Gruenewald, P.J., Ponicki, W., Holder, H.D., and Romelsjö, A. (2006). "Alcohol Prices, Beverage Quality, and the Demand for Alcohol: Quality Substitutions and Price Elasticities." *Alcoholism Clinical and Experimental Research*, Vol. 30, No. 1, pp. 96–105.

Guiso, L., Sapienza, P., and Zingales, L. (2004). "The Role of Social Capital in Financial Development." *American Economic Review*, Vol. 94, pp. 526–557.

Halebsky, M.A. (1987). "Adolescent Alcohol and Substance Abuse: Parent and Peer Effects." *Adolescence*, Vol. 22, pp. 961–967.

Hall, J.H., and Geyser, J.M. (2004). "Are Wine Cooperatives Creating Value?" *Agrekon*, Vol. 43, No. 3, pp. 331–346.

Hall, W. (1996). "Changes in the Public Perceptions of the Health Benefits of Alcohol Use, 1989 to 1994." *Australian and New Zealand Journal of Public Health*, Vol. 20, pp. 93–95.

Hamilton, V., and Hamilton, B.H. (1997). "Alcohol and Earnings: Does Drinking Yield a Wage Premium?" *Canadian Journal of Economics*, Vol. 30, No. 1, pp. 135–151.

Hanf, J.H., and Schweickert, E. (2007a). *Changes in the Wine Chain—Managerial Challenges and Threats for German Wine Co-ops.* AAWE Working Paper No. 7.

Hanf, J.H., and Schweickert, E. (2007b). "How to Deal with Member Heterogeneity—Management Implications." *International Journal of Co-operative Management*, Vol. 3, No. 2, pp. 40–48.

Hannah, L., Roehrdanz, P.R., Ikegami, M., Shepard, A.V., Shaw, M.R., Tabor, G., Zhie, L., Marquet, P.A., and Hijmans, R.J. (2013). "Climate Change, Wine, and Conservation." *PNAS*, Vol. 110, No. 17, pp. 6907–6912. https://doi.org/10.1073/pnas.1210127110.

Hansmann, H.B. (1980). "The Role of Nonprofit Enterprise." *Yale Law Journal*, Vol. 89, pp. 835–901.

Hansmann, H.B. (2012). "Ownership and Organizational Form." In *The Handbook of Organizational Economics*, ed. R. Gibbons and J. Roberts, pp. 891–917. Princeton, NJ: Princeton University Press.

Hanson, B. (1994). "Social Network, Social Support and Heavy Drinking in Elderly Men: A Population Study of Men Born in 1914." *Addiction*, Vol. 89, pp. 725–732.

Harvard School of Public Health (n.d.). "Alcohol: Balancing Risks and Benefits." *The Nutrition Source*. https://www.hsph.harvard.edu/nutritionsource/healthy-drinks/drinks-to-consume-in-moderation/alcohol-full-story/.

Harvey, S.M., and Beckman, L.J. (1986). "Alcohol Consumption, Female Sexual Behavior and Contraceptive Use." *Journal of Studies on Alcohol and Drugs*, Vol. 47, No. 4, pp. 327–332.

Hawkins, J.D., Catalano, R.F., and Miller, J.Y. (1992). "Risk and Protective Factors for Alcohol and Other Drug Problems in Adolescence and Early Adulthood: Implications for Substance Abuse Prevention." *Psychological Bulletin*, Vol. 112, pp. 64–105.

Hay, C. (2010). "The Political Economy of Price and Status Formation in the Bordeaux En Primeur Market: The Role of Wine Critics as Rating Agencies." *Socio-Economic Review*, Vol. 8, No. 4, pp. 685–707.

Hayward, L., Zubrick, S.R., and Silburn, S. (1992). "Blood Alcohol Levels in Suicide Cases." *Journal of Epidemiology and Community Health*, Vol. 46, No. 3, pp. 256–260.

Hazell, P. (1992). "The Appropriate Role of Agricultural Insurance in Developing Countries." *Journal of International Development*, Vol. 4, No. 6, pp. 567–581.

Heath, D.B. (1995). *An Anthropological View of Alcohol and Culture in International Perspective.* In *International Handbook on Alcohol and Culture,* ed. D.B. Heath, pp. 328–347. Westport, CT: Greenwood Press.

Heffetz, O., and Shayo, M. (2009). "How Large Are Non-Budget-Constraint Effects of Prices on Demand?" *American Economic Journal: Applied Economics,* Vol. 1, No.4, pp. 170–199.

Heien, D.M. (1996). "Do Drinkers Earn Less?" *Southern Economic Journal,* Vol. 63, No. 1, pp. 60–68.

Heijbroek, A.M. (2003). "Consequences of the Globalization in the Wine Industry." Presented at the Great Wine Capitals Global Network International Conference, Bilbao, Spain.

Heilbrun, J. (2001). "Empirical Evidence of a Decline in Repertory Among American Opera Companies 1991/2–1997/8." *Journal of Cultural Economics,* Vol. 25, No. 1, pp. 63–72.

Helble, M., and Sato, A. (2011). *Booms and Booze: On the Relationship Between Macroeconomic Conditions and Alcohol Consumption.* Working Paper.

Hellmuth, M.E., Osgood, D.E., Hess, U., Moorhead, A., and Bhojwani, H. (eds.) (2009). *Index Insurance and Climate Risk: Prospects for Development and Disaster Management.* Climate and Society No. 2, International Research Institute for Climate and Society, Columbia University, New York.

Hill, R. (2000). "The Case of the Missing Organizations: Co-operatives and the Textbooks." *Journal of Economic Education,* Vol. 31, No. 3, pp. 281–295.

Hill, T.D., and Angel, R.J. (2005). "Neighborhood Disorder, Psychological Distress, and Heavy Drinking." *Social Science & Medicine,* Vol. 61, No. 5, pp. 965–975.

Hippel, E. von. (1994). "'Sticky Information' and the Locus of Problem Solving: Implications for Innovation." *Management Science,* Vol. 40, No. 4, pp. 429–439.

Hjartåker, A., Meo, M.S., and Weiderpass, E. (2010). "Alcohol and Gynecological Cancers: An Overview." *European Journal of Cancer Prevention,* Vol. 19, No. 1, pp. 1–10.

Hodgson, R.T. (2009). "An Analysis of the Concordance among 13 US Wine Competitions." *Journal of Wine Economics,* Vol. 4, No. 1, pp. 1–9.

Hojman, D.E., and Hunter-Jones, P. (2010). "Wine Tourism: Chilean Wine Regions and Routes." *Journal of Business Research,* Vol. 65, No. 1, pp. 13–21.

Holahan, C.J., Schutte, K.K., Brennan, P.L., Holahan, C.K., Moos, B.S., and Moos, R.H. (2010). "Late-Life Alcohol Consumption and 20-Year Mortality." *Alcoholism: Clinical and Experimental Research,* Vol. 34, No. 11, pp. 1961–1971.

Holt, J.B., Miller, J.W., Naimi, T.S., and Sui, D.Z. (2006). "Religious Affiliation and Alcohol Consumption in the United States." *Geographical Review,* Vol. 96, No. 4, pp. 523–542.

Hoover, H. (1929). "Inaugural Address of Herbert Hoover." *The Avalon Project* (Yale Law School), March 4. https://avalon.law.yale.edu/20th_century/hoover.asp.

Hotelling, H. (1929). "Stability in Competition." *Economic Journal,* Vol. 39, No. 153, pp. 41–57.

Houser, D., and Wooders, J. (2006). "Reputation in Auctions: Theory and Evidence from eBay." *Journal of Economics and Management Strategy,* Vol. 15, No. 2, pp. 353–369.

Howard, M.O., Kivlahan, D., and Walker, R.D. (1997). "Cloninger's Tridimensional Theory of Personality and Psychopathology: Applications to Substance Use Disorders." *Journal of Studies on Alcohol,* Vol. 58, pp. 48–66.

Huerta, M., and Borgonovi, F. (2010). "Education, Alcohol Use and Abuse Among Young Adults in Britain." *Social Science & Medicine*, Vol. 71, No. 1, pp. 143–151.

Hueth, B., Ligon, E., Wolf, S., and Wu, S. (1999). "Incentive Instruments in Fruit and Vegetable Contracts: Input Control, Monitoring, Measuring, and Price Risk." *Review of Agricultural Economics*, Vol. 21, pp. 374–389.

Hull, J.C. (2009). *Options, Futures and Other Derivatives.* Upper Saddle River, NJ: Prentice Hall.

Hummels, D. (2007). "Transportation Costs and International Trade in the Second Era of Globalization." *Journal of Economic Perspectives*, Vol. 21, No. 3, pp. 131–154.

International Center for Alcohol Policies (ICAP) (2009). *Determinants of Drinking.* ICAP Policy Tools Series-Issue Briefings, Washington, DC.

International Co-operative Alliance (ICA) (n.d.a). "Cooperative Identity, Values, and Principles." ICA.coop. https://www.ica.coop/en/cooperatives/cooperative-identity.

International Co-operative Alliance (ICA) (n.d.b). "Cooperatives: Facts and Figures." ICA .coop. http://ica.coop/en/whats-co-op/co-operative-facts-figures.

International Fund for Agricultural Development (IFAD) (2011). *Weather Index-Based Insurance in Agricultural Development—A Technical Guide.* Rome, Italy.

International Organisation of Vine and Wine (OIV) (2017). *International Code of Oenological Practices*, 2017 issue. Paris: OIV. http://www.oiv.int/public/medias/5119/code-2017-en.pdf.

Intesa Sanpaolo Bank (2013). *Economia e Finanza dei Distretti Industriali.* Annual Report No. 6, Milan, Italy.

Italian National Districts Observatory (2013). *IV Rapporto.* Rome, Italy.

Italian National Institute of Statistics (ISTAT) (2001). *8° Censimento Generale dell'Industria e dei Servizi.* Rome, Italy.

Italian National Institute of Statistics (ISTAT) (2013). *L'Uso e l'Abuso di Alcol in Italia-Anno 2012.* Rome, Italy.

Itami, H. (1987). *Mobilizing Invisible Assets.* Cambridge, MA: Harvard University Press.

Jackson, C.K., and Owens, E.G. (2011). "One for the Road: Public Transportation, Alcohol Consumption, and Intoxicated Driving." *Journal of Public Economics*, Vol. 95, pp. 106–121.

Jaeger, D.A., and Storchmann, K. (2011). "Wine Retail Price Dispersion in the United States: Searching for Expensive Wines?" *The American Economic Review—Papers & Proceedings*, Vol. 101, No. 3, pp. 136–141.

Jaeger, E. (1981). "To Save or Savor: The Rate of Return to Storing Wine." *Journal of Political Economy*, Vol. 89, No. 3, pp. 584–592.

Jiménez-Martín, S., Labeaga, J.M., and Vilaplana Prieto, C. (2006). *Further Evidence About Alcohol Consumption and the Business Cycle.* Working Paper N. 2006–06, Fundación de Estudios de Economía Aplicada.

Johnson, F.W., Gruenewald, P.J., Treno, A.J., and Taff, G.A. (1998). "Drinking Over the Life Course Within Gender and Ethnic Groups: A Hyperparametric Analysis." *Journal of Studies on Alcohol*, Vol. 59, No. 5, pp. 568–580.

Johnson, H. (1971). *The World Atlas of Wine.* New York: Simon & Schuster.

Johnson, H. (2009). *Poket Wine Book.* London: Octopus Publishing.

Johnson, N.D., Nye, J., and Franck, R. (2010). *Trade, Taxes, and Terroir*. George Mason University Working Paper No. 10–35.

Jones, D.C., and Pliskin, J.L. (1988). *The Effects of Worker Participation, Employee Ownership and Profit Sharing on Economics Performance: A Partial Review*. Levy Economics Institute Working Paper No. 13.

Jones, G.V., and Storchmann, K.H. (2001). "Wine Market Prices and Investment Under Uncertainty: An Econometric Model for Bordeaux Crus Classés." *Agricultural Economics*, Vol. 26, No. 2, pp. 115–133.

Kaestner, R., and Yarnoff, B. (2011). "Long-Term Effects of Minimum Legal Drinking Age Laws on Adult Alcohol Use and Driving Fatalities." *Journal of Law and Economics*, Vol. 54, No. 2, pp. 365–388.

Kaldor, N. (1961). "Capital Accumulation and Economic Growth." In *The Theory of Capital*, ed. F.A. Lutz and D. Hague, pp. 177–222. London: Palgrave Macmillan.

Kallas, Z., Serra, T., and Gil, J.M. (2010). "Farmers' Objectives as Determinants of Organic Farming Adoption: The Case of Catalonian Vineyard Production." *Agricultural Economics*, Vol. 41, pp. 409–423.

Kalmi, P. (2007). "The Disappearance of Cooperatives from Economics Textbooks." *Cambridge Journal of Economics*, Vol. 31, pp. 625–647.

Kandori, M. (1992). "Social Norms and Community Enforcement." *Review of Economic Studies*, Vol. 59, No. 1, pp. 63–80.

Karafolas, S. (2007). "Wine Roads in Greece: A Cooperation for the Development of Local Tourism in Rural Areas." *Journal of Rural Cooperation*, Vol. 35, No. 1, pp. 71–90.

Karlamangla, A., Zhou, K., Reuben, D., Greendale, G., and Moore, A. (2006). "Longitudinal Trajectories of Heavy Drinking in Adults in the United States of America." *Addiction*, Vol. 101, No. 1, pp. 91–99.

Katz, J. (1997). "Managerial Behaviour and Strategy Choices in Agribusiness Cooperatives." *Agribusiness*, Vol. 13, No. 5, pp. 483–495.

Katz, M., Rosen, H.S., Morgan, W., and Bollino, C.A. (2011). *Microeconomia* (4th ed.). New York: McGraw-Hill.

Kenkel, D.S., and Ribar, D. (1994). *Alcohol Consumption and Young Adults' Socioeconomic Status*. Brookings Papers on Economic Activity: Microeconomics, pp. 119–161.

Kenkel, D.S., and Terza, J.V. (2011). "The Effect of Physician Advice on Alcohol Consumption: Count Regression with an Endogenous Treatment Effect." *Journal of Applied Econometrics*, Vol. 16, pp. 165–184.

Kerr, W.C., Greenfield, T.K., Bond, J., Ye, Y., and Rehm, J. (2009). "Age-Period-Cohort Modelling of Alcohol Volume and Heavy Drinking Days in the US National Alcohol Surveys: Divergence in Younger and Older Adult Trends." *Addiction*, Vol. 104, No. 1, pp. 27–37.

Keser, C. (2003). "Experimental Games for the Design of Reputation Management Systems." *IBM Systems Journal*, Vol. 42, No. 3, pp. 498–506.

Kestilä, L., Martelin, T., Rahkonen, O., Joutsenniemi, K., Pirkola, S., Poikolainen, K., and Koskinen, S. (2008). "Childhood and Current Determinants of Heavy Drinking in Early Adulthood." *Alcohol and Alcoholism*, Vol. 43, No. 4, pp. 460–469.

Klatsky, A.L., Armstrong, M.A., and Friedman, G.D. (1990). "Risk of Cardiovascular Mortality in Alcohol Drinkers, Ex-Drinkers and Nondrinkers." *American Journal of Cardiology*, Vol. 66, pp. 1237–1242.

Klatsky, A.L., Friedman, G.D., Armstrong, M.A., and Kipp, H. (2003). "Wine, Liquor, Beer, and Mortality." *American Journal of Epidemiology*, Vol. 158, No. 6, pp. 585–595.

Klein, B., and Leffer, K. B. (1981). "The Role of Market Forces in Assuring Contractual Performance." *Journal of Political Economy*, Vol. 89, No. 4, pp. 615–641.

Knight, F. (1921). *Risk, Uncertainty, and Profit.* Boston, MA: Houghton Mifflin.

Kollock, P. (1998). "Social Dilemmas: The Anatomy of Cooperation." *Annual Review of Sociology*, Vol. 24, pp. 183–214.

Koppes, L.L., Dekker, J.M., Hendriks, H.F., Bouter, L.M., and Heine, R.J. (2005). "Moderate Alcohol Consumption Lowers the Risk of Type 2 Diabetes: A Meta–Analysis of Prospective Observational Studies." *Diabetes Care*, Vol. 28, pp. 719–725.

Krasker, W.S. (1979). "The Rate of Return to Storing Wines." *Journal of Political Economy*, Vol. 87, No. 6, pp. 1363–1367.

Krebs, S., and Pommerehne, W. (1995). "Politico-Economic Interactions of German Performing Arts Institutions." *Journal of Cultural Economics*, Vol. 19, No. 1, pp. 17–32.

Krekhovets, E., and Leonova, L. (2013). "Alcohol Consumption and Life Satisfaction: Evidence from Russia." *Academic Journal of Interdisciplinary Studies*, Vol. 2, No. 8, pp. 98–105.

Kremer, M., and Levy, D. (2008). "Peer Effects and Alcohol Use Among College Students." *Journal of Economic Perspectives*, Vol. 22, No. 3, pp.189–206.

Kreps, D. (1990). "Corporate Culture and Economic Theory." In *Perspectives on Positive Political Economy*, ed. J.E. Alt and K.A. Shepsle, pp. 90–143. Cambridge, MA: Cambridge University Press.

Krugman, P. (1991). *Geography and Trade.* Cambridge, MA: The MIT Press.

Krugman, P., and Wells, R. (2006). *Microeconomia* (1st ed.). Bologna, Italy: Zanichelli.

Kurz, H., Dietzenbacher, E., and Lager, C. (1998). *Input-Output Analysis.* Cheltenham, UK: Edward Elgar.

Lakonishok, J., Shleifer, A., and Vishny, R.W. (1994). "Contrarian Investment, Extrapolation, and Risk." *Journal of Finance*, Vol. 49, No. 5, pp. 1541–1578.

Landon, S., and Smith, C.E. (1997). "The Use of Quality and Reputation Indicators by Consumer: The Case of Bordeaux Wine." *Journal of Consumer Policy*, Vol. 20, pp. 289–323.

Landon, S., and Smith, C.E. (1998). "Quality Expectation. Reputation and Price." *Southern Economic Journal*, Vol. 64, pp. 628–647.

Lansberg, I.S. (1983). "Managing Human Resources in Family Firms: The Problem of Institutional Overlap." *Organizational Dynamics*, Vol. 12, pp. 39–46.

Larreina, M. (2007). *Detecting a Cluster in a Region Without Complete Statistical Data, Using Input-Output Analysis: The Case of the Rioja Wine Cluster.* CRIEFF Discussion Paper No. 0706.

Larreina, M., Gómez-Bezares, F., and Aguado, R. (2011). "Development Rooted on Riojan Soil: The Wine Cluster and Beyond." *Open Geography Journal*, Vol. 4, pp. 3–15.

Lave, C.A. (1985). "Speeding, Coordination, and the 55 MPH Limit." *American Economic Review*, Vol. 75, No. 5, pp. 1159–1164.

Lecocq, S., Magnac, T., Pichery, M.C., and Visser, M. (2005). "The Impact of Information on Wine Auction Prices: Results of an Experiment." *Annales d'Économie et de Statistique*, No. 77, pp. 37–57.

Lee, J. (2006). "Family Firm Performance: Further Evidence." *Family Business Review*, Vol. 19, No. 2, pp. 103–114.

Lee, J.S., Sudore, R.L., Williams, B.A., Lindquist, K., Chen, H.L., and Covinsky, K.E. (2009). "Functional Limitations, Socioeconomic Status, and All-Cause Mortality in Moderate Alcohol Drinkers." *Journal of the American Geriatrics Society*, Vol. 57, No. 6, pp. 955–962.

Leeuwen, C., and Darriet, P. (2016). "The Impact of Climate Change on Viticulture and Wine Quality." *Journal of Wine Economics*, Vol. 11, No. 1, pp. 150–167.

Leger, A.S., Cochrane, A.L., and Moore, F. (1979). "Factors Associated with Cardiac Mortality in Developed Countries with Particular Reference to the Consumption of Wine." *Lancet*, Vol. 12, No. 1, pp. 1017–1020.

Leibowitz, S.F. (2007). "Overconsumption of Dietary Fat and Alcohol: Mechanisms Involving Lipids and Hypothalamic Peptides." *Physiology and Behavior*, Vol. 91, pp. 513–521.

Leitzmann, M.F., Giovannucci, E.L., Stampfer, M.J, Spiegelman, D., Colditz, G.A., Willett, W.C., and Rimm, E.B. (1999). "Prospective Study of Alcohol Consumption Patterns in Relation to Symptomatic Gallstone Disease in Men." *Alcoholism: Clinical and Experimental Research*, Vol. 23, pp. 835–841.

Leonard, K.E., and Quigley, B.M. (1999). "Drinking and Marital Aggression in Newlyweds: An Event-Based Analysis of Drinking and the Occurrence of Husband Marital Aggression." *Journal of Studies on Alcohol*, Vol. 60, No. 4, p. 537–545.

Lerner, M. (n.d.). "Unintended Consequences." *Prohibition: A Film by Ken Burns and Lynn Novick*. https://www.pbs.org/kenburns/prohibition/unintended-consequences/.

Le Strat, Y., Ramoz, N., Schumann, G., and Gorwood, P. (2008). "Molecular Genetics of Alcohol Dependence and Related Endophenotypes." *Current Genomics*, Vol. 9, pp. 444–451.

Levine, H.J. (1985). "The Birth of American Alcohol Control: Prohibition, the Power Elite, and the Problem of Lawlessness." *Contemporary Drug Problems*, Vol. 12, No. 1, pp. 63–116.

Levinsohn, M. (2006). *The Box: How The Shipping Container Made the World Smaller and the World Economy Bigger*. Princeton, NJ: Princeton University Press.

Levinson, H. (1971). "Conflicts That Plague Family Businesses." *Harvard Business Review*, Vol. 49, pp. 90–98.

Levitt, S.D., and Porter, J. (2001). "How Dangerous Are Drinking Drivers?" *Journal of Political Economy*, Vol. 109, No. 6, pp. 1198–1237.

Lintner, J. (1965). "The Valuation of Risky Assets and the Selection of Risky Investments in Stock Portfolios and Capital Budget." *Review of Economics and Statistics*, Vol. 47, pp. 13–37.

Little, M., Handley, E., Leuthe, E., and Chassin, L. (2009). "The Impact of Parenthood on Alcohol Consumption Trajectories." *Development and Psychopathology*, Vol. 21, pp. 661–682.

Ljunge, M. (2011). "Do Taxes Produce Better Wine?" *Journal of Agricultural & Food Industrial Organization*, Vol. 9, No. 1, pp. 1–14.

Logue, J., and Yates, J. (2005). *Productivity in Cooperatives and Worker-Owned Enterprises: Ownership and Participation Make a Difference!* Paper prepared for the Employment Sector International Labour Office, Geneva.

López, X.A., and Martín, B.G. (2006). "Tourism and Quality Agro-Food Products: An Opportunity for the Spanish Countryside." *Tijdschrift voor Economische en Sociale Geografie*, Vol. 97, No. 2, pp. 166–177.

Los Angeles County Department of Public Health (2001). "Alcohol Consumption and Abuse Among Los Angeles County Adults." *L.A. Health*, December.

Lott, J. (1996). "The Level of Optimal Fines to Prevent Fraud When Reputations Exist and Penalty Clauses Are Unenforceable." *Managerial and Decision Economics*, Vol. 17, No. 4, pp. 363–380.

Lundborg, P. (2006). "Having the Wrong Friends? Peer Effects in Adolescent Substance Use." *Journal of Health Economics*, Vol. 25, pp. 214–233.

Lynch, L., Urban, M., and Sommer, R. (1989). "De-Emphasis on Cooperatives in Introductory Economics Textbooks." *Journal of Agricultural Cooperation*, Vol. 4, pp. 89–92.

Maietta, O.W., and Sena, V. (2008). "Shadow Price of Capital and the Furubotn–Pejovich Effect: Some Empirical Evidence for Italian Wine Cooperatives." *Applied Stochastic Models in Business and Industry*, Vol. 24, pp. 495–505.

Mäkelä, P., and Osterberg, E. (2009). "Weakening of One More Alcohol Control Pillar: A Review of the Effects of the Alcohol Tax Cuts in Finland in 2004." *Addiction*, Vol. 104, pp. 554–563.

Malorgio, G., Hertzberg, A., and Grazia, C. (2008). *Italian Wine Consumer Behaviour and Wineries Responsive Capacity*. Working Paper.

Mandel, B.J. (2009). "Art as an Investment and Conspicuous Consumption Good." *American Economic Review*, Vol. 99, No. 4, pp. 1653–1663.

Manning, W.G., Blumberg, L., and Moulton, L.H. (1995). "The Demand for Alcohol: The Differential Response to Price." *Journal of Health Economics*, Vol. 14, No. 2, pp. 123–148.

Manning, W.G., Keeler, E.B., Newhouse, J.P., Sloss, E.M., and Wasserman, J. (1991). *The Costs of Poor Health Habits*. Cambridge, MA: Harvard University Press.

Mariani, A., Boccia, F., and Napoletano, F. (2006). "Il mercato mondiale del vino: produzione, consumi, scambi e forme di regolamentazione." In *Il mercato del vino, tendenze strutturali e strategie dei concorrenti*, ed. G.P. Cesaretti, R. Green, A. Mariani, and E. Pomarici. Milan, Italy: Franco Angeli.

Markowitz, H. (1952). "Portfolio Selection." *Journal of Finance*, Vol. 7, No. 1, pp. 77–91.

Markowitz, S., Kaestner, R., and Grossman, M. (2005). "An Investigation of the Effects of Alcohol Consumption and Alcohol Policies on Youth Risky Sexual Behaviors." *American Economic Review: Papers & Proceedings*, Vol. 95, No. 2, pp. 263–266.

Marks, D. (2015). *Wine and Economics: Transacting the Elixir of Life*. Cheltenham, UK: Edward Elgar.

Marshall, A. (1890). *Principles of Economics*. London: Macmillan.

Marshall, A., and M.P. Marshall (1879). *The Economics of Industry*. London: Macmillan.

Masih, A., and Masih, R. (1997). "Dynamic Linkages and the Propagation Mechanism Driving Major International Stock Markets: An Analysis of Pre- and Post-Crash Eras." *Quarterly Review of Economics and Finance*, Vol. 37, No. 4, pp. 859–885.

Masset, P., and Henderson, C. (2010). "Wine as an Alternative Asset Class." *Journal of Wine Economics*, Vol. 5, No. 1, pp. 87–118.

Masset, P., Weisskopf, J.P., and Cossutta, M. (2015). "Wine Tasters, Ratings and *en Primeur* Prices." *Journal of Wine Economics*, Vol. 10, No. 1, pp. 75–107.

Masset, P., Weisskopf, J.P., Faye, B., and Le Furb, E. (2016). "Red Obsession: The Ascent of Fine Wine in China." *Emerging Markets Review*, Vol. 29, pp. 200–225.

Massin, S., and Kopp, P. (2010). *Alcohol Consumption and Happiness: An Empirical Analysis Using Russian Panel Data*. Working Paper, Université Paris 1.

McGue, M., Sharma, A., and Benson, P. (1996). "Parent and Sibling Influences on Adolescent Alcohol Use and Misuse: Evidence from a U.S. Adoption Cohort." *Journal of Studies on Alcohol*, Vol. 57, pp. 8–18.

McKee, L.J. (2016). "TTB Proposes New Grape Variety Names." *Wines Vines Analytics*, December 1. https://winesvinesanalytics.com/news/article/177345/TTB-Proposes-New-Grape -Variety-Names.

McQuade, T., Salant, S., and Winfree, J. (2008). *Quality Standard Effects on Goods with Collective Reputation and Multiple Components*. Working Paper.

Mediobanca (2014). *Indagine sul Settore Vinicolo*. Milan, Italy.

Megna, P., and Mueller, D. (1991). "Profit Rates and Intangible Capital." *Review of Economics and Statistics*, Vol. 73, pp. 632–642.

Mei, J., and Moses, M. (2002). "Art as an Investment and the Underperformance of Masterpieces." *American Economic Review*, Vol. 92, No. 5, pp. 1656–1668.

Melnik, M., and Alm, J. (2002). "Does a Seller's Ecommerce Reputation Matter? Evidence from eBay Auctions." *Journal of Industrial Economics*, Vol. 50, No. 3, pp. 337–349.

Meloni, G., and Swinnen, J. (2013). "The Political Economy of European Wine Regulations." *Journal of Wine Economics*, Vol. 8, No. 3, pp. 244–284.

Meloni, G., and Swinnen, J. (2014). "The Rise and Fall of the World's Largest Wine Exporter— And Its Institutional Legacy." *Journal of Wine Economics*, Vol. 9, No. 1, pp. 3–33.

Meloni, G., and Swinnen, J. (2018a). "Trade and Terroir: The Political Economy of the World's First Geographical Indications." *Food Policy*, Vol. 81, pp. 1–20.

Meloni, G., and Swinnen, J. (2018b). "The Political Economy of Regulations and Trade: Wine Trade 1860–1970." *The World Economy*, Vol. 41, No. 6, pp. 1567–1595.

Menzani, T., and Zamagni, V. (2009). "Cooperative Networks in the Italian Economy." *Enterprise and Society*, Vol. 11, No. 1, pp. 98–127.

Mileti, D.S. (1999). *Disaster by Design: A Reassessment of Natural Hazards in the United States*. Washington, DC: Joseph Henry Press.

Milgrom, P., and Roberts, J. (1992). *Economics, Organization and Management*. Englewood Cliffs, NJ: Prentice Hall.

Milhau, J. (1953). "L'Avenir de la Viticulture Française." *Revue Économique*, Vol. 4, No. 5, pp. 700–738.

Minutorizzo, A. (2019). "Competition Policy in the Wine Industry in Europe." *Journal of Wine Economics*, Vol. 14, No. 1, pp. 90–113.

Miron, J.A., and Zwiebel, J. (1991). "Alcohol Consumption During Prohibition." *American Economic Review*, Vol. 81, No. 2, pp. 242–247.

Mollá-Bauza, M.B., Martínez, L.M., Martínez Poveda, A., and Rico Pérez, M. (2005). "Determination of the Surplus that Consumers Are Willing to Pay for an Organic Wine." *Spanish Journal of Agricultural Research*, Vol. 3, No. 1, pp. 43–51.

Monahan, J.L., and Lannutti, P.J. (2000). "Alcohol Myopia Theory, Social Self-Esteem, and Social Interaction." *Human Communication Research*, Vol. 26, No. 2, pp. 175–202.

Moore, A.A., Gould, R., Reuben, D.B., Greendale, G.A., Carter, M.K., Zhou, K., and Karlamangla, A. (2005). "Longitudinal Patterns and Predictors of Alcohol Consumption in the United States." *American Journal of Public Health*, Vol. 95, No. 3, pp. 458–465.

Morrot, G., Brochet, F., and Dubordieu, D. (2001). "The Colors of Odors." *Brain and Language*, Vol. 79, No. 2, pp. 309–320.

Morton, L.M., Zheng, T., Holford, T.R., Holly, E.A., Chiu, B.C., Costantini, A.S., Stagnaro, E., Willett, E.V., Dal Maso, L., Serraino, D., Chang, E.T., Cozen, W., Davis, S., Severson, R.K., Bernstein, L., Mayne, S.T., Dee, F.R., Cerhan, J.R., and Hartge, P. (2005). "Alcohol Consumption and Risk of Non-Hodgkin Lymphoma: A Pooled Analysis." *Lancet Oncology*, Vol. 6, No. 7, pp. 469–476.

Mossakowski, K. (2008). "Is the Duration of Poverty and Unemployment a Risk Factor for Heavy Drinking?" *Social Science & Medicine*, Vol. 67, No. 6, pp. 947–955.

Mossin, J. (1966). "Equilibrium in a Capital Asset Market." *Econometrica*, Vol. 34, pp. 768–783.

Mueller, R.A.E., Sumner, D.A., and Lapsley, J.T. (2006). *Clusters of Grapes and Wine*. Working Paper.

Mukamal, K.J., Conigrave, K.M., Mittleman, M.A., Camargo, C.A., Stampfer, M.J., Willett, W.C., and Rimm, E.B. (2003). "Roles of Drinking Pattern and Type of Alcohol Consumed in Coronary Heart Disease in Men." *New England Journal of Medicine*, Vol. 348, pp. 109–118.

Mullahy, J., and Sindelar, J.L. (1996). "Employment, Unemployment, and Problem Drinking." *Journal of Health Economics*, Vol. 15, No. 4, pp. 409–434.

Murray, C.J.L., and Lopez, A.D. (1996). *The Global Burden of Disease*. Cambridge, MA: Harvard School of Public Health.

Murray, C.J.L., Salomon, J., Mathers, C., and Lopez, A. (2002). *Summary Measures of Population Health: Concepts, Ethics, Measurement and Applications*. Geneva, Switz.: World Health Organization.

Naboum-Grappe, V. (1995). "France." In *International Handbook on Alcohol and Culture*, ed. D.B. Heath, pp. 75–87. Westport, CT: Greenwood Press.

Nash, S.G., McQueen, A., and Bray, J.H. (2005). "Pathways to Adolescent Alcohol Use: Family Environment, Peer Influence, and Parental Expectations." *Journal of Adolescent Health*, Vol. 37, pp. 19–28.

National Institute on Alcohol Abuse and Alcoholism (NIAAA) (2005). *Alcohol Consumption by Children and Adolescents: An Interdisciplinary Overview*. Bethesda, MD: NIAAA.

Nelson, J.P. (1997). "Economic and Demographic Factors in U.S. Alcohol Demand: A Growth-Accounting Analysis." *Empirical Economics*, Vol. 22, No. 1, pp. 83–102.

Nelson, J.P. (1999). "Broadcast Advertising and U.S. Demand for Alcoholic Beverages." *Southern Economic Journal*, Vol. 65, No. 2, pp. 774–790.

Nelson, J.P. (2010). "Alcohol Advertising Bans, Consumption and Control Policies in Seventeen OECD Countries, 1975–2000." *Applied Economics*, Vol. 42, pp. 803–823.

Nelson, J.P. (2014). "Estimating the Price Elasticity of Beer: Meta-Analysis of Data with Heterogeneity, Dependence, and Publication Bias." *Journal of Health Economics*, Vol. 33, pp. 180–187.

Nelson, R. (1991). "Why Do Firms Differ, and How Does it Matter." *Strategic Management Journal*, Vol. 12, pp. 61–74.

Nerlove, M. (1995). "Hedonic Price Functions and the Measurement of Preferences: The Case of Swedish Wine Consumers." *European Economic Review*, Vol. 39, pp. 1697–1716.

Newton, S.K., Gilinsky, A., and Jordan, D. (2015). "Differentiation Strategies and Winery Financial Performance: An Empirical Investigation." *Wine Economics and Policy*, Vol. 4, pp. 88–97.

Nilsson, J. (2001). "Farmer Co-operatives: Organisational Models and their Business Environment." In *The New Mutualism of Public Policy*, ed. J. Birchall, pp. 132–154. New York: Routledge.

Norberg, K., Beirut, L., and Grucza, R. (2009). "Long-Term Effects of Minimum Drinking Age Laws on Past-Year Alcohol and Drug Use Disorders." *Alcoholism: Clinical and Experimental Research*, Vol. 33, pp. 2180–2190.

North, D. (1958). "Ocean Freight Rates and Economic Development 1750–1913." *Journal of Economic History*, Vol. 18, No. 4, pp. 537–555.

Novkovic, S. (2008). "Defining the Co-operative Difference." *Journal of Socio-Economics*, Vol. 37, pp. 2168–2177.

Nunes, P., and Loureiro, M.L. (2012). *Agricultural Landscape, Vineyards and Tourism Flows in Tuscany, Italy: Results from an Applied Economic Study*. AAWE Working Paper No. 103.

O'Boyle, E.J. (2018). *Insurance for Growers of Wine Grapes*. AAWE Working Paper No. 219.

Oczkowski, E. (1994). "A Hedonic Price Function for Australian Premium Table Wine." *Australian Journal of Agricultural Economics*, Vol. 38, No. 1, pp. 93–110.

Oczkowski, E. (2001). "Hedonic Wine Price Functions and Measurement Error." *Economic Record*, Vol. 77, No. 239, pp. 374–382.

Oczkowski, E. (2006). "Modeling Winegrape Prices in Disequilibrium." *Agricultural Economics*, Vol. 34, pp. 97–107.

Oczkowski, E. (2018). "Modelling Prices and the Reputation of Individual Named Wines." *Applied Economics*, Vol. 50, No. 32, pp. 3464–3476.

Okrent, D. (2010). *Last Call: The Rise and Fall of Prohibition*. New York: Scribners and Sons.

Olmos, M.F. (2008). "Why Use Contracts in Viticulture?" *Journal of Wine Research*, Vol. 19, No. 2, pp. 81–93.

Ornstein, S.I., and D. Levy (1983). "Price and Income Elasticities and the Demand for Alcoholic Beverages." In *Recent Developments in Alcoholism* (Vol. 1), ed. M. Galanter. New York: Plenum.

Parker, R. (n.d.). "TWA Rating System." *The Wine Advocate*. https://www.robertparker.com/ratings.

Paroissien, E., and Visser, M. (2018). *The Causal Impact of Medals on Wine Producers Prices, and the Gains from Participating in Contests*. AAWE Working Paper No. 223.

Parry, C.D.H., Plüddemann, A., Steyn, K., Bradshaw, D., Norman, R., and Laubscher, R. (2005). "Alcohol Use in South Africa: Findings of the First Demographic and Health Survey." *Journal of Studies on Alcohol*, Vol. 66, pp. 91–97.

Pascucci, S., Gardebroek, C., and Dries, L. (2012). "Some Like to Join, Others to Deliver: An Econometric Analysis of Farmers' Relationships with Agricultural Co-operatives." *European Review of Agricultural Economics*, Vol. 39, No. 1, pp. 51–74.

Patra, T., and Poshakwale, S.S. (2008). "Long-Run and Short-Run Relationship Between the Main Stock Indexes: Evidence from the Athens Stock Exchange." *Applied Financial Economics*, Vol. 18, No. 17, pp. 1401–1410.

Pauly, M.V. (1968). "The Economics of Moral Hazard." *American Economic Review*, Vol. 58, No. 3, pp. 531–537.

Pearl, R. (1926). *Alcohol and Longevity*. New York: Alfred A. Knopf.

Peele, S., and Brodsky, A. (2000). "Exploring Psychological Benefits Associated with Moderate Alcohol Use: A Necessary Corrective to Assessments of Drinking Outcomes?" *Drug and Alcohol Dependence*, Vol. 15, pp. 305–322.

Pennerstorfer, D., and Weiss, C.R. (2006). *On the Relative Disadvantage of Cooperatives*. Working Paper.

Pennerstorfer, D., and Weiss, C.R. (2012). "On the Relative Disadvantage of Cooperatives: Vertical Product Differentiation in a Mixed Oligopoly." *Journal of Rural Cooperation*, Vol. 40, No. 1, pp. 60–90.

Pennerstorfer, D., and Weiss, C.R. (2013). "Product Quality in the Agri-Food Chain: Do Cooperatives Offer High-Quality Wine?" *European Review of Agricultural Economics*, Vol. 40, No. 1, pp. 143–162.

Pérez-González, F. (2006). "Inherited Control and Firm Performance." *American Economic Review*, Vol. 96, pp. 1559–1588.

Pernanen, K. (1991). *Alcohol in Human Violence*. New York: Guilford.

Pesando, J.E. (1993). "Art as an Investment: The Market for Modern Prints." *American Economic Review*, Vol. 83, No. 5, pp. 1075–1089.

Peterson, C., and Hektner, J.M. (2008). "Change in Drinking Motives from Adolescence Through Emerging Adulthood: Examining the Role of Secondary Motives." Poster session presented at the annual conference of the Society for Prevention Research, San Francisco, California.

Pidd, K.J., Berry, J.G., Roche, A.M., and Harrison, J.E. (2006). "Estimating the Cost of Alcohol-Related Absenteeism in the Australian Workforce: The Importance of Consumption Patterns." *Medical Journal of Australia*, Vol. 185, No. 11, pp. 637–641.

Pierce, J. (2000). "Programmatic Risk-Taking by American Opera Companies." *Journal of Cultural Economics*, Vol. 24, No. 1, pp. 45–63.

Pietilä, A.M., Rantakallio, P., and Läärä, E. (1995). "Background Factors Predicting Non-Response in a Health Survey of Northern Finnish Young Men." *Scandinavian Journal of Social Medicine*, Vol. 23, pp. 129–136.

Pigou, A.C. (1920). *The Economics of Welfare*. London: Macmillan.

Pollner, J.D. (2001). *Catastrophe Risk Management*. World Bank Policy Research Working Paper No. 2560.

Porter, M.E. (1979). "How Competitive Forces Shape Strategy." *Harvard Business Review*, March-April, pp. 1–10.

Porter, M.E. (1980). *Competitive Strategy: Technique for Analyzing Industries and Competitors*. Free Press, New York.

Porter, M.E. (2000). "Location, Competition, and Economic Development: Local Clusters in a Global Economy." *Economic Development Quarterly*, Vol. 14, No. 1, pp. 15–34.

Porter, M.E. (2008). "The Five Competitive Forces that Shape Strategy." *Harvard Business Review*, January, pp. 79–93.

Porter, M.E., and Sölvell, Ö. (2003). *The Australian Wine Cluster*. Boston, MA: Harvard Business School.

Porter, P., and Scully, G. (1987). "Economic Efficiency in Cooperatives." *Journal of Law and Economics*, Vol. 30, pp. 489–512.

Prescott, C.A., and Kendler, K.S. (2001). "Associations Between Marital Status and Alcohol Consumption in a Longitudinal Study of Female Twins." *Journal of Studies on Alcohol*, Vol. 62, pp. 589–604.

Preston, A.E. (1989). "The Nonprofit Worker in a For-Profit World." *Journal of Labour Economics*, Vol. 7, pp. 438–463.

Prial, F.J. (1997). "Bordeaux Again Leads a High-Price Parade." *New York Times*, September 17.

Putnam, R.D. (1993). *Making Democracy Work: Civic Traditions in Modern Italy*. Princeton, NJ: Princeton University Press.

Quinlan, K., Brewer, R., Siegel, P., Sleet, D., Mokdad, A., Shults, R., and Flower, N. (2005). "Alcohol-Impaired Driving Among U.S. Adults, 1993–2002." *American Journal of Preventive Medicine*, Vol. 28, No. 4, pp. 346–350.

Ramchandani, V.A., Bosron, W.F., and Li, T.K. (2001). "Research Advances in Ethanol Metabolism." *Pathologie Biologie*, Vol. 49, No. 9, pp. 676–682.

Rashidkhani, B., Åkesson, A., Lindblad, P., and Wolk, A. (2005). "Alcohol Consumption and Risk of Renal Cell Carcinoma: A Prospective Study of Swedish Women." *International Journal of Cancer*, Vol. 117, No. 5, pp. 848–853.

Ratner, M. (1996). "Investigating the Behaviour and Characteristics of the Madrid Stock Exchange." *Journal of Banking and Finance*, Vol. 20, No. 1, pp. 135–149.

Rees, R. (1985). "The Theory of Principal and Agent—Part I and Part II." *Bulletin of Economic Research*, Vol. 37, No. 1–2, pp. 3–26, 75–95.

Rehm, J., Kanteres, F., and Lachenmeier, D.W. (2010). "Unrecorded Consumption, Quality of Alcohol and Health Consequences." *Drug and Alcohol Review*, Vol. 29, No. 4, pp. 426–436.

Rehm, J., Mathers, C., Popova, S., Thavorncharoensap, M., Teerawattananon, Y., and Patra, J. (2009). "Global Burden of Disease and Injury and Economic Cost Attributable to Alcohol Use and Alcohol-Use Disorders." *Lancet*, Vol. 373, pp. 1–10.

Rehm, J., Room, R., Graham, K., Monteiro, M., Gmel, G., and Sempos, C.T. (2003). "The Relationship of Average Volume of Alcohol Consumption and Patterns of Drinking to Burden of Disease: An Overview." *Addiction*, Vol. 98, No. 9, pp. 1209–1228.

Rehm, J., Room, R., Monteiro, M., Gmel, G., Graham, K., Rehn, N., Sempos, C.T., Frick, U., and Jernigan, D. (2004). "Alcohol Use." In *Comparative Quantification of Health Risks:*

*Global and Regional Burden of Disease Attributable to Selected Major Risk Factors* (Vol 1), ed. M. Ezzati, A.D. Lopez, A. Rodgers, and C.J.L. Murray, pp. 959–1109. Geneva, Switz.: World Health Organization.

Reitlinger, G. (1961). *The Economics of Taste: The Rise and Fall of the Picture Market, 1760–1960*. New York: Holt, Reinhart, and Winston.

Renaud, S.C., Gueguen, R., Siest, G., and Salamon, R. (1999). "Wine, Beer, and Mortality in Middle-Aged Men from Eastern France." *Archives of Internal Medicine*, Vol. 159, pp. 1865–1870.

Renna, F. (2008). "Alcohol Abuse, Alcoholism, and Labor Market Outcomes: Looking for the Missing Link." *Industrial and Labor Relations Review*, Vol. 62, No. 1, pp. 92–103.

Resnick, P., Zeckhauser, R., Swanson, J., and Lockwood, K. (2006). "The Value of Reputation on eBay: A Controlled Experiment." *Experimental Economics*, Vol. 9, pp. 79–101.

Rey, P., and Tirole, J. (2007). "Financing and Access in Cooperatives." *International Journal of Industrial Organization*, Vol. 25, pp. 1061–1088.

Rickard, B.J., Costanigro, M., and Garg, T. (2013). "Economic and Social Implications of Regulating Alcohol Availability in Grocery Stores." *Applied Economic Perspectives and Policy*, Vol. 35, No. 4, pp. 613–633.

Riekhof, G.M., and Sykuta, M.E. (2005). "Politics, Economics, and the Regulation of Direct Interstate Shipping in the Wine Industry." *American Journal of Agricultural Economics*, Vol. 87, No. 2, pp. 439–452.

Rimm, E.B., Klatsky, A., Grobbee, D., and Stampfer, M.J. (1996). "Review of Moderate Alcohol Consumption and Reduced Risk of Coronary Heart Disease: Is the Effect Due to Beer, Wine, or Spirits?" *British Medical Journal*, Vol. 312, pp. 731–736.

Rob, R., and Fishman, A. (2005). "Is Bigger Better? Customer Base Expansion Through Word-of-Mouth Reputation." *Journal of Political Economy*, Vol. 113, No. 5, pp. 1146–1162.

Roberts, P., and Dowling, G. (2002). "Corporate Reputation and Sustained Superior Financial Performance." *Strategic Management Journal*, Vol. 23, No. 12, pp. 1077–1093.

Roberts, P.W., Khaire, M., and Rider, C.I. (2011). "Isolating the Symbolic Implications of Employee Mobility: Price Increases after Hiring Winemakers from Prominent Wineries." *American Economic Review: Papers & Proceedings*, Vol. 101, No. 3, pp. 147–151.

Robinson, J. (1998). *The Oxford Companion to Wine* (2nd ed.). Oxford, UK: Oxford University Press.

Rodgers, B., Windsor, T.D., Anstey, K.J., Dear, K.B., Jorm, A., and Christensen, H. (2005). "Non-Linear Relationships Between Cognitive Function and Alcohol Consumption in Young, Middle-Aged and Older Adults: The PATH Through Life Project." *Addiction*, Vol. 100, No. 9, pp. 1280–1290.

Rogerson, W. (1983). "Reputation and Product Quality." *The Bell Journal of Economics*, Vol. 14, No. 2, pp. 508–516.

Roll, R. (1977). "A Critique of the Asset Pricing Theory's Test—Part I: On Past and Potential Testability of the Theory." *Journal of Financial Economics*, Vol. 4, No. 2, pp. 129–176.

Rose-Ackerman, S. (1996). "Altruism, Nonprofits, and Economic Theory." *Journal of Economic Literature*, Vol. 34, pp. 701–728.

Rosen, S. (1974). "Hedonic Prices and Implicit Markets: Product Differentiation in Pure Competition." *Journal of Political Economy*, Vol. 82, pp. 34–55.

Rota, M., Pasquali, E., Scotti, L., Pelucchi, C., Tramacere, I., Islami, F., Negri, E., Boffetta, P., Bellocco, R., Corrao, G., La Vecchia C., and Bagnardi, V. (2012). "Alcohol Drinking and Epithelial Ovarian Cancer Risk. A Systematic Review and Meta-Analysis." *Gynecologic Oncology*, Vol. 125, No. 3, pp. 758–763.

Rothschild, M., and Stiglitz, J. (1976). "Equilibrium in Competitive Insurance Markets: An Essay on the Economics of Imperfect Information." *Quarterly Journal of Economics*, Vol. 90, No. 4, pp. 629–649.

Ruhm, C.J. (1995). "Economic Conditions and Alcohol Problems." *Journal of Health Economics*, Vol. 14, No. 5, pp. 583–603.

Ruhm, C.J., and Black, W.E. (2002). "Does Drinking Really Decrease in Bad Times?" *Journal of Health Economics*, Vol. 21, No. 4, pp. 659–678.

Rust, R.T., Zahorik, A.J., and Keiningham, T.L. (1995). "Return on Quality (ROQ): Making Service Quality Financially Accountable." *Journal of Marketing*, Vol. 59, No. 2, pp. 58–70.

Saak, A.E. (2012). "Collective Reputation, Social Norms, and Participation." *American Journal of Agricultural Economics*, Vol. 94, No. 3, pp. 763–785.

Sabia, S., Fayosse A., Dumurgier, J., Dugravot, A., Akbaraly, T., Britton, A., Kivimäki, M., and Singh-Manoux, A. (2018). "Alcohol Consumption and Risk of Dementia: 23 Year Follow-Up of Whitehall II Cohort Study." *British Medical Journal*, 362:k2927, pp. 1–11.

Sacks, J.J., Gonzales, K.R., Bouchery, E.E., Tomedi, L.E., and Brewer, R.D. (2015). "2010 National and State Costs of Excessive Alcohol Consumption." *American Journal of Preventive Medicine*, Vol. 49, No. 5, pp. e73–e79.

Saffer, H. (1989). *Alcohol Consumption and Tax Differentials Between Beer, Wine and Spirits.* NBER Working Paper No. 01/1989.

Saffer, H. (1991). "Alcohol Advertising Bans and Alcohol Abuse: An International Perspective." *Journal of Health Economics*, Vol. 10, pp. 65–79.

Sagala, R. (2013). *The Impact of General Public Wine Education Courses on Consumer Perception.* AAWE Working Paper No. 132.

Samokhvalov, A.V., Irving, H., Mohapatra, S., and Rehm, J. (2010). "Alcohol Consumption, Unprovoked Seizures, and Epilepsy: A Systematic Review and Meta-Analysis." *Epilepsia*, Vol. 51, No. 7, pp. 1177–1184.

San Martín, G.J., Brümmer, B., and Troncoso, J.L. (2008). "Determinants of Argentinean Wine Prices in the U.S.." *Journal of Wine Economics*, Vol. 3, No. 1, pp. 72–84.

Sanning, L.W., Shaffer, S., and Sharratt, J.M. (2008). "Bordeaux Wine as a Financial Investment." *Journal of Wine Economics*, Vol. 3, No. 1, pp. 51–71.

Saxenian, A. (1994). *Regional Advantage: Culture and Competition in Silicon Valley and Route 128.* Cambridge, MA: Harvard University Press.

Schamel, G. (2000). *Individual and Collective Reputation Indicators of Wine Quality.* CIES Working Paper No. 10, University of Adelaide, Australia.

Schamel, G., and Anderson, K. (2003). "Wine Quality and Varietal, Regional and Winery Reputations: Hedonic Prices for Australia and New Zealand." *Economic Record*, Vol. 79, No. 246, pp. 357–369.

Schiff, M. (2006). "Living in the Shadow of Terrorism: Psychological Distress and Alcohol Use Among Religious and Non-Religious Adolescents in Jerusalem." *Social Science & Medicine*, Vol. 62, No. 9, pp. 2301–2312.

Schirmer, R. (2012). "Globalization in the New World." *Memorie Geografiche*, Vol. 9, pp. 177–192.

Schmit, T.M., Rickard, B.J., and Taber, J. (2013). "Consumer Valuation of Environmentally Friendly Production Practices in Wines, Considering Asymmetric Information and Sensory Effects." *Journal of Agricultural Economics*, Vol. 64, No. 2, pp. 483–504.

Schneider, A. (2011). "Conservation and Management of Traditional Grape Varieties in Italy." Presented at the Twenty-First International Conference on Grapevine Propagation, July 21–23, Geisenheim, Germany.

Schnohr, C., Højbjerre, L., Riegels, M., Ledet, L., Larsen, T., Schultz-Larsen, K., Petersen, L., Prescott, E., and Grønbaek, M. (2009). "Does Educational Level Influence the Effects of Smoking, Alcohol, Physical Activity, and Obesity on Mortality? A Prospective Population Study." *Scandinavian Journal of Public Health*, Vol. 32, No. 4, pp. 250–256.

Schoenborn, C.A., and Adams, P.E. (2010). "Health Behaviors of Adults: United States, 2005–2007." *Vital and Health Statistics*, Vol. 10, No. 245.

Schroeder, T.C. (1992). "Economies of Scale and Scope for Agricultural Supply and Marketing Cooperatives." *Review of Agricultural Economics*, Vol. 14, No. 1, pp. 93–103.

Schumpeter, J.A. (1911). *The Theory of Economic Development: An Inquiry into Profits, Capital, Credit, Interest, and the Business Cycle.* New Brunswick, NJ: Transaction Publishers, 1983.

Scitovszky, T. (1943). "A Note on Profit Maximisation and Its Implications." *Review of Economics Studies*, Vol. 11, No. 1, pp. 57–60.

Scott, A. (1988). *New Industrial Spaces: Flexible Production Organization and Regional Development in North America and Western Europe.* London: Pion Press.

Scott Morton, F., and Podolny, J. (2002). "Love or Money? The Effects of Owner Motivation in the California Wine Industry." *Journal of Industrial Economics*, Vol. 50, pp. 431–456.

Sexton, R.J. (1990). "Imperfect Competition in Agricultural Markets and the Role of Cooperatives: A Spatial Analysis." *American Journal of Agricultural Economics*, Vol. 72, No. 3, pp. 709–720.

Shapiro, C. (1983). "Premiums for High Quality Products as Returns to Reputations." *Quarterly Journal of Economics*, Vol. 98, No. 4, pp. 659–680.

Sharpe, W.F. (1964). "Capital Asset Prices: A Theory of Market Equilibrium Under Conditions of Risk." *Journal of Finance*, Vol. 19, No. 3, pp. 425–442.

Shepherd, B. (2006). *Costs and Benefits of Protecting Geographical Indications: Some Lessons from the French Wine Sector.* Working Paper, Science Po, Paris.

Shiller, R.J. (1993). *Macro Markets.* New York: Oxford University Press.

Shleifer, A. (2005). "Understanding Regulation." *European Financial Management*, Vol. 11, No. 4, pp. 439–451.

Shrubsole, M.J., Wu, H., Ness, R.M., Shyr, Y., Smalley, W.E., and Zheng, W. (2007). "Alcohol Drinking, Cigarette Smoking, and Risk of Colorectal Adenomatous and Hyperplastic Polyps." *American Journal of Epidemiology*, Vol. 167, No. 9, pp. 1050–1058.

Siegel, M.B., Naimi, T.S., Cremeens, J.L., and Nelson, D.E. (2011). "Alcoholic Beverage Preferences and Associated Drinking Patterns and Risk Behaviors Among High School Youth." *American Journal of Preventive Medicine*, Vol. 40, No. 4, pp. 419–426.

Simpson, J. (2000). "Cooperation and Cooperatives in Southern European Wine Production: The Nature of Successful Institutional Innovation, 1880–1950." *Advances in Agricultural Economy History*, Vol. 1, pp. 95–126.

Simpson, J. (2005). "Cooperation and Conflicts: Institutional Innovation in France's Wine Markets, 1870–1911." *Business History Review*, Vol. 79, pp. 527–558.

Simpson, J. (2009). *Old World versus New World: The Origins of Organizational Diversity in the International Wine Industry, 1850–1914.* Working Papers in Economic History No. 09–01, Universidad Carlos III de Madrid.

Simpson, J. (2011). *Creating Wine: The Emergence of a World Industry, 1840–1914.* Princeton, NJ: Princeton University Press.

Sims, E., and Quintanar, S. (2017). *Analyzing Barrel Purchasing Decisions on Winery Costs.* AAWE Working Paper No. 212.

Singell, L., and Thornton, J. (1997). "Nepotism, Discrimination, and the Persistence of Utility-Maximizing, Owner-Operated Firms." *Southern Economic Journal*, Vol. 63, No. 4, pp. 904–919.

Skees, J.R. (2001). "The Bad Harvest: Crop Insurance Reform has Become a Good Idea Gone Awry." *Regulation: The CATO Review of Business and Government*, Vol. 24, pp. 16–21.

Skees, J.R., and Barnett, B.J. (2004). "Challenges in Government Facilitated Crop Insurance." *China in the Global Economy: Proceedings of the Organisation for Economic Co-operation and Development Workshop on Rural Finance and Credit Infrastructure in China* (October 13–14, Paris). Paris: OECD Publishing.

Skees, J.R., Barnett, B.J., and Collier, B. (2008). "Agriculture Insurance—Background and Context for Climate Adaptation Discussions." Paper presented at the OECD Expert Workshop on Economic Aspects of Adaptation, April 7–8, Paris, France.

Skog, O.J. (2001). "Alcohol Consumption and Mortality Rates from Traffic Accidents, Accidental Falls, and Other Accidents in 14 European Countries." *Addiction*, Vol. 96, Suppl. 1, pp. 49–58.

Skog, O.J., and Elekes, Z. (1993). "Alcohol and the 1950–1990 Hungarian Suicide Trend—Is There a Causal Connection?" *Acta Sociologica*, Vol. 36, pp. 33–46.

Skogen, J.C., Harvey, S.B., Henderson, M., Stordal, E., and Mykletun, A. (2009). "Anxiety and Depression Among Abstainers and Low-Level Alcohol Consumers. The Nord-Trøndelag Health Study." *Addiction*, Vol. 104, pp. 1519–1529.

Smart, R.G. (1988). "Does Alcohol Advertising Affect Overall Consumption? A Review of Empirical Studies." *Journal of Studies on Alcohol*, Vol. 49, No. 4, pp. 314–323.

Soboh, R.A.M.E., Lansink, A.O., Giesen, G., and van Dijk, G. (2009). "Performance Measurement of the Agricultural Marketing Cooperatives: The Gap Between Theory and Practice." *Review of Agricultural Economics*, Vol. 31, No. 3, pp. 446–469.

Solow, R.M. (1995). "But Verify." *The New Republic*, Vol. 213, No. 11, pp. 36–39.

Sridhar, S. (1994). "Managerial Reputation and Internal Reporting." *The Accounting Review*, Vol. 69, No. 2, pp. 343–363.

Stahel, W.R. (2003). "The Role of Insurability and Insurance." *The Geneva Papers on Risk and Insurance*, Vol. 28, No. 3, pp. 374–381.

Stallcup, J. (2005). "Toppling the Wall of Confusion." *Wine Business Monthly*, Vol. 12, No. 4.

Stampfer, M.J., Colditz, G.A., Willett, W.C., Speizer, F.E., and Hennekens, C.H. (1988). "A Prospective Study of Moderate Alcohol Consumption and the Risk of Coronary Disease and Stroke in Women." *New England Journal of Medicine*, Vol. 319, pp. 267–273.

Stein, B.D., Elliott, M.N., Jaycox, L.H., Collins, R.L., Berry, S.H., Klein, D.J., and Schuster, M.A. (2004). "A National Longitudinal Study of the Psychological Consequences of the September 11, 2001 Terrorist Attacks: Reactions, Impairment, and Help-Seeking." *Psychiatry*, Vol. 67, No. 2, pp. 105–117.

Stewart, S.H., Brown, C.G., Devoulyte, K., Theakston, J., and Larsen, S.E. (2006). "Why Do Women with Alcohol Problems Binge Eat? Exploring Connections Between Binge Eating and Heavy Drinking in Women Receiving Treatment for Alcohol Problems." *Journal of Health Psychology*, Vol. 11, pp. 409–425.

Stigler, G.J. (1971). "The Theory of Economic Regulation." *Bell Journal of Economics and Management Science*, Vol. 2, No. 1, pp. 3–21.

Storchmann, K. (2010). "The Economic Impact of Wine Industry on Hotels and Restaurants: Evidence from Washington State." *Journal of Wine Economics*, Vol. 5, No. 1, pp. 164–183.

Storchmann, K. (2012). "Wine Economics." *Journal of Wine Economics*, Vol. 7, No. 1, pp. 1–33.

Storper, M. (1989). "The Transition to Flexible Specialization in the U.S. Film Industry: External Economies, the Division of Labor, and the Crossing of Industrial Divides." *Cambridge Journal of Economics*, Vol. 13, pp. 273–305.

Strachan, J. (2007). "The Colonial Identity of Wine: The Leakey Affair and the Franco-Algerian Order of Things." *Social History of Alcohol and Drugs*, Vol. 21, No. 2, pp. 118–137.

Strand, B.H., and Steiro, A. (2003). "Alcohol Use, Income and Education in Norway 1993–2000." *Tidsskrift for Den Norske Lægeforening*, Vol. 123, pp. 2849–2053.

Sundaresan, S.M. (1989). "Intertemporally Dependent Preferences and the Volatility of Consumption and Wealth." *Review of Financial Studies*, Vol. 2, pp. 73–88.

Tennbakk, B. (1995). "Marketing Cooperatives in Mixed Duopolies." *Journal of Agricultural Economics*, Vol. 46, No. 1, pp. 33–45.

Terzani, T. (2008). *In Asia*. Milan, Italy: Edizioni TEA.

Testa, M., Quigley, B.M., and Leonard, K.E. (2003). "Does Alcohol Make a Difference? Within-Participants Comparison of Incidents of Partner Violence." *Journal of Interpersonal Violence*, Vol. 18, No. 7, pp. 735–743.

Thornton, J. (2013). *American Wine Economics: An Exploration of the U.S. Wine Industry*. Berkeley: University of California Press.

Throsby, D. (1990). "Perception of Quality in Demand for the Theatre." *Journal of Cultural Economics*, Vol. 14, No. 1, pp. 65–82.

Throsby, D. (2001). *Economics and Culture*. Cambridge, UK: Cambridge University Press.

Thun, M.J., Peto, R., Lopez, A.D., Monaco, J.H., Henley, S.J., Heath, C.W., and Doll, R. (1997). "Alcohol Consumption and Mortality Among Middle-Aged and Elderly U.S. Adults." *New England Journal of Medicine*, Vol. 337, pp. 1705–1714.

Tinlot, R. (2001). "Le terroir: un Concept à la Conquête du Monde." *Revue des OEnologues et des Techniques Vitivinicoles et OEnologiques*, Vol. 101, pp. 9–11.

Tirole, J. (1996). "A Theory of Collective Reputations (with Applications to the Persistence of Corruption and to Firm Quality)." *Review of Economic Studies*, Vol. 63, No. 1, pp. 1–22.

Towse, R. (2010). *A Textbook of Cultural Economics*. Cambridge, UK: Cambridge University Press.

Treccani (2012). "Duopsonio." In *Dictionary of Economics and Finance*. Rome, Italy. http://www.treccani.it/enciclopedia/duopsonio_(Dizionario-di-Economia-e-Finanza.

Tremblay, V.J., and Okuyama, K. (2001). "Advertising Restrictions, Competitions, and Alcohol Consumption." *Contemporary Economic Policy*, Vol. 19, No. 3, pp. 313–321.

Turvey, C.G., Weersink, A., and Chiang, S.C. (2006). "Pricing Weather Insurance with a Random Strike Price: The Ontario Ice-Wine Harvest." *American Journal of Agricultural Economics*, Vol. 88, No. 3, pp. 696–709.

Uber and MADD (2015). *Uber and MADD Report*. https://2q72xc49mze8bkcog2f01nlh-wpengine.netdna-ssl.com/wp-content/uploads/2015/01/UberMADD-Report.pdf.

United Nations Educational, Scientific, and Cultural Organization (UNESCO) (1970). Illicit Trafficking of Cultural Property: Convention on the Means of Prohibiting and Preventing the Illicit Import, Export and Transfer of Ownership of Cultural Property. November 17. http://www.unesco.org/new/en/culture/themes/illicit-trafficking-of-cultural-property/1970-convention/text-of-the-convention/.

United Nations, Statistics Division (n.d.). *Demographic Yearbook System*. https://unstats.un.org/unsd/demographic-social/products/dyb/.

Unwin, T. (1991). *Wine and the Vine: An Historical Geography of Viticulture and the Wine Trade*. London: Routledge.

Urrutiaguer, D. (2002). "Quality Judgments and Demand for French Public Theatre." *Journal of Cultural Economics*, Vol. 22, No. 3, pp. 185–202.

US Alcohol and Tobacco Tax and Trade Bureau (2012). "American Viticultural Area (AVA) Manual for Petitioners." Alcohol and Tobacco Tax and Trade Bureau TTB P 5120.4 (09/2012). https://www.ttb.gov/images/pdfs/p51204_ava_manual.pdf.

US Department of Agriculture and US Department of Health and Human Services (2010). *Dietary Guidelines for Americans* (7th ed.). Washington, DC: U.S. Government Printing Office.

US Government Publishing Office (2014). "Title 27: Alcohol, Tobacco Products and Firearms, Part 9—American Viticultural Areas." *Electronic Code of Federal Regulations*. https://www.ecfr.gov/cgi-bin/text-idx?c=ecfr&sid=6bd58158c22269d08137de4a75c5db28&Div5&rgn=&view=text&node=27:1.0.1.1.7&idno=27#_top.

Valette, J, Amadieu, P., and Sentis, P. (2018). "Survival in the French Wine Industry: Cooperatives versus Corporations." *Journal of Wine Economics*, Vol. 13, No. 3, pp. 328–354.

Valliant, P.M. (1995). "Personality, Peer Influence, and Use of Alcohol and Drugs by First-Year University Students." *Psychological Reports*, Vol. 77, pp. 401–402.

Vanek, J. (1970). *The General Theory of Labor-Managed Market Economies*. Ithaca, NY: Cornell University Press.

Van Oers, J.A.M., Bongers, I.M.B, Van De Goor, L.A.M., and Garretsen, H.F.L. (1999). "Alcohol Consumption, Alcohol-Related Problems, Problem Drinking, and Socioeconomic Status." *Alcohol and Alcoholism*, Vol. 34, No. 1, pp. 78–88.

Vaseth, M. (2011). *Wine Wars*. Lanham, MD: Roman & Littlefield.

Vastola, A., and Tanyeri-Abur, A. (2009). *Non-Conventional Viticulture as a Viable System: A Case Study in Italy*. American Association of Wine Economists AAWE Working Paper No. 43.

Vetter, S., Rossegger, A., Rossler, W., Bisson, J., and Endrass, J. (2008). "Exposure to the Tsunami Disaster, PTSD Symptoms and Increased Substance Use—An Internet Based Survey of Male and Female Residents of Switzerland." *BMC Public Health*, Vol. 8, No. 92, pp. 1–6.

Villalonga, B. (2004). "Intangible Resources, Tobin's q, and Sustainability of Performance Differences." *Journal of Economic Behavior & Organization*, Vol. 54, No. 2, pp. 205–230.

Villalonga, B., and Amit, R. (2006). "How Do Family Ownership, Control and Management Affect Firm Value?" *Journal of Financial Economics*, Vol. 80, pp. 385–417.

Vlahov, D., Galea, S., Resnick, H., Ahern, J., Boscarino, J.A., Bucuvalas, M., Gold, J., and Kilpatrick, D. (2002). "Increased Use of Cigarettes, Alcohol, and Marijuana Among Manhattan, New York, Residents after the September 11th Terrorist Attacks." *American Journal of Epidemiology*, Vol. 155, No. 11, pp. 988–996.

Vogel, D. (1995). *Trading Up: Consumer and Environmental Regulation in a Global Economy*. Cambridge, MA: Harvard University Press.

Walter, B.F. (2009). *Reputation and Civil War: Why Separatist Conflicts Are So Violent*. Cambridge, UK: Cambridge University Press.

Warner, L.A., and White, H.R. (2003). "Longitudinal Effects of Age at Onset and First Drinking Situations on Problem Drinking." *Substance Use and Misuse*, Vol. 38, pp. 1983–2016.

Wasserman, D., Varnik, A., and Eklund, G. (1994). "Male Suicides and Alcohol Consumption in the Former USSR." *Acta Psychiatrica Scandinavica*, Vol. 89, No. 5, pp. 306–313.

Waterson, E.J., and Murray-Lyons, I.M. (1990). "Preventing Alcohol Related Birth Damage: A Review." *Social Science & Medicine*, Vol. 30, No. 3, pp. 349–364.

Wechsler, H., Kasey, E.H., Thum, D., and Demone, H.W. (1969). "Alcohol Level and Home Accidents." *Public Health Reports*, Vol. 84, No. 12, pp. 1043–1050.

Weil, R. (1993). "Do Not Invest in Wine, at Least in the U.S. Unless You Plan to Drink it, and Maybe Not Even Then." Paper presented at the Second International Conference of the Vineyard Data Quantification Society, February, Verona, Italy.

Wernerfelt, B. (1988). "Umbrella Branding as a Signal of Product Quality: An Example of Signaling by Posting a Bond." *RAND Journal of Economics*, Vol. 19, No. 3, pp. 458–466.

Westhead, P., and Howorth, C. (2006). "Ownership and Management Issues Associated With Family Firm Performance and Company Objectives." *Family Business Review*, Vol. 19, No. 4, pp. 301–316.

Williams, A.C. (1997). *Risk Management & Insurance* (8th ed.). New York: McGraw-Hill.

Williamson, K., and Wood, E. (2003). *The Dynamics of the South African Wine Industry Cluster: A Basis for Innovation and Competitiveness*. Working Paper, University of Cape Town.

Wilsnack, R.W., Wilsnack, S.C., Kristjanson, A.F., Vogeltanz-Holm, N.D., and Gmel, G. (2009). "Gender and Alcohol Consumption: Patterns from the Multinational GENACIS Project." *Addiction*, Vol. 104, No. 9, pp. 1487–1500.

Wilson, R. (1985). "Reputations in Games and Markets." In *Game-Theoretic Models of Bargaining*, ed. A.E. Roth, pp. 65–84. New York: Cambridge University Press.

Wine Institute and Avalara (n.d.). "Direct-To-Consumer Shipping Laws for Wineries." WineInstitute.org. https://wineinstitute.compliancerules.org/state-map/.

Winetitles Media. (n.d.). "Into the Next Century." https://winetitles.com.au/statistics-2/industry-strategies-plans/into-the-next-century/.

Winfree, J., and McCluskey, J. (2005). "Collective Reputation and Quality." *American Journal of Agricultural Economics*, Vol. 87, No. 1, pp. 206–213.

Winfree, J., McIntosh, C., and Nadreau, T. (2018). "An Economic Model of Wineries and Enotourism." *Wine Economics and Policy*, Vol. 7, No. 2, pp. 88–93.

Wohlgenant, M.K. (2009). *Demand for Wine, Beer, and Spirits: Accounting for On-Premise Consumption*. North Carolina State University Working Paper.

World Bank (2011). *Weather Index Insurance for Agriculture: Guidance for Development Practitioners*. Agriculture and Rural Development Discussion Paper 50, Washington, DC.

World Health Organization (WHO) (2011). *Global Status Report on Alcohol and Health*. Geneva, Switzerland.

World Health Organization (WHO) (2013). *Global Status Report on Road Safety*. Geneva, Switzerland.

World Health Organization (WHO) (2014). *Global Status Report on Alcohol and Health*. Geneva, Switzerland.

World Trade Organization (WTO) (1994). Annex 1C: Trade-Related Aspects of Intellectual Property Rights, Marrakesh Agreement Establishing the World Trade Organization, April 15. https://www.wto.org/english/docs_e/legal_e/27-trips_04b_e.htm#3.

Yermack, D. (2004). "Retention, and Reputation Incentives for Outside Directors." *Journal of Finance*, Vol. 59, No. 5, pp. 2281–2308.

Yörük, B.K., and Yörük, C.E. (2012). "The Impact of Drinking on Psychological Well-Being: Evidence from Minimum Drinking Age Laws." *Social Science & Medicine*, Vol. 75, pp. 1944–1854.

Young, D.J., and Bielinska-Kwapisz, A. (2006). "Alcohol Prices, Consumption, and Traffic Fatalities." *Southern Economic Journal*, Vol. 72, No. 3, pp. 690–703.

Yu, J., Bouamra-Mechemache, Z., and Zago, A. (2018). "What's in a Name? Information, Heterogeneity, and Quality in a Theory of Nested Names." *American Journal of Agricultural Economics*, Vol. 100, No. 1, pp. 286–310.

Zago, A. (2009). "A Nonparametric Analysis of Production Models with Multidimensional Quality." *American Journal of Agricultural Economics*, Vol. 91, No. 3, pp. 751–764.

Zahra, S. (1999). "The Changing Rules of Global Competitiveness in the 21st Century AD." *Academy of Management Executive*, Vol. 13, pp. 36–42.

Zara, C. (2010). "Weather Derivatives in the Wine Industry." *International Journal of Wine Business Research*, Vol. 22, No. 3, pp. 222–237.

Zeckhauser, R. (1970). "Medical Insurance: A Case Study of the Trade-Off Between Risk Spreading and Appropriate Incentives." *Journal of Economic Theory*, Vol. 2, pp. 10–26.

Zubretsky, T.M., and Digirolamo, K.M. (1996). "False Connection Between Adult Domestic Violence and Alcohol." In *Helping Battered Women: New Perspectives and Remedies*, ed. A.R. Roberts and D.C. Dwyer, pp. 222–228. New York: Oxford University Press.

Zylbersztajn, D., and Miele, M. (2005). "Stability of Contracts in the Brazilian Wine Industry." *Revista de Economia e Sociologia Rural*, Vol. 43, No. 2, pp. 353–371.

# Index

Note: Figures, tables, and boxes are indicated by "f" and "t" and "b" respectively, following page numbers.

IGP. *See* Indicazione Geografica Protetta
IGT. *See* Indicazione Geografica Tipica
Imports. *See* Exports and imports
India, 14–15, 245n10
Indicación Geográfica (IG), 178, 179t
Indicazione Geografica Protetta (IGP), 173, 174f, 176–177, 179t
Indicazione Geografica Tipica (IGT), 165t, 174f, 176, 178
  in old classification by country, 175t
  quality correlation to, 66–67, 76t–79t
  requirements for, 174
Industrial districts/clusters
  advantages of, 132–134, 244n19
  in China, 244n19
  defining, 131
  drawbacks of, 134–136
  economies of scale for, 131–132
  innovation advantages for, 133
  input-output methodology for, 136, 244n21
  Italian, 132
  new business creation advantages with, 133
  regulations for, 134, 135, 135f
  structure of, 135–136, 135f
  studies on, 135–136, 135f
  in US, 135–136, 135f
Information asymmetries, 83
  adverse selection as result of, 163, 164
  classification systems role in reducing, 2, 84, 164, 180
  with cooperatives, 129
  defining, 163, 245n1
  experts and guides role in reducing, 63, 84, 166
  in insurance market, 152, 153
  market failures and, 163, 187–188, 212
  moral hazard as result of, 129, 163, 164
  with private companies, 115, 116
  problems and solutions to, 116, 163–165
  quality signals in combating, 164
  reputation role in combating, 164–166, 168
  in resource allocation, 188, 212
Innovations and technologies
  classification hierarchy and, 66, 178
  competition impacted by, 93
  cooperatives adoption of, 119, 130

history of, 32, 33t, 34
industrial clusters advantages in, 133
New World adoption of, 32, 33t, 34, 41–42, 231, 240n27
Old World regulations as barrier to, 32, 34, 40, 42, 66
production costs impacted by New World, 34, 41–42
quality relation to terroir compared with, 65–66
Insurance
  agricultural, for weather risk, 152–154
  against damage, 151–152, 244n6
  index, 153–154, 245n10
  investment in bottles of wine and, 139, 140t–145t, 147
  MPCI, 153
  opportunistic behavior, 152
  risk coverage factors and, 150–153, 244n6
International Cooperative Alliance (ICA), 122, 123b–124b
International Organization of the Vineyard and Wine (OIV), 5, 239n19
Investments and assets, 2. *See also* Profits/profitability; Risk
  benefits with wine compared with financial, 148
  in bottles of wine, 74, 137, 138–139, 140t–145t, 146–149, 244n1
  CAPM for assessing risk in, 137, 146, 147, 159–162
  cointegration model for securities, 149
  costs with wine compared with financial, 147
  deferred returns with vineyard, 36, 62, 65
  diversification strategies and benefits for, 137, 138, 146–147, 156–159, 157f, 158f
  Fama-French's three-factor model for assessing risk of, 137, 146, 147, 161–162
  intangible, 95–96, 241n13
  liquidity with wine compared with financial, 148
  portfolio selection, conditions for optimal, 159–160, 159f